Electricity and basic electronics

by

Stephen R. Matt
Assistant Professor
Industrial Arts Department
University of Georgia
Athens, Georgia

South Holland, Illinois
THE GOODHEART-WILLCOX COMPANY, INC.
Publishers

Library of Congress Cataloging in Publication Data

Matt, Stephen R.
  Electricity and basic electronics.

  Includes index.
  1. Electric engineering. 2. Electronics. I. Title.
TK146.M376   621.3   79—6346
ISBN 0—87006—285—9

INTRODUCTION

ELECTRICITY AND BASIC ELECTRONICS is intended to reflect the growing and expanding interest in electricity and electronics. Here is a first course that teaches the basic fundamentals. It is a starting point for a career in this most important area.

This text is clearly written and profusely illustrated, making it easy for beginning students to understand. Chapters are organized so that content is presented in the logical order of use. They are laid out in a sequence that facilitates learning. Each new concept builds on the information learned in the previous chapter. Safe work habits are stressed throughout the text.

Using this text, students will find it easy to learn fundamentals and apply them. Important details are clarified with use of extra color. Numerous proven projects are included with construction procedures. These not only help motivate students but build interest and confidence.

Suggested activities at the end of each chapter range from simple to demanding. These activities may be used to challenge the more advanced students or to diversify the program. An effort has been made to relate all activities and projects to actual applications in home and industry.

Stephen R. Matt

CONTENTS

PROJECTS TO MAKE

SAFETY PRECAUTIONS FOR THE ELECTRICITY-ELECTRONICS SHOP

There is always an element of danger when working with electricity. Observe all safety rules that concern each project and be particularly careful not to contact any live wire or terminal regardless of whether it is connected to either a low voltage or a high voltage. Projects do not specify dangerous voltage levels. However, keep in mind at all times that it is possible to experience a surprising electric shock under certain circumstances. Even a normal healthy person can be injured or seriously hurt by the shock or what happens as a result of it. Do not fool around. The lab is no place for horseplay.

Chapter 1
LEARNING AND APPLYING THE FUNDAMENTALS

Electricity is truly amazing. To better appreciate its benefits, think for a minute what life would be like without electricity. Can you imagine no lights, no radio or stereo, no television or motion pictures, not being able to drive a car?

Very few things operate or exist without some help from electricity. It helps heat your home in winter, air condition it in summer. The clothes you wear are the products of electric sewing machines and the giant, electrically powered machinery that made the cloth.

Newspapers, magazines and books are printed on presses operated by electricity. How about all the electrical devices used in hospitals, laboratories and research centers? Consider all of the electrical gear designed into a rocket, Fig. 1-1. What a dull world this would be without electricity and the benefits we obtain from its use.

USE OF ELECTRICITY

Using electricity is a lot like driving a car. It is not necessary for you to be able to repair anything that goes wrong in order to enjoy using these conveniences. To turn on a light at home, for example, all that you have to do is flick a switch. To start a car, you just turn the switch with an ignition key.

If the light in your home does not come on, or if the car fails to start, you may have to call in a specialist to make the repair. Normally, however, you do *not* have to be an expert service technician to use and enjoy the benefits of the light or the car.

SAFETY

The more you know about a given subject, the easier it will be to perform your own service work if a problem arises. However, a word of warning is given to do-it-yourselfers. Many people who know a lot about a given subject are injured because they get careless. They get that way when they believe that, "It can only happen to someone else, never to me."

Respect electricity, then, and it will serve you well. Treat it carelessly, and one day you may receive the shock of your life.

Fig. 1-1. Space vehicles carry complex electrical equipment. (NASA)

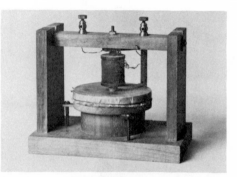

Fig. 1-2. Alexander Graham Bell's first telephone. (Western Electric Co.)

Since Benjamin Franklin flew his now famous kite and key, and Alexander Graham Bell invented the telephone, Fig. 1-2, many important discoveries have been made in the electrical field.

CAUTION: In attempting to duplicate Franklin's kite flying experiment in a thunderstorm, some people have been killed or seriously injured. As you will see in studying this text, there are safer ways of experimenting with electricity.

Over the years, electrical discoveries have led to many things which have made our work easier and life more comfortable. In order to understand today's remarkable advances, it will be necessary to examine many of the early discoveries. Each one has served as a building block in the house of electrical knowledge that people have constructed so far.

LEARNING OPPORTUNITY

ELECTRICITY AND BASIC ELECTRONICS is designed to provide an explanation of theory and, through the many projects, offer practical application of the principles. It also will serve as a foundation for future studies leading to a career in electricity/electronics or in a related field.

To avoid confusion and make learning easier, each new electrical or electronic term and/or unit will be explained as much as possible when it is introduced. In some cases, the explanation will be quite brief. However, you also will be told which chapter will give complete details on that particular subject area. This method of treating new ideas will help you build a strong understanding of the subject with only the facts that are necessary at the time.

BASIC CONCEPT OF A CIRCUIT

How do you turn on a light? Earlier mention was made of flicking on a switch. What did that action do to result in the light coming on? Understanding this action and reaction will bring you a long way along the path of knowledge about electricity.

First, look at Fig. 1-3. It shows a battery, wires and a test lamp (light bulb and socket). One wire connects the test lamp to the battery. The other wire is connected to the lamp, but not to the battery. Since a wire is disconnected, the test lamp is out. However, when both lamp wires are connected to the battery, Fig. 1-4, the

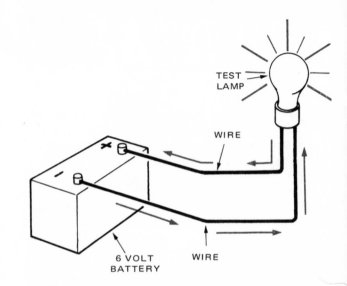

Fig. 1-4. This circuit has the same parts as in Fig. 1-3. The test lamp is on because the wire is connected to the battery.

lamp will burn brightly. This happens because electricity is flowing through the wires in a "completed circuit."

Any completed circuit requires three things: First, it needs a source of electricity, or energy (in this case, the battery). Second, it needs a load, or something which uses that energy. The load in Figs. 1-3 and 1-4 is the light bulb. Third, it needs a path to get the energy from source to load and back. As you may have guessed, this path is formed by the wires.

There is more to electricity than a simple circuit, but studying circuit makeup is a strong first step toward understanding the subject.

ELECTRONS

As you study these examples of electric circuits in greater detail, you will need to apply physics (matter and energy, and their interaction). Suppose you have a

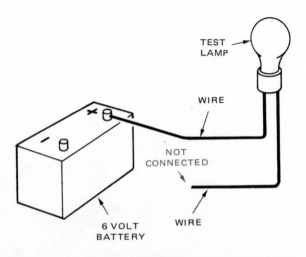

Fig. 1-3. A battery, test lamp and wires are needed to form a simple circuit. However, since a wire is disconnected, this circuit is not completed.

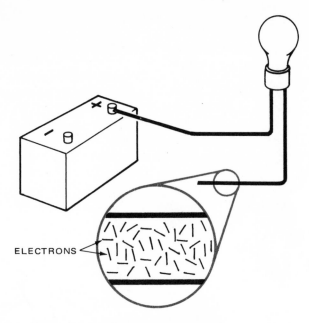

called "electrons" are moving at random (without direction or pattern). These electrons are so small that millions upon millions would fit on the point of a pin.

When the lamp wires are connected, Fig. 1-6, the electrons flow through the wires from the battery to the lamp and back to the battery. How fast do they move? *Electrons move at the speed of light, which is about 186,000 miles per second.*

Notice, also, that when the circuit is completed, the electrons flow through the wires from negative (−) to positive (+). This important fact will be explained in detail in Chapter 3.

The use of a switch was mentioned earlier in this chapter. A switch makes it easy to open or close a circuit, Fig. 1-7. Closing the switch allows the electrons to move or flow. When the switch is opened, none of the electrons can jump across the gap. This is called an "open circuit."

Fig. 1-5. Using a very powerful microscope to look inside the wires of Fig. 1-3 shows that the electrons are not flowing.

very powerful microscope and can take a look at what is happening inside the wires, both before and after they are connected to the battery. Note that before connecting the wires, Fig. 1-5, extremely small particles

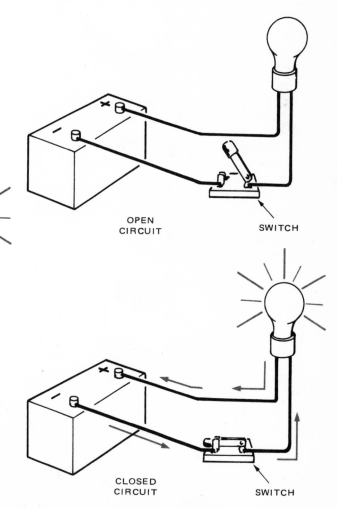

OPEN CIRCUIT · SWITCH

CLOSED CIRCUIT · SWITCH

Fig. 1-7. A switch installed in the simple circuit allows easier opening and closing of the circuit.

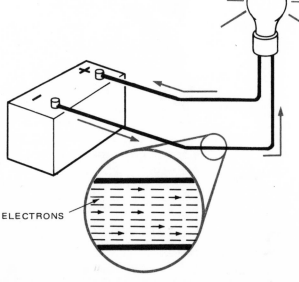

Fig. 1-6. Using a very powerful microscope to look inside the wires of Fig. 1-4 shows that the electrons are flowing (moving rapidly in one direction).

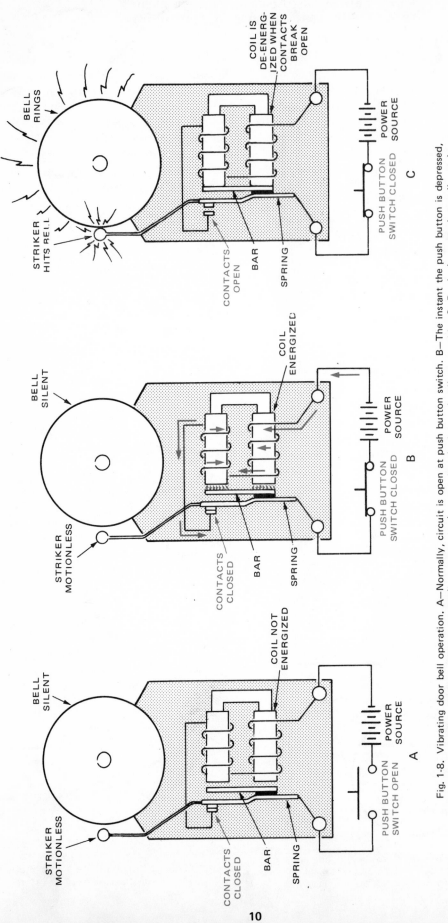

Fig. 1-8. Vibrating door bell operation. A—Normally, circuit is open at push button switch. B—The instant the push button is depressed, circuit is completed and coil is energized (has magnetism), attracting the bar and striker assembly. C—Bar is drawn against coil, causing striker to ring bell. However, movement of bar causes contacts to break and circuit is opened. Coil is de-engerized (lost its magnetism) and spring pulls bar and striker away from coil. This, in turn, completes circuit at contacts and striker hits the bell again.

Basically, a switch provides a simple and convenient way of opening or closing the circuit without having to disconnect and reconnect wires to the battery. A push button can be connected to the circuit to do the same thing at your house or apartment. See A in Fig. 1-8. A person at your door will press the push button, as in B and C, allowing electrons to flow to the door bell. Releasing the push button, as in A in Fig. 1-8, stops the flow of electrons and the sound. As long as the push button is depressed, the circuit repeatedly opens and closes at the contacts (as in B and C), causing the door bell to ring.

Several other methods used to open and close circuits will be explained as you progress in your knowledge of electricity. At the same time, you will learn what is happening electrically, as you go through this book.

BATTERIES

You probably have noticed that the batteries pictured in some of the illustrations are labeled 6 volts. Your flashlight uses two or more 1 1/2 volt cells, like the one shown in Fig. 1-9. These are *not* batteries.

Fig. 1-9. The familiar flashlight battery is not a battery. Proper terminology for the item shown is a "1 1/2 volt cell." (Burgess Battery Div., Gould, Inc.)

When two or more cells are combined, they form a battery. Another type of battery, Fig. 1-10, is used in cars, golf carts and electric vans.

Although most people refer to cells as "batteries," this is *not* correct. Proper names or terminology are very important in the study of electricity and electronics. Imagine your surprise if you asked somebody for a screwdriver and they gave you a hacksaw. Use the proper name for any particular item or component (part) and save yourself a lot of trouble later.

Fig. 1-10. Pictured is a 12 volt battery for automotive use. It is an assembly of six 2 volt cells.

SYMBOLS

Take a closer look at the drawing of the power source, wires and load, Fig. 1-4. Note that the elements of the circuit are drawn in some detail. Now that you are learning more about electricity and electronics, it is necessary to use some other method of showing circuits. Otherwise, drawings will be difficult to make and take a long time to complete.

Therefore, rather than picturing each element, symbols should be used. For example, instead of drawing a cell, use the symbol for a cell, Fig. 1-11.

Fig. 1-11. Symbols are used to identify electrical elements or components in circuit drawings. The symbol for a cell is shown.

Since a battery is a combination of two or more cells, its symbol should be drawn as a group of cell symbols, as shown in Fig. 1-12. Note that the long line in the battery symbol is the positive terminal; the short line is the negative terminal.

Fig. 1-12. The symbol for a battery is a combination of several cell symbols.

When using symbols, a wire is simply drawn as a line. The symbol for "load," on the other hand, will not always be the same. It depends on what element or component you are using for a load. For example, Figs. 1-13 and 1-14 show common loads (speaker, lamp and bell) along with their symbols. There are many others.

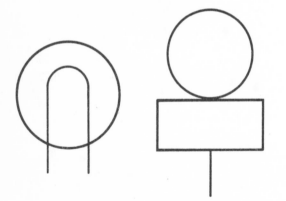

Fig. 1-13. A speaker and its symbol are shown.   (Electro Voice)

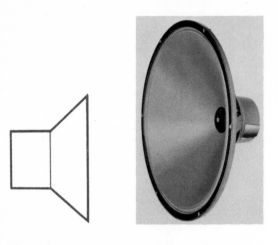

Fig. 1-14. Symbols pictured are: Left. Lamp. Right. Door Bell.

I.C. SAYS:

"Hi! My name is I.C. I will be around from time to time to help you understand new words, phrases and formulas. My body is a printed circuit board with integrated circuits attached. Can you recognize the rest of my parts? You will learn the function of these parts and their symbols in your study of electricity and basic electronics."

Knowing symbols will allow us to redraw some of the circuits we saw earlier in this chapter. Fig. 1-16, would be another way of drawing the circuit shown in Fig. 1-4, using the battery, wires and lamp. If you want

Fig. 1-15. I.C. stands for "integrated circuit." It also is the name of this helpful character made of various parts and symbols.

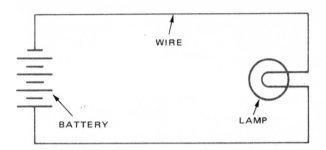

Fig. 1-16. Use of symbols permits us to simplify the pictorial drawing of electrical circuits. Compare this schematic with the drawing in Fig. 1-5.

to add another lamp to the circuit, just put a large dot where the wires connect to one another. Fig. 1-17 shows how this is done.

If two circuits are shown near each other, or if the wires do not connect, leave off the dot, Fig. 1-18. This tells the person reading the drawing that the wires just pass over each other but are not connected.

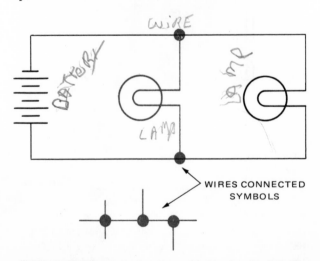

Fig. 1-17. Another lamp has been added to the circuit in Fig. 1-16. Note that a large dot is the symbol for "wires connected."

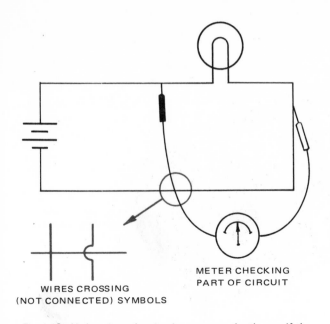

WIRES CROSSING
(NOT CONNECTED) SYMBOLS

METER CHECKING
PART OF CIRCUIT

Fig. 1-18. If the wires of a circuit are near each other, or if the wires cross but do not connect, either one of the two symbols shown may be used.

Drawings using symbols are called "schematics" by most people working in the electrical field. Fig. 1-19 shows what a finished industrial schematic looks like. The television service technician would be lost without a schematic for a TV set brought in for repairs.

Not all television receivers have the same circuits, and even the same brand varies as the models change from year to year. The TV technician usually has several hundered TV schematics. Some companies print

them to sell. Can you imagine the trouble it would be to check the circuits of the assembly shown in Fig. 1-20 *without* a schematic?

PROBES

As you go through this text, you will be shown things you can build. Many of these projects will be illustrated with circuit drawings and photographs to give you a good idea what the finished job looks like.

Some of the projects you will be asked to build require the use of probes to make tests. You can buy a set of probes or make your own, Fig. 1-21. The probes you can make are just nails with wires soldered to them. Then, the whole thing must be cast in plastic to insulate the tips. This is done so you can hold the probes in your hand without fear of electric shock.

Some people use plastic tubing for insulation instead of casting in plastic. Simple ideas like this make projects more economical to build than if you bought the materials from an electronics store. Also, it is more fun to "build your own," and you really can learn how it works. Then, too, if something happens to your project, you will have an easier time fixing it.

ELECTRONS

Earlier, we told you a few facts about electrons. If you used a powerful electron microscope, you would be able to see that everything around you is made up of electrons. These microscopes, however, are much too expensive for a school to buy. Therefore, we will try to

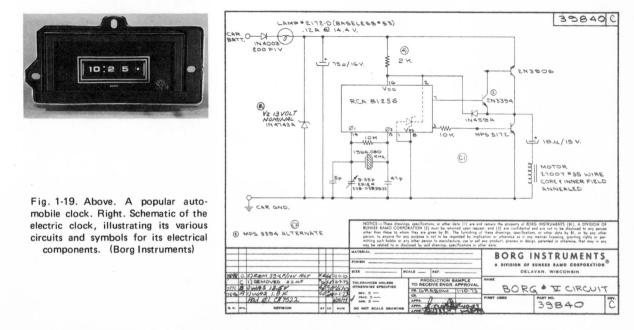

Fig. 1-19. Above. A popular automobile clock. Right. Schematic of the electric clock, illustrating its various circuits and symbols for its electrical components. (Borg Instruments)

Fig. 1-20. This complex assembly contains printed circuits and dozens of electrical and electronic components. (Fisher Radio)

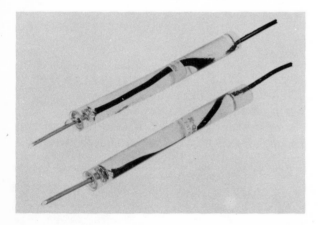

Fig. 1-21. Probe used to perform circuit tests made from nails soldered to wires and cast in plastic. (Project by Rico Righetti)

explain by text and illustration what electrons are and how they act.

One of the simplest atoms is the helium atom, Fig. 1-22. The helium atom has two electrons (negatively charged particles) in orbit around the center of the atom, known as its nucleus. The nucleus is made up of neutrons (particles which have no charge) and protons (particles which have a positive charge). The number of neutrons, protons and electrons varies, depending on which atom you are studying. Some atoms have 50 to 100 electrons in orbit around the nucleus.

Think of the nucleus of an atom as our planet Earth, and the electrons as two satellites orbiting around it,

Fig. 1-23. If a satellite speeds up or slows down, its orbit will change. If you want the satellite to orbit farther from the planet, you must give it more energy, usually by firing a rocket to increase the satellite's speed. To lower its orbit, you would fire rockets that retard (hold back) the forward speed of the satellite.

You can do similar things to electrons to make their orbits change. For example, you can take away some of

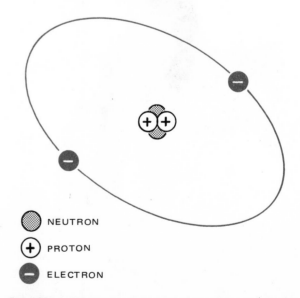

NEUTRON

⊕ PROTON

⊖ ELECTRON

Fig. 1-22. A helium atom is made up of two electrons (negative charge) and a nucleus composed of neutrons (no charge) and protons (positive charge).

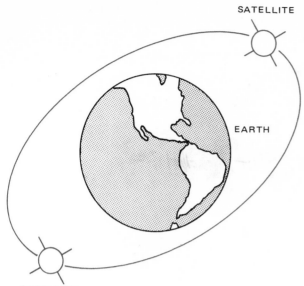

Fig. 1-23. The parts of a helium atom can be compared with the Earth and two satellites orbiting around it. The Earth represents the nucleus and the satellites are the electrons in orbit.

their energy and make them drop to a lower energy level. If you want to shift the electrons to a higher level, you must give them more energy.

As you move away from the simple atoms, such as helium, you get into others that are more complex. Think again in terms of our Earth. You now have a large number of satellites in orbit, but they are not all at the same altitude nor in the same orbit. The more

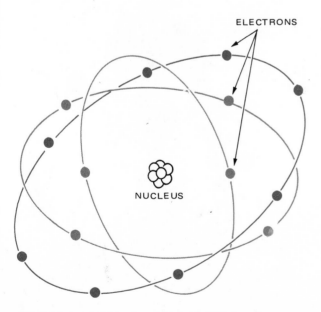

Fig. 1-24. In studying the makeup of more complex atoms, note that electrons orbit around nucleus at different levels.

complex atoms, Fig. 1-24, have many more orbiting electrons at different energy levels.

In Chapter 17, you will find that most electronic components depend on electrons to work properly. How we are able to move electrons or shift them around means a great deal to us. For example, electronic components such as diodes and transistors cannot work if we are not able to move those electrons.

PROJECT TESTER AND FUSE SAVER

Do you want to save yourself a lot of trouble after building a project? Put together the project tester shown in Fig. 1-25, then plug it into a wall receptacle.

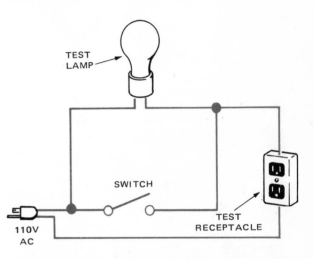

Fig. 1-25. Project tester and fuse saver. First, obtain electrical parts and wire as shown. Then, assemble the unit by making the necessary connections and proceed with the tests as outlined in the text.

The project tester is just what its name implies. It will serve as an advance guard against project failure when you put the project into operation for the first time.

You can check a new 110 volt project for short circuits (unwanted electron flow) or opens by plugging it into the test receptacle with the switch open. The tester will prevent blown fuses and avoid the possibility of burning up a circuit that has taken hours to build.

Here is how the tester works: First, be sure the tester switch is open. Then, screw in a test lamp of about the same wattage as the project you are building. Next, plug the power lead of your tester into a 110 volt receptacle. Then, plug the lead of your 110 volt project into the test receptacle.

Here are the possible results:
1. With a short, the test lamp will burn brightly.
2. Without a short, the test lamp will be dim.
3. With an open circuit, the test lamp will not light.

Once you establish that there is no short or open, close the tester switch. This bypasses the test lamp and lets you try your project under normal operating conditions.

TEST YOUR KNOWLEDGE

1. The three requirements for a complete circuit are _Load_, _Source_ and _____.
2. Electron flow usually means:
   a. Flow from positive to negative.
   b. Flow of positive charges.
   c. Flow of negative and positive charges.
   d. Flow of negative charges.
3. Electrons come from the _Neg._ post of a battery and go back to the _Pos._ post.

4. One way to stop the flow of electrons is to _open_ the circuit.
5. Electrons travel at:
   a. The speed of light.
   b. The speed of sound.
   c. 100 miles per hour.
   d. 1000 feet per second.
6. An electrical schematic is _____.
7. Draw the symbols for a cell and for a battery.

SOMETHING TO THINK ABOUT

1. Why are there so many different symbols for an electrical load?
2. Why is a load so important to an electrical circuit?

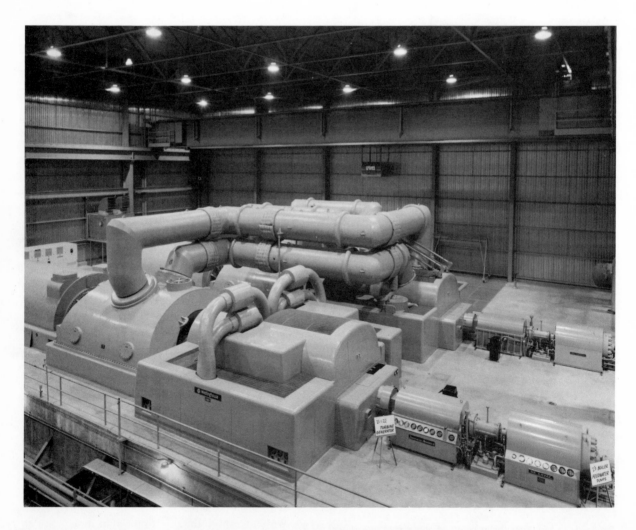

This electric power generating plant is a prime example of "using magnetism to generate electricity." These huge generators are a major source of power in the Morro Bay, California, area.

Chapter 2
SOURCES OF ELECTRICITY

After reading the first chapter, you probably have several questions about how electricity is produced. This chapter is designed to provide the answers.

There are many ways and means of making, converting or generating electricity. Basically, they fall into five general areas as follows:

1. Chemical energy.
2. Light.
3. Pressure.
4. Heat.
5. Magnetism.

These five methods of generating electricity will be explained in some detail in the order given. Specific examples and related projects will be included.

ELECTRICITY FROM CHEMICAL ENERGY

Chemical energy means that electricity is generated through the use of chemicals and their reaction with other materials. The simple cell mentioned earlier is a good example of this means of producing electricity. If you were to cut one of these cells in half, you would see something like Fig. 2-1.

This type of cell is often called a "dry cell," although the chemicals used (electrolyte-material that provides the path for electricity inside a cell) are in the form of a paste. The paste is a mixture of salammoniac, manganese dioxide and carbon with water. These chemicals react with the zinc can and the carbon rod to produce electricity. When the zinc can is finally eaten away, the cell can no longer produce electricity. It is called a "dead cell."

Older dry cells would swell up as the zinc can dissolved, and flashlights were ruined because electrolyte leaked through the case. Today, most companies making dry cells have solved this problem by building the outer section of the case with different materials.

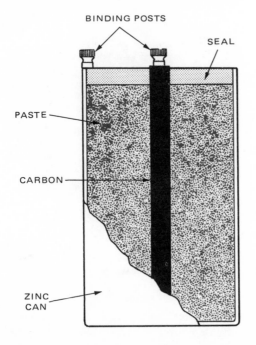

Fig. 2-1. In a dry cell, chemical energy produces electricity. The paste (electrolyte) in the cell reacts with the zinc can and the carbon rod.

Fig. 2-2. A dc powered portable light is a common application of a dry cell. (Burgess Battery Div., Gould, Inc.)

Dry cells and batteries produce a stream of electrons that flow from negative (–) to positive (+). Remember, this was mentioned in Chapter 1. This flow of electrons in the electrical circuit is called direct current (often abbreviated to dc).

How many things can you think of that get their power from a dc source? Flashlights, emergency lamps or lanterns, Fig. 2-2, portable radios and children's toys use dc as a source of electricity.

CONTINUITY TESTER PROJECT

Figs. 2-3 and 2-4 illustrate a handy item you can make for yourself. It is called a "continuity tester." With it, you can test for broken wires or burned out (open) fuses. Another good use is finding the matching wire ends in a wiring harness (a group of wires routed together).

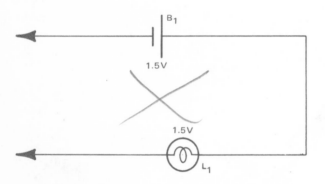

Fig. 2-3. This schematic for a shop-built continuity tester includes: B_1 — One 1.5 volt dry cell. L_1 — One 1.5 volt test lamp.

To make the continuity tester, connect leads to a 1.5 volt dry cell and a 1.5 volt test lamp, Fig. 2-3. Keep the circuit small. In that way, you can put the dry cell and

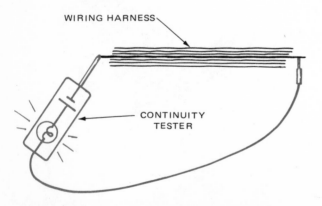

Fig. 2-4. The continuity tester can be used to trace the flow of electricity in individual wires in a wiring harness.

lamp in almost anything you want to, such as a plastic tube or toy hotdog.

To use the continuity tester, touch the leads to the item under test. If the light comes on, the item has continuity (will conduct electricity). If the light stays off, the item lacks continuity (will not conduct electricity).

You cannot use this continuity tester to check out certain things. The coil in a transformer or a large inductor, for example, offers too much resistance for the low voltage tester. A dc cell of 1.5 volts cannot do the job. Later, you will find why this is so.

If you want to test a wiring harness, Fig. 2-4, hook one lead of the tester to one end of a wire. To find the other end of the wire, touch each of the other wire ends with the other test lead. When the light comes on, you have continuity. Now you have both ends of the same wire. Sometimes, one end of the wiring harness is hard to reach. For this reason, wires are often color coded (each wire has a specific color of insulation).

STORAGE BATTERIES

Cells may contain electrolytes other than the paste type. In Fig. 2-5, for example, you see several cells combined in a storage battery like the one installed in an automobile.

Fig. 2-5. A car's storage battery is made up of six cells filled to a prescribed level with electrolyte.
(ESB Brands, Inc.)

A storage battery uses an acid as its electrolyte. This is an advantage over the dry cell, because the car battery can be recharged over and over again. When an inexpensive dry cell goes dead, it is thrown away and replaced with a new one. The storage battery used in a car, however, is fairly expensive. To save the cost of constantly replacing batteries, they are designed so they can be recharged.

In the case of a car battery, the alternator (generator) does the charging. However, it is possible to drain (discharge) the battery of its charge. If, for example, you leave the headlights on, it may be necessary to use a battery charger, Fig. 2-6, to recharge the battery. In some cases, you will see mechanics or

Fig. 2-6. A portable battery charger serves to keep the storage battery fully charged (high specific gravity).

car owners use two heavy "jumper cables" to start an engine from the battery in a second car. Once the engine in the first car is started, the alternator will begin to recharge the battery.

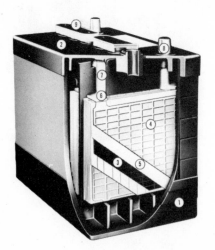

Fig. 2-7. Cutaway battery reveals relationship of parts. 1—Case. 2—Cover. 3—Positive plates. 4—Negative plates. 5—Separators. 6—Cell. 7—Cell connector. 8—Terminal post. 9—Vent cap.

Consider, however, that car batteries do not last forever. When they get old, it is harder to keep them fully charged. This is especially true if the owner does not take care of the battery, and it is allowed to discharge repeatedly. When this happens, the life of the battery is shortened.

Older batteries or those that have been misused do not hold their charge because chemical action is poor. As the battery is used, the electrolyte begins to react with the metal plates. See 3 and 4 in Fig. 2-7. This causes them to be eaten or dissolved like the zinc can in the dry cell. The major difference, however, is that the metal in a car battery can be made to deposit itself back on its plates. This is done through a chemical action known as charging the battery. This battery charging process is explained in full detail in most chemistry books.

BUZZER PROJECT

Fig. 2-8 illustrates a circuit for a buzzer you can build. This circuit will be seen as part of other circuits which are to follow. In some cases, the buzzer is replaced with a light, or you can use a bell or siren if you wish.

The parts needed to complete this project include a 1.5 volt cell, a 1.5 volt buzzer, a single-pole single-throw switch and sufficient wire to make the connections shown in Fig. 2-8.

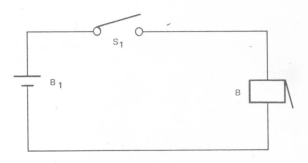

Fig. 2-8. This schematic for a buzzer includes: B—One 1.5 volt buzzer. B_1—One 1.5 volt dry cell. S_1—SPST switch.

HYDROMETER TEST

The acid or electrolyte in a battery also can be used to tell you whether or not the battery is fully charged. You may recall seeing a service station attendant checking the battery with a device called a "hydrometer," Fig. 2-9.

First, the hydrometer bulb must be squeezed and released to fill the hydrometer tube with acid from the

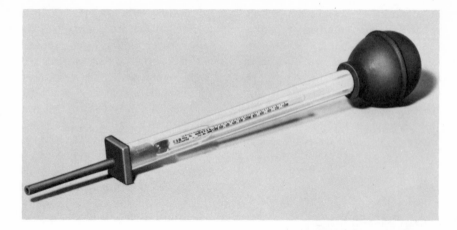

Fig. 2-9. A hydrometer is an instrument used to measure specific gravity of electrolyte in each cell to determine state of charge of the battery.

battery. Then, by reading the scale on the float inside, it is possible to tell how much of a charge the battery has available. Many hydrometers are marked to tell you whether the battery is fully charged, half-charged or discharged. This is possible because the acid actually changes in strength. This strength is measured in terms of "specific gravity."

Specific gravity is a comparison of the density of the acid with the density of water. Water has a specific gravity of 1.000. As the battery discharges, the specific gravity of the acid gets closer to the 1.000 mark. In some colder sections of the country, this loss of charge could result in a battery freezing in the winter. Note the table in Fig. 2-10.

FREEZING POINT OF BATTERY ELECTROLYTE

	SPECIFIC GRAVITY	FREEZING POINT
FULLY CHARGED	1.275	−85 DEG. F (−65 C)
	1.250	−62 DEG. F (−52 C)
HALF-CHARGED	1.225	−35 DEG. F (−37 C)
	1.200	−16 DEG. F (−27 C)
	1.175	− 4 DEG. F (−20 C)
DISCHARGED (WATER)	1.150	+ 5 DEG. F (−15 C)
	1.000	+32 DEG. F (0 C)

Fig. 2-10. Chart shows freezing points of battery electrolyte at various specific gravity readings.

ELECTRICITY FROM LIGHT ENERGY

Light energy is another means of producing electricity. In recent years, more and more research has been involved with changing or converting light energy into electricity.

Light energy becomes especially important in space travel. Figs. 2-11 and 2-12 show solar cells being used in two different satellites. In space, liquid chemicals like the electrolyte used in batteries would be of limited

Fig. 2-11. This spacecraft is part of the Earth Resources Technology Satellite program designed to gather environmental information and relay it to a ground facility. Solar cells convert light energy into electricity.
(NASA)

use. However, the sun is a huge source of power. By using solar cells, it is possible to convert the sun's rays into enough electrical energy to power all the electronic devices in the satellite.

Fig. 2-12. Telestar satellite uses solar cells to convert the sun's rays into electrical energy while orbiting in space. (Western Electric Co.)

PHOTOELECTRIC CELL

Light can be used to produce effects other than storing electricity for later use. For example, a portable light meter or a similar device that senses light inside of a camera is a valuable aid when taking photographs.

In any situation, the film must have just the right amount of exposure to light to produce good photographs. Too much light or too little light will give poor results. By using a light metering device, you are able to know just how much of an adjustment to make to your camera shutter and lens controls for the best possible photographs. Fig. 2-13 will help you to understand why this is so.

Some cameras automatically control the various adjustments. Rays of light strike a photoelectric cell (which measures the amount of light that hits it). If there is a lot of light, a signal is sent to a diaphragm (variable size opening) which makes the opening smaller. If there is a small amount of light, the signal will make the diaphragm opening larger.

Other devices use the photoelectric cell to convert light energy into electric energy. For example, have you ever seen a rainstorm occur during the daytime? The next time that happens, watch the streetlights. In some cities, as the sky grows darker, the lights will come on automatically through the use of light sensing devices. The sensor is a photoelectric cell mounted on top of the light assembly.

FIBER OPTICS LAMP PROJECT

You, too, can make a fiber optics lamp. You will need an old can, a light bulb, a single-pole single-throw switch, fiber optics tubing and some wire, Figs. 2-14 and 2-15. To dress up the project, install contact paper on the outside of the can.

The size of the bulb you use will depend on how big and how fancy you wish to get. However, your best bet is to stay down in wattage. This will prevent the bulb

Fig. 2-14. The base of this fiber optics lamps is made from an old can covered with contact paper. (Project by Allan Witherspoon)

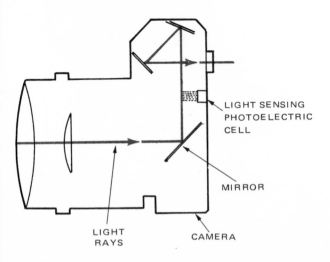

Fig. 2-13. A built-in light meter uses a photoelectric cell to detect the amount of light that will enter the camera.

from overheating the inside of the can. If you decide to paint the outside of the can, use a smaller bulb to avoid burning your design. A base can be cut from wood, so the can does not mar your table. For further protection, glue a felt material on the bottom of the wood base.

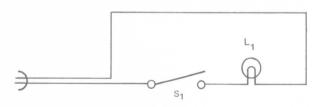

Fig. 2-15. This schematic for the fiber optics lamp includes: L_1—One 110 volt ac lamp. S_1—One SPST switch and one rubber cap with line cord.

ELECTRICITY BY PRESSURE

Pressure is still another method used to generate electricity for many of the things you see and hear around you every day. Piezo is the Greek name for pressure. Therefore, piezoelectricity, or the piezoelectric effect, is the name often mentioned when talking about the effect of electricity from pressure.

A common application of pressure being used to generate electricity is shown in Fig. 2-16. It is a microphone used with a citizens band radio. Another popular use is the tone arm which you place on your favorite record. The telephone, one of the great conveniences of the modern household, also depends on pressure to generate electricity.

All of these devices use crystalline materials, like quartz, in their makeup. When a force is exerted on these materials, they bend or become distorted and electrons are forced to move. Fig. 2-17 illustrates how this is done.

Fig. 2-16. A citizens band radio microphone is a good example of electricity being generated by pressure.
(E.F. Johnson Co.)

Another recent use of quartz is shown in Fig. 2-18. Here it is being used in a wristwatch to aid in keeping accurate time. Fig. 2-19 shows both the mechanical and electric parts of the watch, and its main sections are identified. The quartz crystal is pulsing at 32,768 times per second (Hz) to give off electrical signals. The abbreviation Hz stands for hertz, which will be explained in Chapter 10.

The item in Fig. 2-19 called an "integrated circuit module" is another way of saying the circuits being used are miniaturized (extremely small). Integrated circuits will be described in some detail when you get to Chapter 17. The important thing at this point is to note that quartz is being used to give off an electrical signal, which is what the piezoelectric effect is all about.

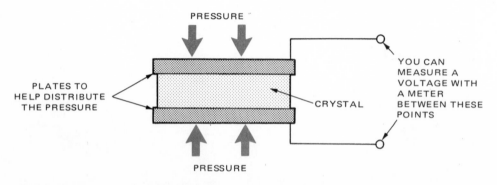

Fig. 2-17. Electricity by pressure can be demonstrated by placing quartz crystal between pressure plates. When pressure is applied, electricity flows.

Fig. 2-18. Another use of quartz crystal is in the "tuning fork" wristwatch. Pressure created by the pulsing of the crystal produces electrical signals. (Bulova Watch Co., Inc.)

ELECTRICITY BY HEAT

Heat also can be used to generate electricity. Many schools have an area called a foundry, set aside for casting hot metals into many different shapes. Other schools have furnaces used for heat treating metals at a specific temperature. Still others have a kiln where ceramic objects are fired. At home, ovens are used to cook our food and to bake cakes and pies. How, then, can we tell the temperatures at which certain things melt and others cook? All this is possible because of thermoelectricity or electricity from heat.

Many of the temperatures needed to handle the tasks mentioned are measured with a device called a thermocouple, Fig. 2-20. A thermocouple consists of two dissimilar metals, such as copper and zinc or constantan and iron joined tightly at one end. To check output voltage, connect the other two ends of the dissimilar metals to a sensitive meter, such as a galvanometer.

A galvanometer is a meter that can sense a very small number of electrons flowing through a circuit. As the electrons flow, the needle on the meter will begin to move. If the flow of electrons increases, the needle will

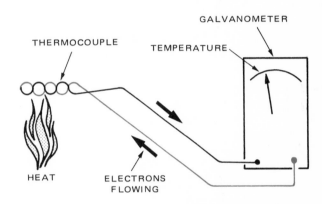

Fig. 2-20. A thermocouple will generate electricity if heat is applied to the connection of two dissimilar metals. If the other two ends of the metals are connected to a meter, the meter needle will indicate the temperature of the thermocouple at the joined end.

Fig. 2-19. "Tuning fork" wristwatch features a quartz power cell. A-Integrated circuit module. B—Electromagnetic coils. C—Tuning fork. D—Power cell nest. (Bulova Watch Co., Inc.)

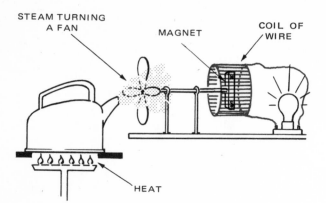

STEAM TURNING
A FAN

MAGNET

COIL OF
WIRE

HEAT

Fig. 2-21. The principle of the steam turbine powered generator is illustrated. Heat changes water to steam, which forces the fan (turbine wheel) to turn. The turbine output shaft rotates the generator to produce electricity.

move farther across the meter face. To relate this to the thermocouple, consider that when the joined end is heated, the needle of the galvanometer will move, Fig. 2-20. The scale of the meter can then be marked with degrees of temperature.

ELECTRICITY BY MAGNETISM

Using magnetism to generate electricity is the best known and most often used method of all. Yet, for the average person, magnetism probably is the least understood source of electricity.

Everyone knows that the power company supplies electricity to their homes. The company produces this electricity with a device called a generator. To produce electricity the generator has to be turned in a rotary motion, Fig. 2-21. Power companies operate generators in several ways:

1. By means of waterfalls and water under pressure from giant dams or reservoirs, Figs. 2-22 to 2-25.
2. By means of steam produced by burning coal or oil, Fig. 2-26.
3. By means of steam from nuclear reactors, Figs. 2-27 and 2-28.

We will go into motors and generators in greater detail when we get to Chapter 12. At this time, just bear in mind that the spinning generator will create a magnetic field which is used to give us electricity.

MAGNETISM

All material is either magnetic or nonmagnetic. This characteristic is based on whether or not the material will be attracted to a strong magnet placed near it. Most ferrous metals (those that have iron or steel as part of their makeup) are magnetic. Stainless steel is an exception to this fact because, usually, it is not magnetic.

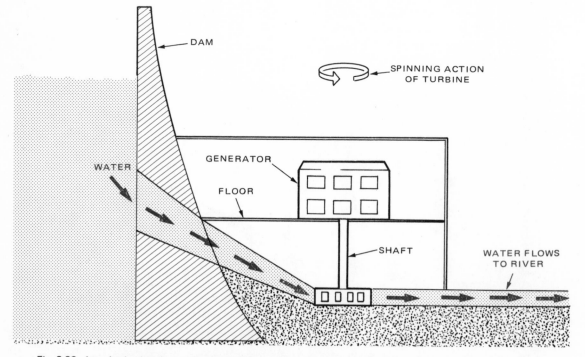

DAM

SPINNING ACTION
OF TURBINE

WATER

GENERATOR

FLOOR

SHAFT

WATER FLOWS
TO RIVER

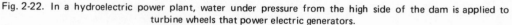

Fig. 2-22. In a hydroelectric power plant, water under pressure from the high side of the dam is applied to turbine wheels that power electric generators.

Fig. 2-23. The difference in water level on high and low sides of the dam creates the "head" (water pressure) needed to spin the turbines.
(Boeing Electronics)

Fig. 2-24. Turbine shaft drives a generator in the hydroelectric power plant. The greater the water pressure on the turbine wheel, the higher the electrical output of the generator.

Fig. 2-25. Electricity generated by magnetism is the method employed by huge generators installed in a power station.

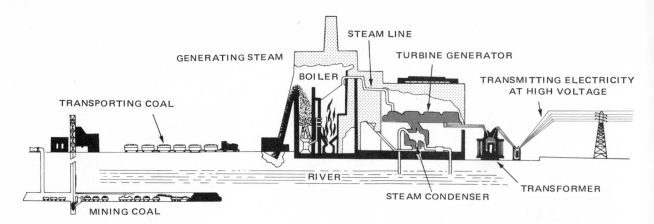

Fig. 2-26. In a steam turbine generator application, coal fires the boiler that generates steam which, in turn, drives the turbine generator that produces electricity.

Fig. 2-27. Nuclear power plants are the modern approach to producing electricity. In this plant, workers are loading fuel cells in the nuclear reactor.   (Consumers Power Co.)

Sources of Electricity

Fig. 2-28. Nuclear power complex includes auxiliary building, fuel building and reactor building, at left, which houses a free-standing steel containment and nuclear reactor. (Consumers Power Co.)

Most nonferrous metals are not magnetic. Nonferrous metals are metals other than iron. Copper, brass and aluminum are examples of these nonmagnetic materials. An easy way to find out whether or not a metal is magnetic is to place a magnet on it. If it is attracted to the magnet, it *is* magnetic.

I.C. SAYS:

"If you look at a magnet attracting a piece of steel, you can see the movement of the steel toward the magnet. However, you cannot see the magnetic field because it is invisible."

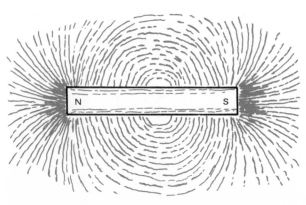

Fig. 2-29. In this magnetic field experiment, first cover a bar magnet with a sheet of paper, plastic or glass. Then sprinkle on the iron filings and tap gently. The iron filings will form the pattern shown between the poles of the magnet.

There *are* some simple ways that make it possible for you to observe a magnetic field. One way is to place a piece of paper, plastic or glass over a bar magnet. Then, obtain some fine steel particles by filing a steel rod or bar. Next, lightly cover the material with filings, sprinkling them right over the magnet, and gently tap it. You will see the filings fall into a pattern of attraction between the north and south poles of the magnet, Fig. 2-29. Do not use particles that are too large or too heavy. If you use a horseshoe magnet, you will see a magnetic field like the one illustrated in Fig. 2-30.

The lines which make up the magnetic field are

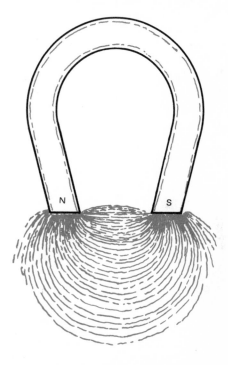

Fig. 2-30. Iron filings on a paper, plastic or glass placed over a horseshoe magnet reveals the pattern of lines of force around and between the poles of the magnet.

called lines of flux or lines of force. Figs. 2-31 and 2-32 reveal that the lines of force start from one end of the magnet and are attracted to the other end. The two ends of the magnet are called poles. One is known as the north pole, the other is the south pole. The lines of force are thought to leave the north pole, enter the magnet at the south pole, and complete the path through the magnet to the north pole.

PRINCIPLES OF MAGNETISM

In the 1800s, an Englishman named Michael Faraday first performed the lines of force experiment with iron filings. He also found that when two magnets are used

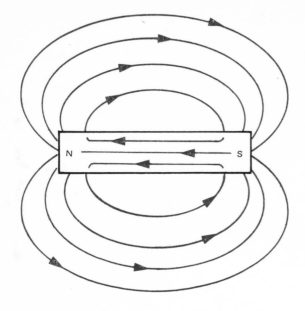

Fig. 2-31. The magnetic lines of force around a bar magnet follow the pattern of iron filings shown in the experiment in Fig. 2-29.

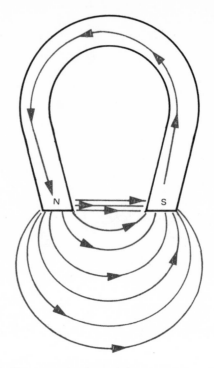

Fig. 2-32. The magnetic lines of force around a horseshoe magnet are similar to the pattern of the iron filings in Fig. 2-30.

in the experiment, you will get the results shown in Figs. 2-33 and 2-34. From this discovery, he was able to state two basic principles of magnetism:

1. Like poles repel.
2. Unlike poles attract.

Whether the two poles are both north or both south, they *will* repel one another, Fig. 2-33. Unlike poles, on the other hand, will always attract each other, Fig. 2-34.

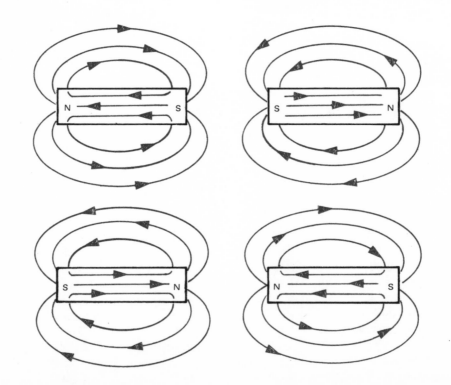

Fig. 2-33. The opposing lines of force of these two bar magnets show that like poles repel each other, whether they are aligned north to north or south to south.

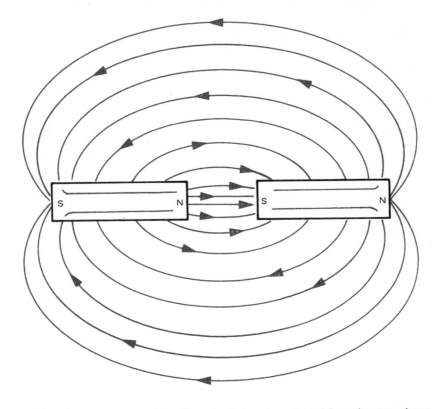

Fig. 2-34. Two bar magnets with unlike poles facing, show lines of force that merge into a pattern of attraction. Note that the lines of force always go from north to south.

I.C. SAYS:

"When you are working with magnets or experimenting with magnetism, the important facts to remember are: like poles repel and unlike poles attract."

MAGNETIC DECLINATION

The north pole of a compass needle is really the north seeking pole, Fig. 2-35. As an experiment, hold a compass level and watch the needle. You will find that the pole of the needle labeled N (north) will align itself in the general direction of the north geographic pole. However, unlike poles attract. Therefore, the compass needle has to be pointing at a south magnetic pole, the earth's "magnetic north" pole.

This north pole/south pole relationship is confusing to most people, so today's accepted terminology is "magnetic north" because it attracts the north seeking pole of the compass needle. Fig. 2-36 will help clarify this point.

From this explanation, it can be seen that the Earth

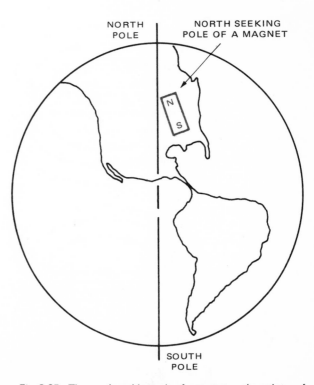

Fig. 2-35. The north seeking pole of a magnet or the pointer of a compass always points in the general direction of the north geographic pole.

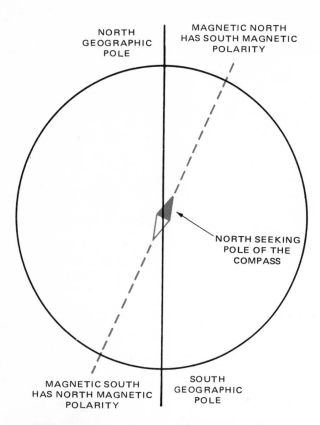

Fig. 2-36. The north seeking pole of the compass needle points at the south magnetic pole, which is termed "magnetic north."

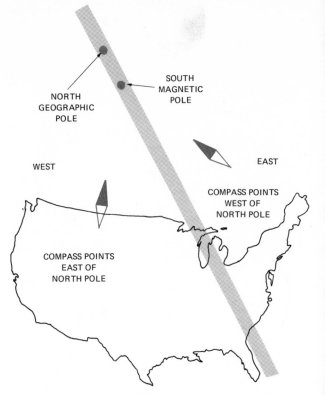

Fig. 2-37. The north geographic pole and "magnetic north" are in similar but separate locations. The compass needle points at both poles only in the narrow band shown in color.

is a giant magnet. The major problem with this theory, is that the two magnetic poles do not line up with the two geographic poles. Depending on where you live on the earth, the magnetic pole may be either east or west of the true north pole. In the United States, people living in the narrow band shown in Fig. 2-37 will have their compasses point toward the geographic and magnetic poles with the same reading. People living to the west of that band will have their compasses point east of the true north pole.

The difference between the geographic and magnetic poles is called "magnetic declination." It is another way of saying that a difference exists between the two poles. Magnetic declination is an important fact that airplane pilots must know. If they fail to make corrections when reading their navigational instruments, it is possible that they will not find the airfield when it is time to land their planes.

BENDING LINES OF FORCE

Earlier in this chapter, we placed paper over a magnet and sprinkled iron filings on it to show magnetic lines of force. Paper is an insulator. An insulator is a material that does not allow electrons to flow through it easily. Bakelite, glass, air, paper and mica are insulators.

Although insulators block the flow of electrons, they do not block the flow of the magnetic lines of force. This is why you are able to see the lines of force formed on the paper. As you saw, you can use plastic, glass or other insulators over the magnet and still perform the experiment with iron filings.

Fig. 2-38 shows how magnetic lines of force can pass right through most nonmagnetic materials. However, there are ways of bending the lines of force so that they can take another path through the magnetic material.

DEMONSTRATION OF MAGNETISM

To demonstrate how electricity is generated through magnetism, perform this experiment using a galvanometer. First, take a coil of wire and attach the two ends to the terminals of the galvanometer, Fig. 2-39. Next, take a bar magnet and quickly insert it into the coil while watching the meter pointer move. Quickly remove the magnet and note that pointer movement is in the opposite direction. The meter pointer moves because the magnetic lines of force cut through the copper coil even though the wire is insulated.

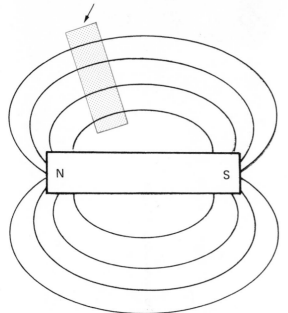

NONMAGNETIC MATERIAL:
PAPER, GLASS, PLASTIC, COPPER OR ALUMINUM

N   S

Fig. 2-38. Magnetic lines of force are unaffected by insulators and most nonmagnetic materials.

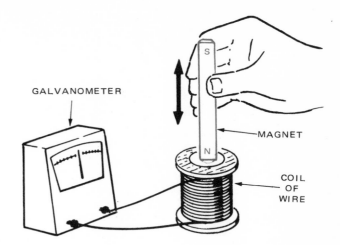

GALVANOMETER

MAGNET

COIL
OF
WIRE

Fig. 2-39. Moving a magnet into and out of a coil of wire will cause the needle of a galvanometer to move, first in one direction, then in the opposite direction.

This "cutting" action of the magnetic lines of force generates a small movement of electrons which, in turn, causes the pointer of the meter to move. Different amounts of electrons can be made to move by:

1. Changing the size of the magnet.

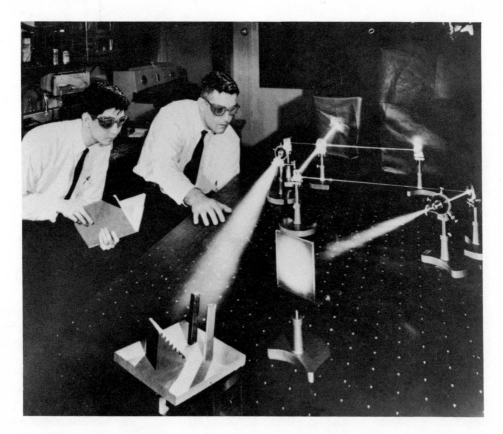

Fig. 2-40. Laser beams created by electricity are used to make holograms, which are three dimensional pictures made without a camera. (Western Electric Co.)

2. Using coils with different numbers of copper wire turns.

3. Changing the speed with which the magnet is moved into the coil.

You can see these differences by watching how far the pointer moves on the meter. It is also possible in this experiment to move the pointer by holding the magnet still and moving the coils of copper wire over the magnet. Relative movement between the magnetic field and the conductor causes the pointer to move.

Later, we will see how all these ideas are used to build motors and generators to do specific jobs.

USES OF ELECTRICITY

Five methods are commonly used to generate electricity: chemical energy, light, pressure, heat and magnetism. It also is possible to reverse this procedure and use electricity to create these same five things.

In Fig. 2-40, for example, laser light beams are being used to make holograms. A laser (Light Amplification by Stimulated Emission of Radiation) is a highly concentrated or focused beam of light. The hologram being produced is a three dimensional picture. In Fig. 2-41, for example, you can see the front, side and back views of various chess pieces.

White light, Fig. 2-42, also can be produced through the use of electricity. Special tungsten bulbs are used in this application.

One of the more specialized uses of electricity is

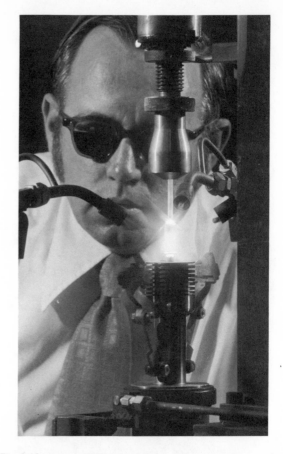

Fig. 2-42. In an application where electricity is used to produce light, tungsten halogen bulbs make white lights for a manufacturing process. (Guide Div., General Motors Corp.)

Fig. 2-41. These holograms were produced by laser beams. Chess pieces appear in three different perspectives on each hologram plate.

Fig. 2-43. In another factory operation, an electrical discharge machine is shown producing an automobile grille.

Fig. 2-45. Ultrasonic welding of two dissimilar plastics also illustrates electricity being used to produce heat for a manufacturing operation.

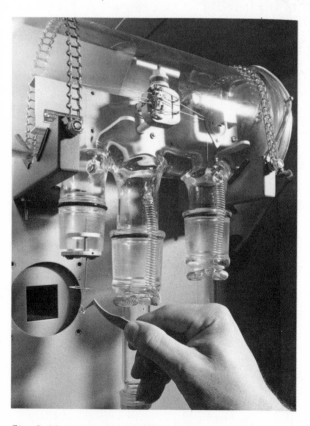

Fig. 2-44. Laboratory thermal analysis equipment is an example of electricity being used to generate heat. (Guide Div., General Motors Corp.)

shown in Fig. 2-43, where an electrical discharge machine is pictured making plastic grilles for automobiles. Two other uses include thermal analysis equipment, Fig. 2-44 and the welding of two different types of plastic through ultrasonics, Fig. 2-45. There are many other examples all around you: telegraph, telephone, buzzers, soldering guns and other devices that are powered by electricity.

TEST YOUR KNOWLEDGE

1. The basic law of magnetism says that _____ poles repel and _____ poles attract.
2. The north pole of a compass needle is really the north seeking pole. When held level, the compass will always point at:
   a. The north geographic pole.
   b. Magnetic north, which has south magnetic polarity.
   c. The south geographic pole.
   d. Magnetic south, which has north magnetic polarity.
3. Name two uses for a thermocouple.
4. One of the materials used to produce electricity by pressure is _quartz._
5. Which of the following is a type of chemical energy used to produce electricity?

a. Nuclear power plants.

b. Burning coal to produce heat.

c. The storage battery in a car.

d. Lights in your house.

6. A thermocouple is made from ___TWO___ dissimilar metals.

7. The magnetic flux lines come out of the ___South___ pole and go into the ___North___ pole of a permanent magnet.

8. A ___Hydrometer___ is used to measure the specific gravity of the acid used in a storage battery.

9. Name five ways to produce electricity.

   Chemical Heat magnetism Light Pressure

10. Why would a battery freeze up in the winter if it is discharged?

SOMETHING TO THINK ABOUT

1. Find the magnetic declination for your city.
2. Why is a keeper bar used on a magnet?
3. Where are the poles on a ring magnet?
4. Why does striking a magnet a sharp blow demagnetize it?
5. Why does a magnet grow weaker with age?

Flexible printed circuits are used to conduct electricity in telephone handsets and switchboard equipment. (Western Electric Co.)

Chapter 3
CONDUCTORS AND INSULATORS

As discussed in Chapter 1, electrons must be provided with a path from the power source to the load. This path usually is in the form of a wire, although it could be any kind of a conductor of electricity. *A conductor is any material that easily passes an electric current.*

Fig. 3-1. Printed circuit boards utilize printed wiring to conduct electricity between parts of the circuit.

In some cases, the path for the electrons is provided by printed wiring on specially made "printed circuit boards," or "PC boards." See Fig. 3-1. Many radios and television sets have circuits of this type, Fig. 3-2.

Fig. 3-3 presents a close-up of two typical PC boards. Note that many of the parts are connected with thin strips of copper or silver. This is printed wiring. While it is used extensively in certain applications, many electric circuits still require wire.

Fig. 3-2. Electric circuits in some television sets include printed circuit boards   (Heath Co.)

WIRE

Wire used in electric circuits can be made from many different materials. Common conductors are shown in the table in Fig. 3-4. When the electrons can flow easily in the material, it is called a good conductor. Those in the table are ranked by their ability to conduct. If it is difficult to pass an electric current through a material, it is known as an "insulator."

Some good conductors of electricity are copper, silver and aluminum. Most wire is made from copper because it *is* a good conductor and can be purchased at a reasonable price. Silver is a better conductor of electricity than copper, but higher cost and poorer availability usually prevent it from being used as wire.

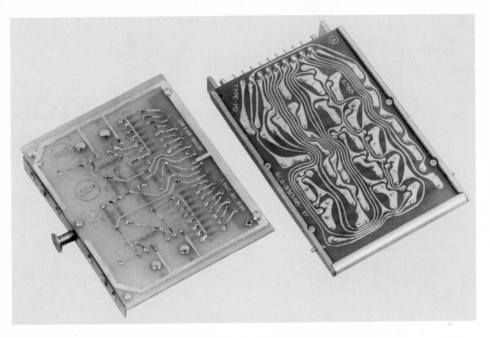

Fig. 3-3. Close-up of printed circuit boards reveals printed wiring usually made of copper or silver.

There *are* times when silver is used to make wire, such as the printed wire mentioned earlier. However, much more use is made of copper wire, while aluminum wire is used for high voltage lines that carry electricity long distances from the power plant to the users.

SOLID VS STRANDED WIRE

If you examine various pieces of wire, you will see how they differ. One piece may be a solid single wire. Another may be made of many fine strands of wire in the form of a braid, Fig. 3-5. Solid wire does not flex as readily as stranded wire, so it is used in places where, normally, the wire does not move. Stranded wire is used in applications where the wire has to flex.

Wire is used in thousands of different electric circuits. Solid wire, for example, completes the circuit to your lights, washer, dryer and in most circuits routed behind the plaster or dry wall in your home. Stranded wire will be found in lamp cords, toaster cords, and in the heavy-duty cord connected to the iron used to press your clothes. All of these electric cords are moved

CONDUCTOR	APPROXIMATE RESISTANCE*	TYPICAL APPLICATION
SILVER	9.8	ELECTRONIC CIRCUITS
COPPER	10.35	SOLID AND STRANDED WIRE
GOLD	14.6	SPECIALIZED ELECTRICAL INSTRUMENTS
ALUMINUM	17.0	LIGHTWEIGHT WIRE
NICKEL	52.0	RADIO TUBES
STEEL	100	TELEPHONE LINES
CONSTANTAN	295	HEAT MEASURING DEVICES
NICHROME	676	HEATER ELEMENTS
CARBON	2000 +	AUTOMOBILE WIRING

*Resistance is given in ohms per circular mil-foot at 25 deg. on Celsius scale. These terms will be explained as you progress through this and the next two chapters.

Fig. 3-4. Electrical conductors are listed in order of rising resistance. Typical applications are given for each conductor.

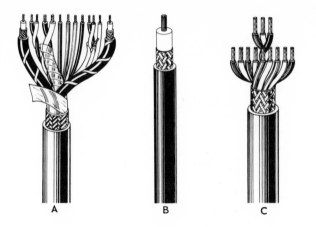

Fig. 3-5. Stranded wire with braided covering is used in communications field. A—TV camera extension cable. B—Mobile TV camera coaxial cable. C—Shielded remote control cable. (Belden Corp.)

frequently, and the stranded wire is what makes the moves possible. If solid wire was used, it would break due to the constant flexing.

WIRE COVERING

Wire is made in many different ways. In addition to being either solid or stranded, wire can be coated with varnish, left bare or covered with insulating material, and with a braided covering.

Coated wire usually is referred to as "magnet wire." It can be found in electromagnets, coils for transformers (see Chapter 11) and in the antennas of portable radios. Generally, the varnish coating must be removed when this type of wire is to be soldered.

Fig. 3-6. Data transmission cable contains 66 pairs of color-coded wires, built into a plastic jacket.

However, there is a coated wire available which does not require the protective coating to be stripped off before the wire can be soldered.

INSULATED WIRE

Insulated wire is used for lamp cords, house wiring and telephone cables. *By definition, a cable is an assembly of two or more wires inside a common covering.* Cables used in telephone circuits have color-coded insulation on the individual wires for ease in connecting or tracing wires. See Fig. 3-6.

Since many people are color blind, it is important to remember that certain electrical jobs require working with color-coded wires. Color coding is one reason why some companies insist on giving a color blindness test when they are interviewing people to fill new job openings. It would create a problem if color blind persons were hired as telephone installers or TV camera installers, because some cables have 20 or more color-coded conductors inside.

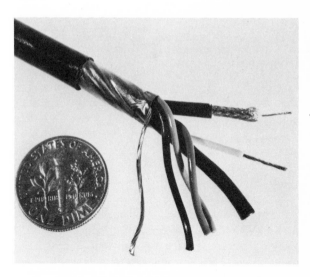

Fig. 3-7. This 6-conductor TV camera cable includes two coaxial cables with braided covering to shield against interference.

Wire with braided covering, Fig. 3-7, is used extensively in communications work. Applications include lead-in cables that connect antennas to TV sets, leads for electrical instruments and transmission line cables. This type of cable is used because the braided covering reduces interference problems in radio and television reception. It is much more expensive than other wire, but it is used because of its special "shielding" properties. For this reason, wire with braided covering is commonly called "shielded cable."

A coaxial cable, B in Fig. 3-5, utilizes two conductors having the same axis. The outer, braided conductor shields magnetic fields from the inner conductor.

STRIPPING WIRE

When working with insulated wire for electric circuits, it is necessary to remove the insulation before it can be soldered. Stripping usually is done with wire

Fig. 3-8. A wire stripper is a specialty hand tool used to remove insulation from an insulated wire.

strippers, Fig. 3-8, or it can be done with a knife. Use care, however, to insure that the wire is not nicked or strands broken off during the stripping, Fig. 3-9. When this happens, the size of the wire is reduced. This, in turn, reduces the number of electrons that can flow through the wire.

Fig. 3-10 shows what happens when broken strands reduce the size of a wire. In view A, for the sake of comparison, water hoses of two different sizes are shown. Note that as the size of the hose increases, the amount of water passing through also increases. The reverse is also true. When the size of the hose decreases, the amount of water also decreases.

Electron flow through two wires is shown in view B

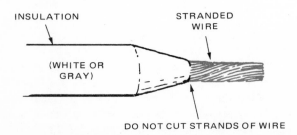

Fig. 3-9. Sketch shows an insulated wire that has been properly stripped by means of a knife.

in Fig. 3-10. One wire is whole and undamaged. The other is smaller because many strands are broken. *Note that the smaller the wire, the fewer electrons that can pass through.*

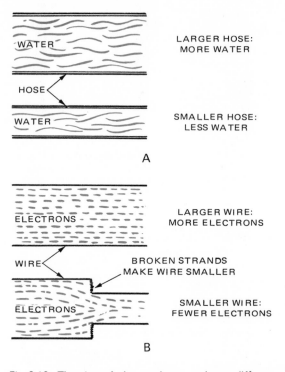

Fig. 3-10. The size of the conductor makes a difference in volume of flow, whether the product is water or electricity. A—Large and small hoses. B—Whole and damaged wires.

WIRE SIZE

Fig. 3-11 lists various sizes of wire. Note that the size is given by a number, ranging from 0000 to 40. This numbering system is known as the American Wire Gage (AWG). It is used to give specific information about the various sizes of wire.

Study the chart and you will see that the larger the wire number, the smaller the diameter of the wire. It is important to select the proper size of wire for the job. Otherwise, failure is possible and a fire might result.

I. C. SAYS:

"NEVER USE WATER ON AN ELECTRICAL FIRE. Use the correct type of fire extinguisher or common baking soda. Also be sure to disconnect the source of electricity powering the device that caused the fire."

DIMENSIONS AND RESISTANCES OF COPPER WIRE

GAGE NO. AWG OR B & S	DIAMETER OF BARE WIRE IN INCHES	DIAMETER IN MILS	AREA IN CIRCULAR MILS	RESISTANCE FEET PER ONE OHM AT 68 DEG. F
0000	.460	460.0	211600	20400
000	.410	409.6	167800	16180
00	.365	364.8	133100	12830
0	.325	324.9	105500	10180
1	.289	289.3	83690	8070
2	.258	257.6	66370	6400
3	.229	229.4	52640	5075
4	.204	204.3	41740	4025
5	.182	181.9	33100	3192
6	.162	162.0	26250	2531
7	.144	144.3	20820	2007
8	.129	128.5	16510	1592
9	.114	114.4	13090	1262
10	.102	101.9	10380	1001
11	.091	90.74	8234	794
12	.081	80.81	6530	629.6
13	.072	71.96	5178	499.3
14	.064	64.08	4107	396.0
15	.057	57.07	3257	314.0
16	.051	50.82	2583	249.0
17	.045	45.26	2048	197.5
18	.040	40.30	1624	156.5
19	.036	35.89	1288	124.2
20	.032	31.96	1022	98.5
21	.028	28.46	810.1	78.11
22	.025	25.35	642.4	61.95
23	.023	22.57	509.5	49.13
24	.020	20.10	404.0	38.96
25	.018	17.90	320.4	30.90
26	.016	15.94	254.1	24.50
27	.014	14.20	201.5	19.43
28	.013	12.64	159.8	15.41
29	.011	11.26	126.7	12.22
30	.010	10.03	100.5	9.691
31	.009	8.928	79.7	7.685
32	.008	7.950	63.21	6.095
33	.007	7.080	50.13	4.833
34	.006	6.305	39.75	3.833
35	.006	5.615	31.52	3.040
36	.005	5.000	25.00	2.411
37	.004	4.453	19.83	1.912
38	.004	3.965	15.72	1.516
39	.004	3.531	12.47	1.202
40	.003	3.145	9.888	0.9534

Fig. 3-11. Copper wire is charted by gage number, by diameter in inches and in mils and by area in circular mils. Resistance is shown in feet per ohm at 68 deg. F (20 deg. C).

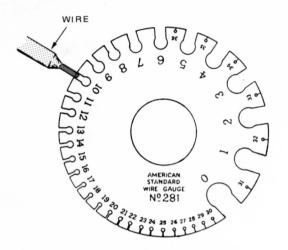

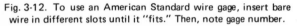

Fig. 3-12. To use an American Standard wire gage, insert bare wire in different slots until it "fits." Then, note gage number.

Other problems that result from the use of undersized wire are:
1. Toasters do not brown the bread.
2. Electric motors run too slow.

Fig. 3-12 shows how a wire gage is used to determine the size of the wire. The gage is easy to use. First remove the insulation from the wire. Then, hold the gage firmly and place the wire being measured between the slots. Check the size, then use the round hole at the bottom of the slot for easy removal of the wire. The numbers on the outside of the gage are the same numbers found in the table in Fig. 3-11.

INTERCOM PROJECT

You can build a simple intercom for use in your home. Use two permanent magnet speakers about 4 in. in diameter. Get speakers with large magnets. Place the speakers in separate rooms and connect them with two wires, Fig. 3-13.

Take turns either talking or listening. Both parties cannot speak at once. You will not be blasted by the volume from this simple intercom setup, but the results will surprise you. Note, however, that this intercom will not work in a noisy building.

AMPERES/AMPS

Earlier in this chapter, we stressed the importance of wire size as it relates to the number of electrons that can flow through the wire. Since the number of electrons is *so large,* when speaking in electrical terms, the word amperes or amps is used instead. A one amp reading, in fact, is another way of stating that

6,240,000,000,000,000,000 (6.24 x 10^{18}) electrons are flowing past a given point in one second, Fig. 3-14. The term "amp" is explained in detail in Chapter 5.

Changing the 6,240,000,000,000,000,000 electrons to 6.24 x 10^{18} is called "scientific notation." (See

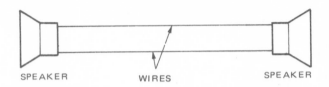

Fig. 3-13. This simple intercom project consist of two 4 in. diameter permanent magnet speakers and connecting wires.

reference section.) It is a way people working on electrical circuits can shorten the numbers to make them easier to use. The given value for one amp is a "standard" issued by the National Bureau of Standards.

You probably have seen the term "amps" printed on motors, electric frypans and other appliances around your home. There are two reasons why this is important. First, for safety reasons, each of the various

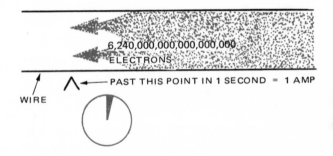

Fig. 3-14. The term amperes, or amps, is used to simplify, and it refers to how many electrons are flowing through a circuit.

appliances must have the proper size lead wire to prevent it from burning up. Second, by keeping a total of the number of amps used by each appliance every hour, the power company can send you the bill for the amount of money owed them for using the electricity.

Remember, the larger the wire, the smaller the AWG number, and the higher the number of amps that the wire will carry without overheating.

RESISTANCE WIRE

If you watch a toaster while it is in operation, you will notice that it gets so hot that the heating element inside turns red. If you look closely after it cools off, you will see that the part that was red is also made from wire. *Actually, this type of wire is made so that it will heat up. It is called "resistance wire."* It is used inside light bulbs, in heating elements for hair dryers and in electric space heaters.

Because resistance wire is designed to heat up, it is not made from copper. Instead, it is made from special

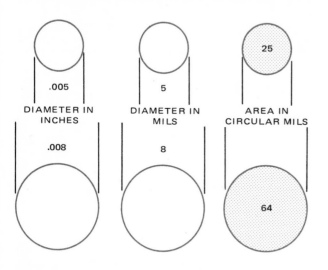

Fig. 3-15. A given size of wire is expressed in inches, mils and circular mils.

metals with names like Nichrome and Constantan. Although most wire is round, resistance wire can also be purchased in a ribbon shape. Ribbon shape resistance wire is what is commonly used in toasters.

CIRCULAR MILS

Round wire is measured in "circular mils." A mil is one thousandth of an inch (.001 in.). A circular mil is the cross-sectional area of a wire that has a diameter of one mil. Note in Fig. 3-15 that as the cross-sectional area (shaded area) gets larger, the number of circular mils is increased by the square of the diameter.

Circular mils are easier to visualize through an example. When you square a number, it means you must multiply the number by itself. Number 5 squared would be 5 x 5, which is equal to 25. Therefore, if you had a wire 5 mils in diameter, Fig. 3-15, the cross-sectional area of the wire would contain 25 circular mils. The second wire shown in Fig. 3-15 has a diameter of 8 mils. Its area then, would be 64 circular mils.

There is a good reason for converting the diameter of wire into circular mils. Working with the cross-sectional area of the wire makes it possible for you to select the right size of wire to safely carry the prescribed number of amps in the circuit.

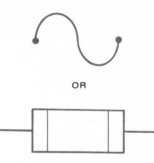

Fig. 3-16. Wiring diagrams and schematics use symbols to identify parts of circuits. Two different fuse symbols are shown.

FUSES

Most electric circuits are equipped with a safety device called a "fuse." It is designed to prevent fires caused by overloading the circuit with too many amps. Fuse symbols most commonly found on schematics are shown in Fig. 3-16.

A fuse is a conductor of electricity containing a thin element made of aluminum and other metals. When too many amps pass through the fuse, the thin element will overheat, then melt. This opens the circuit and stops

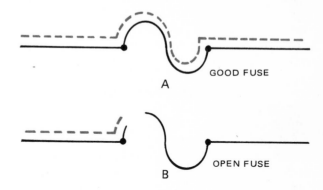

Fig. 3-17. The most popular fuse symbol is used to indicate a good fuse. An open fuse is shown with a break in the symbol. A—Electrons can flow. B—Electrons are stopped.

the electrons from flowing, Fig. 3-17. Fuses range in size from 1/100 amp to heavy-duty units able to carry several hundred amps.

The reason for using a fuse in the circuit is to keep the wires from burning up or having other parts of the circuit damaged. When the element does melt most people say, "The fuse has blown." When it does blow, the bad fuse must be replaced with a new one. It also is very important to find out *why* the fuse blew. Something must be wrong with the circuit, or the circuit is being overloaded.

FUSE TESTER PROJECT

Build the fuse tester shown in Figs. 3-18 and 3-19 by obtaining the following materials and assembling them according to the construction procedure given.

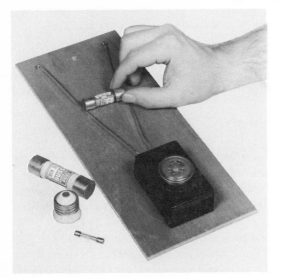

Fig. 3-18. Shop-built fuse tester emits a buzzing sound if the fuse is good. (Project by Allan Witherspoon)

No.	Item
1	1.5 to 3 volt dc alarm buzzer.
1	1 in. x 3 in. x 2 in. chassis box.
1	Penlight battery (cell).
1	1/4 in. x 6 in. x 14 in. plywood.
4	1/2 in. screw eyes.
1	Battery holder.
2	1 ft. lengths of stranded copper wire (uncoated).

CONSTRUCTION PROCEDURE

1. Cut a hole in chassis box and mount buzzer.
2. Install battery and mount chassis box at bottom of plywood.
3. Position screw eyes on plywood in a triangular shape.
4. Connect wires on screw eyes as shown in diagram.
5. Drill a 1/4 in. hole in plywood for hanging fuse tester on the wall.

To use the tester, Fig. 3-18, pick up a fuse and place it between the two wires on the plywood. Move the fuse inward until it makes contact with both wires. The buzzer will sound off if the fuse is good. If it fails to buzz, the fuse is blown.

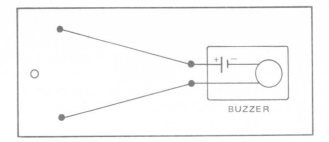

Fig. 3-19. Fuse tester schematic shows the arrangement of parts. See the text for construction details.

BLOWN FUSES

Blown fuses are shown in Fig. 3-20. The blackened section is the result of a short circuit. This condition will result if a large number of amps are surging through the circuit. When this happens, the thin metal strip is heated to a very high temperature and turns into a vapor, which discolors the fuse window.

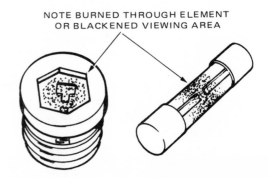

Fig. 3-20. When fuses of the snap-in or screw-in type blow, the element vaporizes.

The term, "short circuit" is explained in Fig. 3-21. Notice that a short circuit bypasses the load. This allows an extremely high number of amps to flow in the circuit, which is what the fuse is meant to stop.

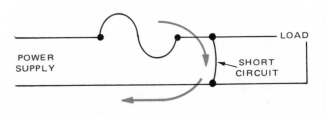

Fig. 3-21. A short circuit occurs when two bare wires come into contact and bypass the load, allowing a large number of amps to take a shortcut.

SLOW BLOWING FUSES

Some circuits normally have a large number of amps flowing for short periods of time. You probably have seen some lights in your home go dim when certain appliances (such as the refrigerator) first turn on. If you have an electric saw, you may notice the same dimming effect when the saw is turned on, or if the saw slows down while cutting a large piece of wood. At times like this, you do not want the fuse to blow if the overload is only going to last a few seconds. To prevent this, a slow blowing fuse, Fig. 3-22, is made by some manufac-

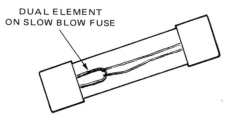

Fig. 3-22. Slow blowing fuses generally have a dual element to help withstand a brief overload.

turers. Note that the dual element pictured is heavier at one end. Some types of slow blowing fuses are designed to withstand an overload for 12 to 25 seconds without blowing.

Fig. 3-23 illustrates some of the many fuses that are

Fig. 3-23. Fuses are designed for every electrical protection need. (Bussmann Mfg. Div., McGraw-Edison Co.)

manufactured. Fig. 3-24 pictures a fuse holder with a fuse in place to protect the circuit from being overloaded. This unit is designed for easy removal and replacement of the blown fuse. Many automobiles are equipped with this type of fuse holder.

Fig. 3-24. Fuse holders usually are placed for easy access and designed for quick fuse replacement.

The electric circuits in many older homes are protected with fuses. In most newer homes, however, the circuits are protected with circuit breakers, Fig. 3-25. A circuit breaker has a distinct advantage over a fuse. When an overload occurs, the circuit breaker "trips," and the circuit is opened. *Unlike a fuse, the circuit breaker does not have to be replaced after it trips.* It works like a switch. If an overload trips it, all you have to do is turn it off, then back on and the circuit is completed again.

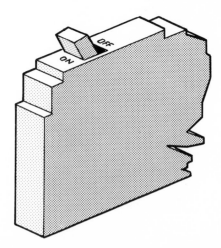

Fig. 3-25. A circuit breaker will open automatically when a current overload flows through it.

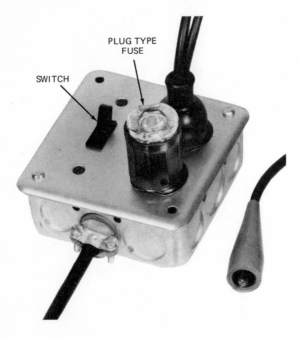

Fig. 3-26. This portable receptacle is fitted with a toggle switch and plug type fuse.

SWITCHES

As mentioned in Chapter 1, switches are used to open and close circuits without having to go through the inconvenience of removing and attaching wires. Fig.

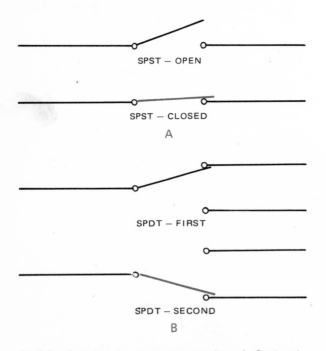

Fig. 3-27. Symbols for single-pole switches. A—Single-pole single-throw switch in open/closed position. B—Single-pole double-throw switch in first/second positions.

3-26 shows one of the common types of switches, along with a plug type fuse. The purpose of the glass top is to help prevent shocks when you change a blown fuse.

The symbols for various switches are shown in Figs. 3-27 and 3-28. The letters SPST stand for single-pole

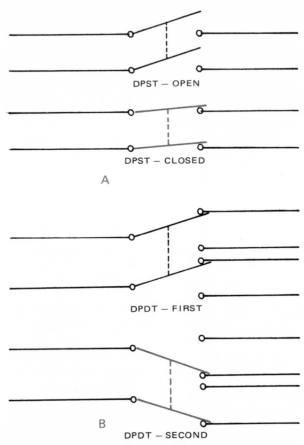

Fig. 3-28. Symbols for double-pole switches. A—Double-pole single-throw switch in open/closed positions. B—Double-pole double-throw switch in first/second positions.

single-throw; SPDT is single-pole double-throw; DPST is double-pole single-throw; DPDT is double-pole double-throw. To make it easier to understand, each switch is shown in both the normal position and in the second position after the switch is thrown.

Switches are wired into circuits so that one or two circuits are completed by closing (throwing) the switch. When the circuit is open, no electrons can flow and the switch is said to be OFF. When the circuit is closed, electrons can flow and the circuit is said to be ON.

The number of switches available ranges in the hundreds. To mention a few, there are: slide switches, rotary switches (Fig. 3-29), plunger switches, push button switches (Fig. 3-30) and roller arm switches.

Fig. 3-29. Rotary switches come in many styles and shapes, including these open frame types. (Centralab Electronics Div. of Globe-Union Inc.)

Some switches will indicate that a piece of magnetic material is close to it. These are known as "magnet actuated proximity switches."

AUTO BURGLAR ALARM PROJECT

You can put a switch to good use in building an auto burglar alarm. First look at the horn circuit on your car, Fig. 3-31. When you push the horn button or press the horn ring, the horn "blows." When you release the pressure, the blowing stops. You can use the same sound to scare away a car thief. Do this by wiring a switch into the horn circuit as shown in Fig. 3-32.

When the switch is installed and the car is in use, keep the burglar alarm switch in the open position. When you leave your car, close the switch and hide it where only you can find it. As soon as someone starts the engine, the horn will start blowing and frighten the would-be thief away.

When *you* want to start the engine of your car, turn off the alarm switch. Always keep this in mind or the blasting sound of the horn will surprise you, too.

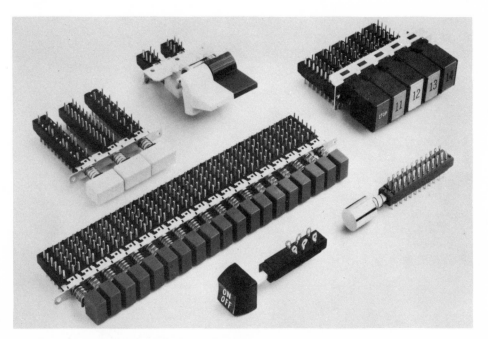

Fig. 3-30. Lighted and unlighted push button switches are used in many applications such as car radios, telephone, intercoms and elevators in buildings. (Centralab Electronics Div. of Globe-Union Inc.)

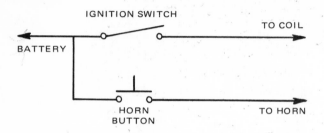

Fig. 3-31. This schematic shows a simplified circuit for the horn on a passenger car.

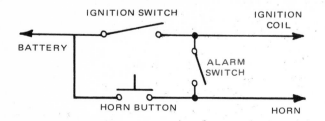

Fig. 3-32. Install a burglar alarm switch into the horn circuit as shown. When the ignition switch is turned on (closed), the alarm will sound off.

INSULATORS

As mentioned earlier, insulation is a material used to cover wire. Wires are insulated to keep them from touching each other or contacting other conductors. If they did touch, it would cause a bright spark called an "arc." Anytime there is an arc, there is danger that a fire will start or a fuse will blow. Therefore, for protection and safety, wire is manufactured with an insulating cover molded on it.

Insulation comes in white, black, red, green, blue, yellow and many other colors, combinations of colors and striped patterns. Insulation also prevents any person touching the wire from getting a shock. Imagine plugging in an uninsulated radio lead and receiving a shock or bad burn. Usually, the plug is molded from an insulating material, which offers a high resistance to the flow of electrons.

The best insulators are rubber, shellac, glass, mica, plastics and paper. In Fig. 3-33, the plastic ends (connectors) on the wires in the instrument panel module (unit) are insulators. The electrical tape and plastic shields are also forms of insulators. When a large number of wires are taped together, or encased in protective tubing, the assembly is called a "wiring harness." Fig. 3-34 shows a few of the many insulators you will find being used in conjunction with electric circuits.

SOLDERING

Many times, it will be necessary to connect wires or attach them to an electrical part. Usually, this is done by soldering the connection. *Solder is a metal mixture used to connect metal parts or components.* Generally, the mixture is made of tin and lead. For example, a mixture of 60 percent tin and 40 percent lead is known

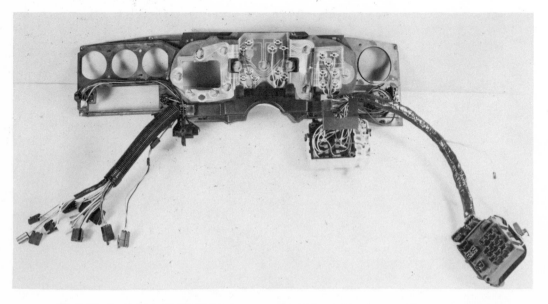

Fig. 3-33. Car instrument panel circuits are combined in a wiring harness. They are protected from damage by arcing by various types of insulators.
(Packard Electric Div., General Motors Corp.)

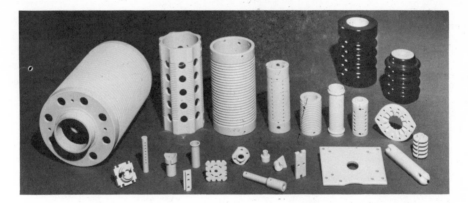

Fig. 3-34. Insulators for electric circuits come in many shapes and forms. Insulators shown are termed "structural ceramics" by professionals in the field.
(Centralab Electronics Div. of Globe-Union Inc.)

as 60/40 solder. This 60/40 mixture is used by most people because it melts at a lower temperature than other types (70/30 or 40/60).

I. C. SAYS:

"Here is a word of warning. Solder will melt at 370 deg. F or higher. At these temperatures, you can get a bad burn. Do not let any of the melted solder touch your skin. Also, do not touch the soldering iron or gun used to melt the solder."

Typical soldering irons are shown in Figs. 3-35 and 3-36. The proper way to use a soldering iron is shown in Fig. 3-37. By using it this way, you can avoid some serious burns.

Fig. 3-35. This soldering iron is a 25 watt pencil model.
(Heath/Weller)

STEPS IN SOLDERING

To make a satisfactory solder joint, several important steps are necessary. First, strip the insulation or coating from the ends of the wire to be soldered. Next, make a strong mechanical connection to insure good electrical flow after solder has been placed on the

joint. If the joint is not mechanically tight, you will get poor electrical flow because solder is not as good a conductor as copper.

Fig. 3-36. This soldering iron kit provides a built-in heat source. Note heat control knob. (Heath/General Electric)

Then, use a soldering iron Fig. 3-37, to heat the mechanical connection on one side while applying the solder from the other side. Do not put the solder directly on the soldering iron; a poor solder joint may result. By heating one side, the connection will get hot enough to melt the solder on the other side. Use only enough solder to flow over all the surfaces and wires at the connection. Too much solder will result in poor appearance, and solder may drip into places where it is not wanted.

When enough solder has melted on the connection, remove the iron. Do not move the parts being soldered until the solder has cooled. If you accidentally move the parts while the connection is cooling, apply the iron

Fig. 3-37. The correct soldering technique is to heat one side of the connection while applying solder from the other side.

and reheat the joint. This technique will insure a good solder connection. The finished soldering joint should be completely covered by the flowing solder, bright and smooth, with all wires firmly held in place. Study the soldering sequence in Figs. 3-38 through 3-41.

Fig. 3-38. The first step in soldering is to heat the joint.

Fig. 3-39. The second step in soldering is to apply the solder.

Fig. 3-40. The third step in soldering is to remove the iron and allow the solder to cool.

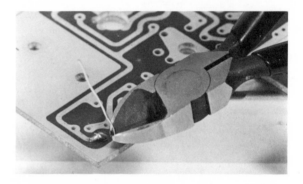

Fig. 3-41. The fourth and final step in soldering is to cut off excess wire.

If the solder is dull looking or if it has lumps or small beads of solder, the job is poorly done. This is known as a "cold solder joint," which is the mark of someone who has not learned the correct way to solder. When you know how, even printed wiring, Fig. 3-42 can be repaired with solder. Small breaks or cracks can be fixed by heating the bad spot and applying a small amount of solder.

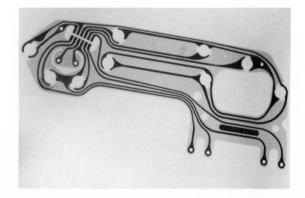

Fig. 3-42. Small breaks in a printed wiring circuit can be soldered. The circuit shown is for a car instrument panel. (Packard Electric Div., General Motors Corp.)

TYPES OF SOLDER

Most hardware and building supply stores carry two types of solder, rosin core and acid cord. *Use only rosin core solder when doing electrical work.* Acid core solder is used for sheet metal work. If it is used for electrical work, it will eat through connections and damage parts.

Rosin core solder means that the solder has a small opening in the center, filled with special material called "flux." Flux is a general term for materials used to remove impurities (called "oxides") from the surfaces being joined by the solder. The clean surfaces permit the solder to stick to the connection when it reaches the correct temperature.

HEAT SINKS

Items like transistors, diodes and even some resistors (parts to be described in detail later) can be damaged from the heat created by the soldering iron. Therefore, when soldering near delicate electrical parts, protect them with a heat sink. A heat sink is a better conductor of heat than the electrical part, so it will pick up the heat and stop it from going to the nearby part.

The heat sink shown in Fig. 3-43 is made from an alligator clip and a piece of copper. One end of the

Fig. 3-43. A satisfactory heat sink can be made from an alligator clip and a piece of copper. (Project by Ron Simons)

copper is cut to fit the clip. The other end is broad and flat to give off absorbed heat. This heat sink is easy to solder together and convenient to store for use when needed. When preparing to solder electrical work, place the heat sink between the electrical part to be protected and the joint to be soldered, Fig. 3-44. Needle nose pliers and other metal objects also can be used as heat sinks.

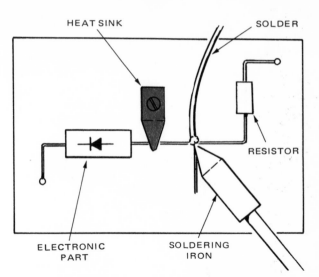

Fig. 3-44. The heat sink is a means of attracting and giving off heat to protect delicate electrical parts from heat damage.

POOR SOLDER JOINTS

Check these common reasons why some solder joints are unsatisfactory:
1. Not heating connection to proper temperature so solder can flow.
2. Applying solder to iron instead of on joint opposite iron.
3. Using too much solder.

The only way to learn to solder is to do it. Learning to solder is like learning to ride a bicycle. In both cases, the only way to learn is by trying.

PROTECTING THE WIRE

Cables and wires have to be protected in some job locations. Often, where there is danger of having the

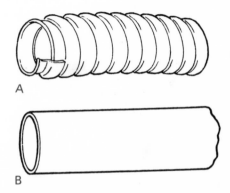

Fig. 3-45. Wires between switches, receptacles and electrical loads must be installed in protective tubing. A—Flexible steel conduit. B—Thinwall conduit.

insulation worn away, cut or scraped, the wire is protected by placing it in a hollow pipe called "conduit." In some applications, it is protected by flexible steel conduit, or "greenfield." Both types of protection are shown in Fig. 3-45.

Another covering used to protect wire is called "spaghetti," Fig. 3-46. It is a hollow insulating material

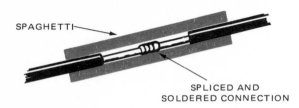

Fig. 3-46. "Spaghetti" is a common term used for a hollow insulating material used to cover spliced and soldered wires.

similar to a plastic straw. When splicing two pieces of wire, remove the insulation from both ends. Slide a piece of spaghetti over one of the ends. Then, connect the two pieces of wire and solder them together. Next, slide the spaghetti over the completed connection so that it completely covers the bare ends of both wires. This method of protection will prevent bare wire from coming into contact with other wires or metal parts, causing an arc or giving someone an electric shock.

Another material, called "heat shrink," also can be used to protect wires. It looks much like spaghetti. With heat shrink, however, heat is applied to it after assembly and the material actually gets smaller in size. This shrinking action makes the covering fit more tightly on the wire.

ELECTRICAL CODES

Guidelines for proper routing of wires, and other safe practices, vary from city to city, depending on local electrical codes. The National Electrical Code contains the general rules and regulations that contractors, builders and manufacturers must follow. Local codes supplement (add to) National Code requirements. These codes have been established to protect those people who are not familiar with some of the dangers of electricity. City and state building inspectors visit new buildings under construction to make sure that codes have been followed.

If you are becoming more interested in the electrical field, get yourself a small set of tools, Fig. 3-47, and a heat sink for project work and other supervised activities. With the kit shown, you should be able to do the soldering and handle electrical work involved with small nuts, screws and other fasteners.

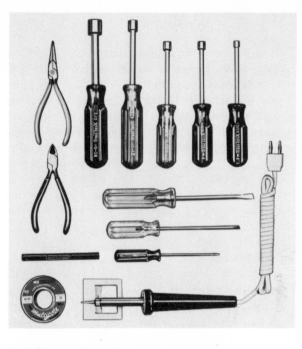

Fig. 3-47. A basic set of tools for simple electrical work should include pliers, screwdrivers, nut drivers, soldering iron and a supply of solder.   (Heath Co.)

TEST YOUR KNOWLEDGE

1. The purpose of soldering in electrical circuits is to _____. *metal parts or compone*
2. Rosin-core solder should be used for all electrical work. True or False?
3. What two precautions should you observe when soldering?
4. Three good materials for conducting electricity are *Copper, Silver* and *Aluminin*
5. Name three commonly used insulator materials.
6. Which size wire is larger, No. 12 or No. 16?
7. Describe how a wire should fit into the slot on the American Wire Gage to determine its size.
8. Name two tools used to remove insulation from wires. *Knife, wire stripper*
9. How is a circuit breaker reset after it has opened the circuit?
10. What kind of wire is used in heating elements?
11. List three nonmagnetic materials.
12. Acid core solder should not be used for _____ work.
13. Draw the schematic for a circuit to show how a single-pole, single-throw switch can be used to control a 6 volt light bulb connected to a storage battery.
14. Draw a circuit for a continuity tester.

SOMETHING TO DO

1. Why are the ends of stranded wire twisted together and tinned before it is used?
2. List the different mixtures of solder and find the temperature where each will melt.
3. Many companies use a method of soldering which does not use soldering irons or guns. Find out what they call this type of soldering and how it is done.
4. There are many different types of insulation used on wire. Make a list of the different kinds and their colors.
5. How is insulation put on wire?
6. How is the rosin flux put into the core of the solder? How do the companies make the solder with a hollow center?
7. List the different types of fuses you can find, then determine how much current they will carry without blowing.
8. How is stranded wire made?
9. Find out what types of tests are given to see if a person is color blind.
10. What is flux and how is it made?
11. Where does copper come from? Make a map of the different places where it is found.

This cable winding operation calls for the use of giant spools rotated by large electric motors. (Western Electric Co.)

Chapter 4

RESISTORS AND CAPACITORS

In Chapter 3, you studied electrical conductors and insulators. You found that good conductors are made from various materials that allow electrons to flow easily. Conductors are used to form a path or circuit for electrons to flow from source to load and back again.

Insulators, on the other hand, are *very* poor conductors of electricity. They are made from materials that offer strong resistance to the flow of electrons. Because of this non-conducting characteristic, insulators are used to contain the flow of electricity in planned circuits and to prevent severe electrical shock when you touch wires, plugs or switches.

RESISTORS

Conductors, then, allow electrons to flow. Insulators do not. In between these two extremes, we have other materials that conduct electricity but make it difficult for electrons to flow. These are called "resistors." One familiar type of resistor is a resistance wire, which offers so much resistance that it turns bright red and gives off enough heat to make toast, heat rooms and dry people's hair.

The resistance of a given material to electron flow is based on a comparison with copper. See the table in Fig. 3-4 in the last chapter, to review this comparison. Silver, for example, is listed higher on the table than copper because it has lower resistance and is a better conductor. Aluminum, nichrome and carbon have higher resistance, so they are poorer conductors.

Since carbon resists the flow of electrons much more than copper, carbon is often used in making resistors. The material, size and makeup of the resistor determines its resistance value, which is measured in ohms.

Resistors can be purchased in any one of hundreds of different values, ranging from about 1 ohm to over 22 million ohms. Fig. 4-1 shows some of the common types of resistors that are used in electrical circuits.

RESISTOR SYMBOLS

The symbols for a resistor (or resistance) are shown in Fig. 4-2. Do not confuse the second symbol in this illustration with the symbol for a fuse. Note that there are no lines in the box in the resistor symbol as there are in the fuse symbol.

Fig. 4-1. Many resistors are used in this electrical circuit. Note identifying stripes on body of each resistor. (Heath Co.)

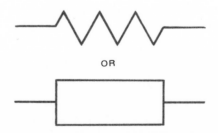

OR

Fig. 4-2. These two symbols are the most popular resistor symbols used in schematics.

In some schematics, you will see what looks like an upside down horseshoe (Ω) being used instead of the word "ohm." This symbol is the Greek letter "omega." In other circuits, you may find the letter "R" being used. This letter helps locate the place to put the resistor on the circuit board when you assemble the kit. In Fig. 4-3, for example, note R 29.

Later in this text, we will see still other symbols and letters being used in drawings to represent some common electrical components.

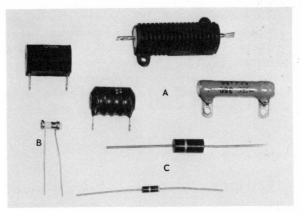

Fig. 4-4. Fixed resistors. A—Wirewound. B—Thin Film. C—Composition.

and you will find that it is made of a carbon material and a "binder" that holds the carbon together. Note also that the wire leads at either end of the resistor do not go all the way through. They stop soon after entering the body of the resistor. The different numerical values of resistors depend on the amounts of carbon and binder used in their construction.

Carbon resistors come in many different sizes. Those most commonly used range from approximately 3/32 in. to about 5/16 in. in diameter. (Wirewound and thin film types will have still different sizes.)

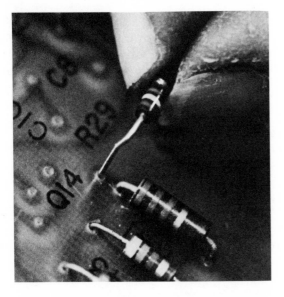

Fig. 4-3. Installing resistors on a coded circuit board. (Acme Electric Corp.)

TYPES OF RESISTORS

Resistors are of two types, either fixed or variable. A fixed resistor has one specific value, such as 150 ohms. A variable resistor has a range of values or it can be adjusted to the resistance needed for a specific circuit.

FIXED RESISTORS

Fixed resistors vary in makeup. They may be molded composition, metal film, carbon film or wirewound. Fig. 4-4 shows various types of fixed resistors that are available. The big difference between resistors lies in the heat they create and the heat they can safely get rid of to the air around them. Therefore, the amount of heat in the area where they are being used and the humidity of the area, are two typical problems which a designer must consider when drawing a circuit that uses resistors.

A molded composition resistor is often referred to as a carbon resistor. Break apart a composition resistor

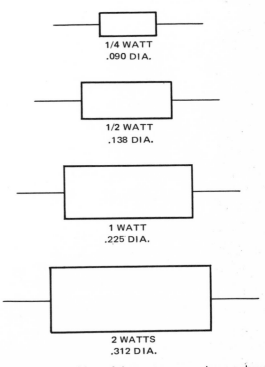

Fig. 4-5. Resistors of four of the most common sizes are shown by wattage and diameter.

The different sizes of carbon resistors, Fig. 4-5, also tell the amount of heat that a given resistor can transfer or dissipate to the outside air. Carbon resistors are rated at 1/8, 1/4, 1/2, 1 and 2 watts.

The term "watts" will be described in detail in Chapter 5. For the time being, just think of this term as the amount of heat the resistor can stand without being destroyed.

THIN FILM RESISTORS

Thin film resistors are sort of a compromise between carbon resistors and wirewound resistors. The primary differences are in size and cost. Fig. 4-6 shows how thin film resistors are constructed. The substrate shown is an insulating material something like ceramic.

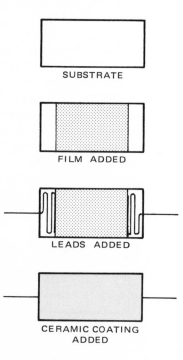

Fig. 4-6. Four steps in the construction of thin film resistors.

WIREWOUND RESISTORS

Wirewound resistors, Fig. 4-7, are built in a different manner. They are constructed by winding a long length of wire around a core. The length of wire, its diameter, and material from which it is made determine the value of the resistor.

After the core is wound, a ceramic insulating coating is applied to the body of the resistor. This coating protects the wires when the resistor is put to use. Each end of the winding is attached to a terminal used to install the resistor in the circuit.

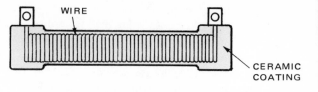

Fig. 4-7. A wirewound resistor is made up of a core wound with a continuous piece of wire and coated with ceramic insulation.

Wirewound resistors are used in circuits where large amounts of heat must be dissipated (given off) to the surrounding air. As a result they usually are made much larger than carbon resistors. Some wirewound resistors are over 1 in. in diameter and 10 in. long. Power ratings range from about 3 watts to over 200 watts; much higher than carbon resistors. For this reason, they are much more expensive than the smaller carbon types.

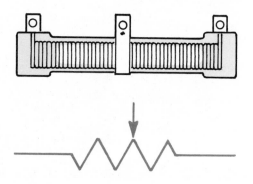

Fig. 4-8. An adjustable wirewound resistor and its symbol.

ADJUSTABLE WIREWOUND RESISTORS

Some wirewound resistors are adjustable. This type of resistor is not completely covered with the ceramic material that protects the windings. As you can see in Fig. 4-8, these resistors have a small portion of one side exposed. A slider arm is mounted across the exposed portion as a means of adjustment.

To make a resistance adjustment, move the slider arm along the resistor, then clamp it in a given position. By changing the slider arm location, different amounts of resistance can be obtained. This type of resistor is not designed to be adjusted very often. If repeated adjustments were made, the resistor would soon wear out from the action of the slider arm on the exposed copper.

VARIABLE RESISTORS

Variable resistors, Fig. 4-9, are used in applications where it is necessary to make frequent adjustments of resistance values. Some common usages are as follows:

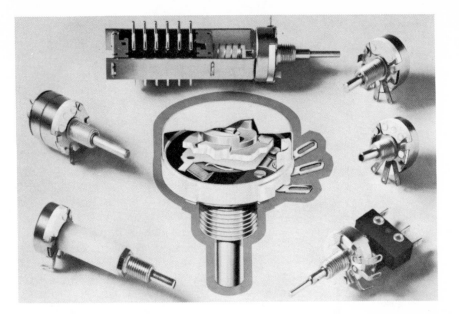

Fig. 4-9. Variable resistors fill a variety of electrical needs. (Centralab Electronics Div. of Globe-Union Inc.)

1. When you adjust the volume of your stereo or television set.
2. When you change the speed of your car windshield wipers.
3. When you dim lights in your house.
4. When you change the speed of variable speed drill in your home workshop.

In each case, you are adjusting a variable resistor. Remember, too, that this type of resistor will do many other things for you, depending on how it is used.

POTENTIOMETER

Variable resistors are also known as potentiometers, rheostats and trimmer resistors, Fig. 4-10. These names refer to some of the many uses found for variable resistors. A potentiometer, for example, allows the electrons to have two paths: one through part of the resistance, then to the load; the second path goes only through the resistance, Fig. 4-11. The rheostat allows only one path.

Fig. 4-10. Potentiometers and trimmer resistors are popular forms of variable resistors.

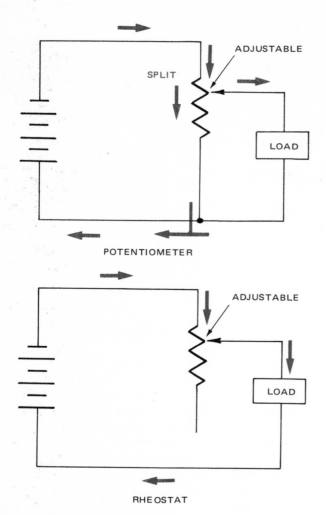

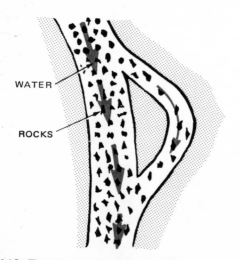

When there is more than one path for electrons, you have what is called a "parallel circuit." This will be explained more fully when you get to Chapter 7. At this point, you should know that if more than one path is present, some of the electrons will travel down each path. The amount depends on the size of the path and how much resistance is in the path, Fig. 4-12.

VARIABLE NEON VOLTMETER PROJECT

Here is a small voltmeter, Fig. 4-13, which you can use to find the exact circuit voltage. After you have built the circuit (inset) and mounted it in a box, you can calibrate the instrument. Connect the test leads to a variable power supply. *Do not* exceed 600 volts. Turn the dial on the potentiometer to the point where the light goes out. Mark this voltage value on the front of

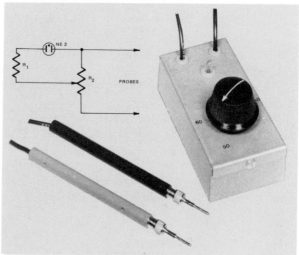

Fig. 4-13. A variable neon voltmeter is pictured. Note the inset schematic showing circuit. NE 2-Bulb. R_1-200K resistor. R_2-300K potentiometer. (Project by Allan Witherspoon)

your voltmeter box. Keep changing the variable power supply and making this test until you have all of the values you want marked on the front of your voltmeter. Be sure to use grommets where the wires come through the metal box.

RHEOSTAT

The rheostat is a device that just adds resistance to the circuit. It serves to put greater restriction in the path of the electrons, Fig. 4-14. The further the electrons must travel through the rheostat's resistance before getting to the load, the more resistance the electrons must overcome. So, the more resistance you

Fig. 4-11. Arrows indicate path of electron movement in two similar circuits. Note there are two paths in circuit with potentiometer and one path in circuit with rheostat.

Fig. 4-12. The principle of splitting the path of electron flow is illustrated by water flowing in the mainstream and in a branch. Rocks show resistance to flow.

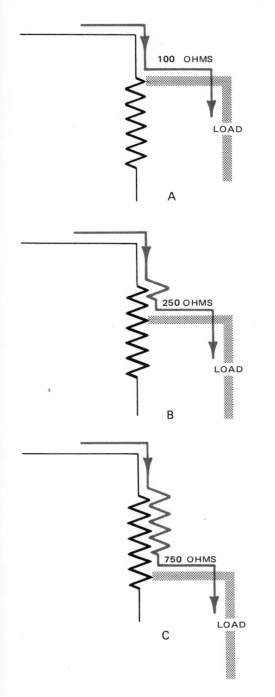

Fig. 4-14. A rheostat adds resistance to the circuit. A—Easy flow of electrons. B—More difficult flow of electrons. C—Very difficult flow of electrons.

want in the circuit, the further you have to turn the movable arm of the variable resistor.

TRIMMER RESISTORS

A trimmer resistor, Fig. 4-15, is a variable resistor that is similar to a potentiometer in construction. It is

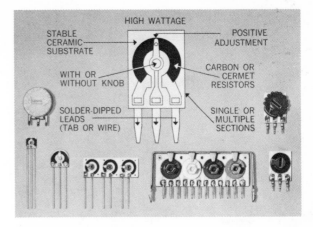

Fig. 4-15. Trimmer resistors used in fine tuning applications. (Centralab Electronics Div. of Globe-Union Inc.)

used to make very small changes in the resistance of a circuit. Fine tuning of certain circuits in television sets, stereos and other communication devices are typical applications of trimmer resistors.

VALUES

Resistors are included in most circuits as a part of the design. They come in many different shapes and sizes. Generally, to help you identify a given resistor, a number is printed on its body, telling you its value. You will see numbers like 5 ohms or 450 ohms. Values such as 3.8k ohms or 3 meg ohms are a short method of labeling the size of the resistor without having to write out all the zeros.

The letter k stands for kilo, which is the same as *thousand*. A 3.8k resistor, for example, would have a value of 3800 ohms. Meg is another way of writing *million*. Therefore, a 3 meg resistor would have the value of 3,000,000 or 3 million ohms.

Pursuing this method of marking resistors further, what is the value of a 2.7k resistor? This would be 2700 ohms. How about 7.9 meg? This resistor would have a value of 7,900,000 ohms or 7.9 million ohms. The term meg and kilo are very commonly used with resistors and for sizing other electrical components, which you will learn about later.

COLOR CODE

Some resistors are color coded with rings or bands to indicate their size, Fig. 4-16. Most resistors have four bands, while others have five. At this point, an explanation of how to read these bands is in order. First, note in Fig. 4-16, that the bands are closer to one end of the resistor. Therefore, when you read its value,

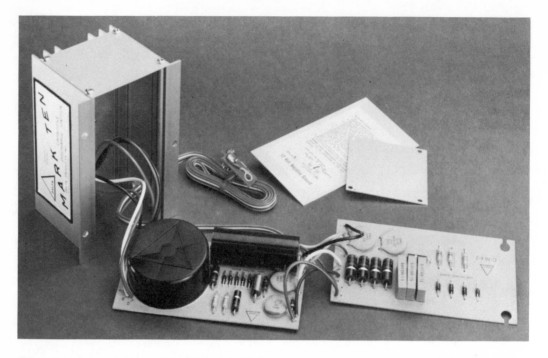

Fig. 4-16. An automotive capacitive discharge ignition system uses a number of resistors. (Delta Products, Inc.)

always place the resistor so the bands are to your left, Fig. 4-17. The value placed on the resistor depends on the order in which you read the colors.

I.C.SAYS:

"When using a carbon resistor, it may be soldered into the circuit in either direction. We will discover later that many other components can be mounted only one way. If they are installed any other way, they might be destroyed."

Next, we will get down to the business of *reading* the code. Basically, each color represents a number. The colors used for the first three bands are:

COLOR	NUMBER IT REPRESENTS
Black	0
Brown	1
Red	2
Orange	3
Yellow	4
Green	5
Blue	6
Violet	7
Gray	8
White	9

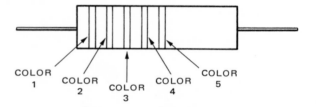

Fig. 4-17. The right way to read the color bands on a resistor is to position the resistor with the bands at your left.

To make a reading, list each of the colors used in the three bands. Then, list their respective numbers as follows.

Red – 2   Green – 5   Orange – 3

Write the first two numbers in order. In this case, the numbers would be 25. The third color tells you the number of zeros to add. Therefore, we would have a resistor with a value of 25 with 3 zeros or 25,000 ohms.

If the color bands are brown, yellow and red, write the code as follows:

Brown – 1   Yellow – 4 ₒₒ   Red – 2

This indicates a resistor value of 14 with 2 zeros or 1400 ohms.

If the color bands are red, violet and blue, write the code as follows:

Red – 2   Violet – 7   Blue – 6

This indicates a resistor value of 27 with 6 zeros or 27,000,000 ohms.

The values of these three resistors could be written as follows:

$$25,000 \text{ ohms} = 25 \text{ k ohms}$$
$$1400 \text{ ohms} = 1.4 \text{ k ohms}$$
$$27,000,000 \text{ ohms} = 27 \text{ meg ohms}$$

The fourth band, if used, will either be gold or silver. This band tells how close the rated value of the resistor is to the actual value. Usually, the resistor is coded to some percentage, (range of values) specified by the manufacturer. This is known as its tolerance.

The reason for having a tolerance on resistors is because of problems in controlling the exact amount of carbon and binder used to make each resistor. When making a "batch," a very small change in amounts called for will result in a group of resistors that are not all the same exact value. Therefore, the manufacturer will give their value within a range or tolerance from one value.

Possible resistor tolerances are:

Gold	Plus or minus 5 percent ($\pm$ 5%)
Silver	Plus or minus 10 percent ($\pm$ 10%)
No 4th band	Plus or minus 20 percent ($\pm$ 20%)

Using our previous examples, assume that the 1400 ohm resistor had a gold fourth band. Its value and tolerance would be:

1400 ohms plus or minus 5 percent ($\pm$ 5%)
5 percent of 1400 ohms is
1400 ohms x .05 = 70 ohms

Therefore, this resistor would have a value of 1400 $\pm$ 70 ohms. This would mean that if you measured the resistor with a meter that showed values in ohms, the value could go as high as 1470 ohms or as low as 1330 ohms and still be within the manufacturer's tolerance. This is possible because:

1400 ohms + 70 ohms = 1470 ohms
1400 ohms − 70 ohms = 1330 ohms

A 25,000 ohm resistor with a silver fourth band would be:

25,000 ohms plus or minus 10 percent ($\pm$ 10%)
10 percent of 25,000 ohms is
25,000 ohms x .10 = 2500 ohms

Therefore, its range would be:

25,000 ohms $\pm$ 2500 ohms or from
27,500 ohms to 22,500 ohms

A 27,000,000 ohm resistor with no fourth band would be:

27,000,000 ohms plus or minus 20 percent ($\pm$ 20%)
20 percent of 27,000,000 ohms is
27,000,000 ohms x .20 = 5,400,000 ohms

Therefore, its range would be:

27,000,000 ohms $\pm$ 5,400,000 ohms or from
32,400,000 ohms to 21,600,000 ohms.

Another way of stating this range would be, "Between 32.4 meg ohms and 21.6 meg ohms." This is the manufacturer's tolerance for a 27 meg ohm resistor with no fourth band.

Occasionally, you will find a gold or silver color on the third band of a carbon resistor. When you do, write down the first two numbers based on the color code, then divide by 10 for a gold band or by 100 for a silver band. For example, if you have a resistor with stripes of red, green and gold, you would have:

Red − 2   Green − 5   Gold − Divide by 10

This resistor would have a value of 25 divided by 10, which equals 2.5 ohms.

A resistor with colors of red, green and a silver third band would be:

Red − 2   Green − 5   Silver - Divide by 100

This resistor would have a value of 25 divided by 100 which equals .25 ohms. Resistors of this size will not be found in many circuits.

FIFTH BAND

Earlier, we mentioned the possibility of a resistor having a fifth band. When such a resistor is found, it also will be color coded. The fifth band is used to predict the percentage of failure per thousand hours of use. It is based on tests which the manufacturer has conducted on similar resistors. The coding is:

Brown	−	1.0 percent
Red	−	0.1 percent
Orange	−	0.01 percent
Yellow	−	0.001 percent

For example, if a group of resistors all had a brown fifth band, only one out of each one hundred (1.0 percent) would be likely to fail during one thousand hours of use. Reading a resistor is similar to learning to multiply. The more you do it, the easier it becomes. To sum up the entire resistor code, look at Fig. 4-18.

SHELF LIFE

Resistors tend to heat up when they are used in a circuit. Therefore, when reusing a resistor, make sure it

RESISTOR VALUES

COLOR	1st & 2nd BAND DIGITS	3rd BAND MULTIPLY BY	4th BAND TOLERANCE ± %	5th BAND FAILURE RATE LEVEL %/ THOUSAND HOURS USE
BLACK	0	1	—	
BROWN	1	10	—	1.0
RED	2	100	—	0.1
ORANGE	3	1000	—	0.01
YELLOW	4	10000	—	0.001
GREEN	5	100000	—	
BLUE	6	1000000	—	
VIOLET	7	10000000	—	
GRAY	8	100000000	—	
WHITE	9	1000000000	—	
GOLD	—	0.1	5	
SILVER	—	0.01	10	
NO COLOR	—	—	20	

Fig. 4-18. Table shows resistor color code and tolerances.

is still good. This can be done by using an ohmmeter to measure the resistance. This type of measuring instrument will be described in detail when you get to Chapter 8.

Resistors can be stored for long periods of time without going bad. If they have been on a shelf and never used in a circuit, they can be considered as good as new. Some electrical devices should be used as soon as possible after they have been purchased. Batteries, for example, will age when stored on a shelf.

TOLERANCES

Some popular resistors are made in large quantities to cover the full range for each tolerance. If you want to buy resistors with a 20 percent tolerance, you can cover the range with fewer sizes.

Note in Fig. 4-19 that a 10 ohm resistor with 20 percent tolerance would range from 8 to 12 ohms; a 15 ohm resistor, 12 to 18 ohms; a 22 ohm resistor, 17.6 to 26.4 ohms. Because of the overlap in their ranges, it is not necessary to get resistors with values in between the ranges.

When a 10 percent tolerance is required, the number of resistors needed to fill the range will increase. See

20% TOLERANCE RESISTORS

RESISTOR SIZE IN OHMS	TOLERANCE RANGE
10	8 — 12
15	12 — 18
22	17.6 — 26.4
33	22.4 — 39.6
47	37.6 — 56.4
68	54.4 — 81.6
100	80 — 120
150	120 — 264
330	224 — 396
470	376 — 564
680	544 — 816
1000	800 — 1200

Fig. 4-19. Table specifies resistor sizes in ohms and 20 percent tolerance range.

10% TOLERANCE RESISTORS

SIZE IN OHMS	TOLERANCE
10	9 — 11
12	10.8 — 13.2
15	13.5 — 16.5
18	16.2 — 19.8
22	19.8 — 24.2
27	24.3 — 29.7
33	29.7 — 36.3
39	35.1 — 42.9
47	42.3 — 51.7
56	50.4 — 61.6
68	61.2 — 74.8
82	73.8 — 90.2
100	90 — 110
120	108 — 132

Fig. 4-20. Table covers resistor sizes in ohms and 10 percent tolerance range.

Fig. 4-20. When a 5 percent tolerance is required, Fig. 4-21, the range gets smaller and still more resistors are needed to fill the overlap.

Note also in Fig. 4-19 that all 20 percent tolerance resistors are included in the 10 percent tolerance group. Both of these groups are also included in the 5 percent tolerance group. Bear in mind, too, that manufacturers charge more for resistors they must produce in limited quantities. By keeping all of these tolerances and ranges in mind, it is possible to save the expense of buying all different sizes of resistors.

THERMISTORS

One other type of resistor called a "thermistor" deserves special mention. See Fig. 4-22. Thermistors have come into popular usage in electric circuits of heating and cooling systems. The name, thermistor, was derived from the phrase THERMally sensitive resISTOR.

Most thermistors differ from conventional resistors in how they react to temperature changes. A resistor starts to heat up when electrons begin to flow in the

SIZE IN OHMS	TOLERANCE RANGE
10	9.5 — 10.50
11	10.45 — 11.55
12	11.40 — 12.60
13	12.35 — 13.65
15	14.25 — 15.75
16	15.20 — 16.80
18	11.10 — 18.90
20	19.00 — 21.00
22	20.90 — 23.10
24	22.80 — 25.20
27	25.65 — 28.35
30	28.50 — 31.50
33	31.35 — 34.65
36	34.20 — 37.80
39	37.05 — 40.95
43	40.85 — 45.15
47	44.65 — 49.35
51	48.45 — 53.55
56	53.20 — 58.80
62	58.90 — 65.10
68	64.60 — 71.40
75	71.25 — 78.75
82	77.90 — 86.10
91	86.45 — 95.55
100	95.00 — 105.00
110	104.50 — 115.50

Fig. 4-21. Table shows resistor sizes in ohms and 5 percent tolerance range.

Fig. 4-22. Thermistors are thermally sensitive resistors. Since thermistors can detect slight temperature changes, they often are used as safety switches. (Fenwal Electronics, Inc., Div. of Walter Kidde & Co., Inc.)

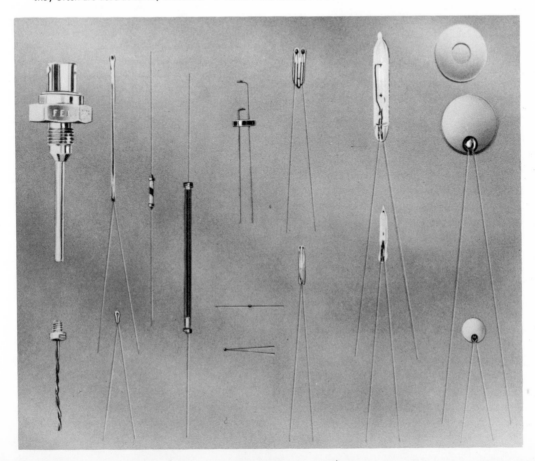

circuit. As the temperature rises, resistance increases in an effort to reduce the amount of moving electrons so the resistor can cool off. In contrast, most thermistors will have an opposite reaction. As the temperature rises, resistance decreases. As mentioned, most thermistors react this way; others behave just like resistors.

Actually, there are two different kinds of thermistors, the negative temperature coefficient (NTC) type and the positive temperature coefficient (PTC) type. The NTC type *decreases* in resistance as the temperature rises. The PTC type, like a conventional resistor, *increases* in resistance as the temperature rises.

Since thermistors have this special ability to react to temperature changes, they are used to measure temperatures, protect electric motors from overheating, sense liquid levels, Fig. 4-23, or serve as safety switches in fire or frost alarms.

Fig. 4-23. In another application, thermistors can be used to sense liquid levels.

The frost alarm application is very important to people who are concerned with the danger of frost on fruits and vegetables. A killing frost could destroy hundreds of thousands of dollars worth of food in just one night. Thermistors are used because they are able to detect small differences in temperature.

There are two terminals, Fig. 4-24, on most thermistors, permitting them to be installed into circuits in the same way as resistors. To get a better understanding of the use of resistors and thermistors, read Chapter 5. It is designed to clear up the kinds to be used, their wattage value, the type of circuits where they are found, and the value in ohms for each one.

CAPACITORS

Another key part found in many electric circuits is a "capacitor," Fig. 4-25. The basic job of a capacitor is to

store electrons until they are needed by the circuit in which the capacitor is installed. Although many different symbols for a capacitor are in current use, the symbol shown in Fig. 4-26 is a popular choice.

Fig. 4-24. Thermistors generally have two leads or terminals for use in connecting this type of resistor into the circuit.

CAPACITOR ACTION

To understand how a capacitor works in a circuit, try relating it to a water jug placed in a cooler in an office. First, it must be filled. Next, people take water by the cupful until the jug is empty. Then, it must be refilled.

With the capacitor, the sequence is the same. First, it must be charged with electrons. Next, it must give up the electrons as needed. Then, the capacitor must be recharged so it can give off more electrons when called on to do so.

In Chapter 9, we will go into details of capacitor operation for some circuits. At this point, however, all you need to know is that a capacitor can be charged and discharged.

CAPACITOR CONSTRUCTION

A capacitor is made up of two plates and an insulator. The plates are made from material that can be charged, such as aluminum foil. In order to hold this

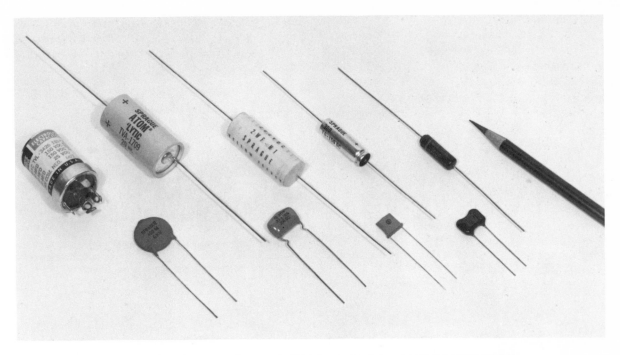

Fig. 4-25. Capacitors store electrons, holding the electrical charge until the circuit needs it. (Sprague Electric Co.)

charge, an insulator (dielectric) must be placed between the plates, Fig. 4-27. If the plates touch one another, the charge will be lost. These losses, in turn, will appear as heat in the capacitor. The more rapidly we charge and discharge the capacitor, the more heat we will create. If these heat gains are not held to a reasonable level, the capacitor will be destroyed.

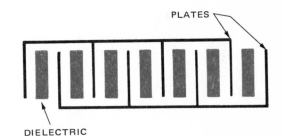

Fig. 4-27. The dielectric in a capacitor is the insulator placed between the plates that store the electrical charge.

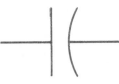

Fig. 4-26. This simple sketch depicts the most commonly used symbol for a capacitor.

TYPES OF CAPACITORS

To hold down the heat, capacitors are designed to do specific jobs. Because of this, capacitors are given names like "tantalum electrolytic," shown in Fig. 4-28, "film capacitors" in Fig. 4-29, "ceramic capacitors" in Fig. 4-30, and "mica capacitors" in Fig. 4-31.

Mica capacitors are used in applications where it is necessary to charge and discharge the capacitor billions of times in a second.

Fig. 4-28. Tantalum electrolytic capacitors utilize paper soaked with electrolyte as insulator between tantalum plates or sheets. (Sprague Electric Co.)

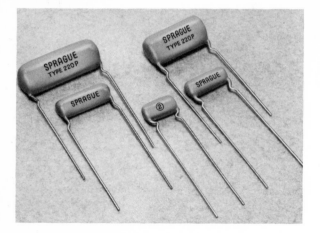

Fig. 4-29. A film capacitor usually is made up of two conductors separated by a plastic film.

Fig. 4-31. Mica capacitators are made of foil segments with sheets of mica serving as insulators. These capacitors are used in high current transmitting applications. (Sprague Electric Co.)

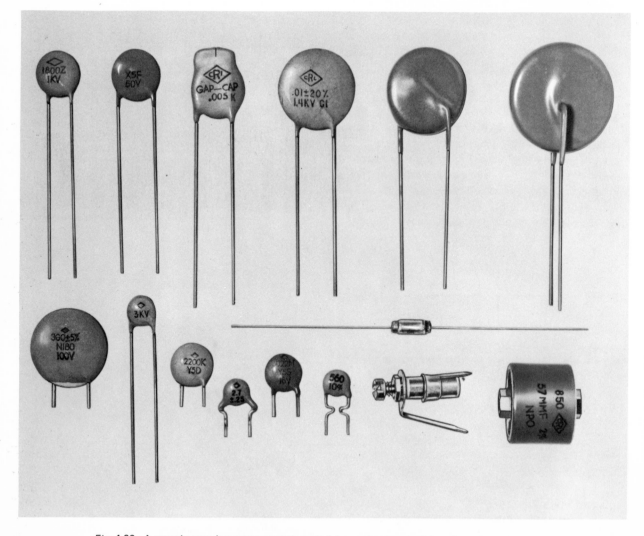

Fig. 4-30. A ceramic capacitor uses a ceramic material as an insulator between the two conductors.
(Centralab Electronics Div. of Globe-Union Inc.)

FLASHER TOY PROJECT

If you are good at tracing, or if you have a friend who can draw well, here is a project you can make for a little brother or sister. It is a framed picture of a rabbit in a flower bed backed by flashing lights. See Fig. 4-32.

After you have completed the painting, drill small holes for the five NE 2 bulbs. The lamps can be placed anywhere to get the effect you want. Then, hook up the circuit so that the lamps will flash. By picking various sizes of resistors, you can make the lamps flash at different times.

Note where the line cord has been hooked up in Fig. 4-32. As a further precaution, drill a hole through the wooden frame and tie the end of the cord in a knot so that it cannot be yanked from the circuit.

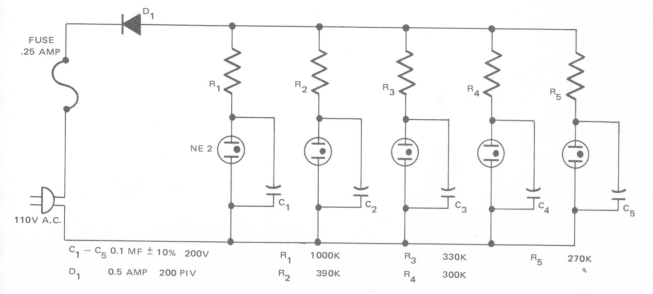

Fig. 4-32. Front view, back view and the schematic for building the flasher toy described in the text. In the parts list, C is for capacitor, R is for resistor, D is for diode, NE 2 for the bulbs. (Project by Don Winchell)

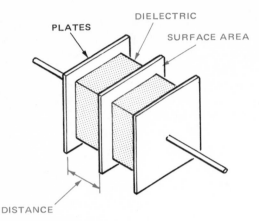

Fig. 4-33. The three factors which determine capacitance are: plate surface area (total of all plates); distance between plates; value of dielectric material (see Fig. 4-35).

FACTORS AFFECTING CAPACITANCE

At this point, you probably are asking yourself "What are the important things I need to know about capacitors?" There are three important factors that determine the capacitance of a capacitor, Fig. 4-33. These are:
1. Surface area of the plates.
2. Distance between the plates.
3. Insulating material used between the plates. This insulating material is usually known as "dielectric."

SURFACE AREA

To get a better understanding of capacitance, look at some examples. First, look at the surface area of the plates, while keeping the distance between the plates and dielectric equal for both. The capacitor with the most plate surface area will have the greatest capacitance, Fig. 4-34. This is because the capacitor would have more space to store an electrical charge. *Increase the surface area and you will increase the capacitance.* Remember though, this is only true if everything else is kept equal and the plates are perfectly parallel to each other.

DISTANCE BETWEEN PLATES

The second thing which affects our capacitance is the distance between the plates. Again, we will keep the dielectric and surface areas constant. If the two charged plates are moved apart (double the space between them), the strength of the capacitor's charge will be lowered. In effect, the strength is inversely proportional to the distance between them.

Suppose, for example, the distance between the plates is increased by one thousandth of an inch (.001 in.). Will the capacitance increase or decrease? Since the strength is inversely proportional, capacitance will get weaker. *If the distance between the plates is reduced, capacitance will increase.*

DIELECTRIC

The third affect on capacitance concerns the type of dielectric (insulating material) used in the capacitor. Remember from Chapter 3 that an insulator resists the flow of electrons. See Fig. 4-35 for dielectrics used in capacitors. Approximate values are given.

Fig. 4-34. Plate area determines the difference in capacitance of these two aluminum electrolytic capacitors. Note that one at left is marked 100000 mfd; right 56000 mfd.

DIELECTRIC CONSTANTS OF CAPACITOR MATERIALS

MATERIAL	DIELECTRIC CONSTANT
AIR	1.0
ALUMINUM OXIDE	10.0
BEESWAX	3.0
CAMBRIC (VARNISHED)	4.0
CELLULOID	4.0
GLASS (PYREX)	4.2
GLASS (WINDOW)	7.6
MICA (CLEAR INDIA)	7.5
MYLAR	3.0
PAPER (KRAFT)	4.0
PORCELAIN	6.2
QUARTZ	5.0
TANTALUM OXIDE	11.0
VACUUM	1.0

Fig. 4-35. Dielectric materials are listed at left; dielectric constants for figuring capacitance are given at right.

FORMULA

To better understand the three major factors that affect capacitance, examine the following formula:

$$C = 0.2235 \frac{KA}{d} \; (N\text{-}1)$$

Where

C is the capacitance in picofarads (pF).
K is the dielectric constant.
A is the area of one plate in square inches.
d is the distance between the plates in inches.
N is the number of plates.

Example:

What is the capacitance of four 1 in. square plates separated by quartz which is 0.001 in. thick?

$$C = 0.2235 \frac{KA}{d} \; (N\text{-}1)$$

$$C = 0.2235 \frac{5 \times 1 \text{ sq. in.}}{0.001 \text{ in}} \; (4\text{-}1)$$

$$C = 0.2235 \frac{5}{.001} \; (3)$$

$$C = 0.2235 \times 5000 \; (3)$$

$$C = 3352.5 \text{ pF}$$

If we double the distance between the plates by increasing the thickness of the quartz to 0.002 in. thick, we have:

$$C = 0.2235 \frac{KA}{d} \; (N\text{-}1)$$

$$C = 0.2235 \frac{5 \times 1 \text{ sq. in.}}{.002} \; (4\text{-}1)$$

$$C = 0.2235 \frac{5}{.002} \; (3)$$

$$C = 0.2235 \times 2500 \; (3)$$

$$C = 1676.25 \text{ pF}$$

As you can see, doubling the distance cut the capacitance to one-half of the previous example.

TERMS

Note that the capacitance in the example is given in picofarads. The original discoverer of the capacitor was Pieter van Musschenbrock, but the basic unit of capacitance was named "farad" after Michael Faraday.

A capacitor has a capacitance of one farad when, with one volt applied to its plates, it stores one coulomb of charge on each of its plates. A coulomb is equal to 6.24×10^{18} electrons. You may be familiar with the term "volt," having seen it marked on batteries. We will cover what a volt is and how it works with other units in the next chapter.

The farad is a very large unit and is not used much. Most capacitors are rated picofarads and microfarads. A microfarad is one-millionth of a farad, while a picofarad is one-millionth of a microfarad. The chart which follows will give you an idea of how this works.

1 farad (F)	= 1,000,000 microfarads (μF)
1 microfarad (μF)	= 1,000,000 picofarads (pF)
1 picofarad (pF)	= .000001 microfarad (μF)
1 microfarad (μF)	= .000001 farad (F)
1 picofarad (pF)	= .000 000 000 001 farad (F)

It is common to use the term microfarads down to about 0.0001 μF. The term picofarads is used up to about 1000 pF (.001 μF). Even though a crossover point exists, the idea is to try to use the smallest amount of numbers.

DISCHARGING CAPACITORS

At this point, you probably are saying to yourself that these are pretty small numbers. As small as they may seem, they carry enough electrical charge to kill.

I.C. SAYS:

"CAUTION: Before you pick up any capacitor, make sure it has been discharged."

Use care when working with electrical circuits that include capacitors. Always assume that any capacitor is "charged." Therefore, use some special means of discharging it before attempting to handle it. Fig. 4-38 shows a discharging device that is safe and easy to use, simply by touching a probe to each capacitor terminal.

Some people make the mistake of using a screwdriver or a short section of insulated wire. Two things can result. First, you could ruin the capacitor by discharging it this way. Second, if the capacitor is very large, it is possible to cause an arc. In some cases, such as the capacitors in an electronic flash for a camera, you could even melt the end of the screwdriver.

CAPACITOR DISCHARGER PROJECT

Capacitor discharging devices can be useful when working on radios, television sets and other electrical circuits involving capacitors. However, the right kind of equipment is needed to do the job properly. Fig. 4-36

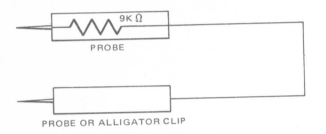

Fig. 4-36. Schematic for a capacitor discharger shows a two probe setup. An alligator clip can be substituted for plain probe.

gives details on how to build a capacitor discharger. Fig. 4-37 reveals a resistor built into the probe and how the boot fits over the alligator clip.

Parts needed to build this capacitor discharger include: two probes or one probe and an alligator clip; one 9K 1 watt resistor (or larger, depending on the capacitor to be discharged); three feet of 14 AWG wire.

Building procedures are as follows:

1. Obtain parts and check resistor ratings.
2. Drill out the inside of one probe to accomodate the resistor or resistors.

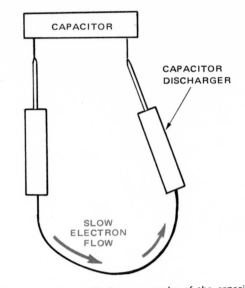

Fig. 4-38. The resistor built into one probe of the capacitor discharger slows the flow of electrons to avoid a damaging sudden discharge of the capacitor.

3. Strip the wire on both ends and make good mechanical joints. Connect the wire to one end of the resistor and connect the other end of the resistor to the metal part of the probe. Connect the other end of the wire to the second probe or alligator clip, if used.
4. Screw the handles on the metal part of the probes. To use the capacitor discharger, attach the alligator clip (or probe) to one terminal of the capacitor and

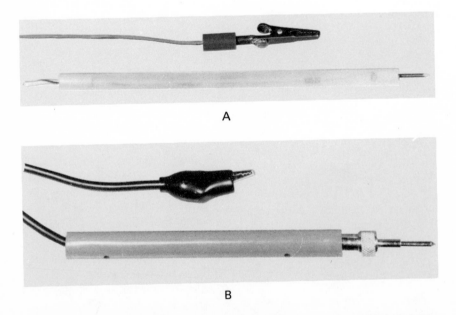

Fig. 4-37. Two probe and alligator clip setups are shown for use with capacitor discharger. Note insulating boot on the lower clip.

touch a probe to the other terminal for at least five seconds. See Fig. 4-38. This will give the capacitor time to discharge slowly.

The principle of operation of the discharger is simple. The wire acts as a path for the electrons to flow from one set of capacitor plates to the second set of plates. The resistor slows up the discharging process and stops arcing by avoiding a sudden buildup of heat.

ELECTROLYTIC CAPACITORS

Electrolytic capacitors deserve special attention. They are made in a different way, Fig. 4-39, and provide more capacitance for their size than any other type of capacitors.

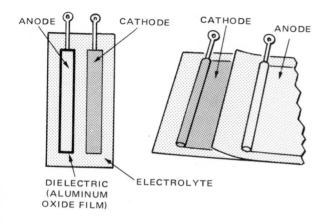

Fig. 4-39. In manufacturing an electrolytic capacitor, one aluminized sheet is coated with a layer of oxide. As a result, the capacitor becomes "polarized" (has a positive side and a negative side).

Fig. 4-40. In use, electrolytic capacitors must be installed positive to positive and negative to negative. Ends of capacitors are marked with a + and −. (Fisher Radio)

Instead of having its plates separated by a dielectric, the electrolytic capacitor consists of two aluminum sheets separated by a layer of paper soaked with a liquid chemical called "electrolyte." All three sheets are rolled together, sealed in a container and dc voltage is applied between the two aluminized sheets.

The charge of electricity causes a thin oxide layer to be formed on one of the aluminized sheets. As a result, the capacitor is said to be "polarized." This means that one side of the capacitor is positive; the other side is negative. See Fig. 4-40. The oxide acts as the dielectric. It has a high resistance to the flow of electrons in one direction and very low resistance in the opposite direction.

I.C. SAYS:

"Electrolytic capacitors are marked with a positive (+) end and a negative (−) end. If they are not connected into circuits positive to positive and negative to negative, the oxide film will be ruined. In some cases, the capacitor may explode."

ELECTROLYTICS DEFORM

Electrolytic capacitor will slowly deform (decrease in capacitance) if it is not used for awhile. If not used, then given a sudden surge of electrons, it may be ruined. This surge can be caused by just installing the electrolytic capacitor in a circuit. When electrolytics are purchased from a store or taken off a shelf after being stored for a long time, they must be charged slowly.

Many capacitors have voltage ratings marked on their outside cover. See Fig. 4-41. The electrolytic capacitor shown is marked WVDC, which stands for "Working

Fig. 4-41. Many capacitors are labeled with a voltage rating in addition to capacitance. This electrolytic capacitor is marked 150 volts. (Sprague Electric Co.)

Voltage Direct Current." To reform an electrolytic, hook up a circuit shown in Fig. 4-42. It is important that the voltage output of the power source is *lower* than the voltage marked on the outside of the capacitor.

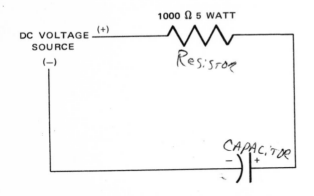

Fig. 4-42. This schematic shows an electrolytic capacitor being reformed (charged) to build up its capacitance. Note that it is installed positive to positive and negative to negative.

HOLDING A CHARGE

Remember that capacitors can hold a charge for many hours and, in some cases, for several days. Before you handle any capacitor, make sure it is discharged.

SUBSTITUTING CAPACITORS

When repairing a circuit or building a new one, install capacitors of the specified voltage rating. If you cannot find a capacitor of the correct voltage rating, substitute one that has a higher voltage rating. If you install one with a lower rating, you are asking for trouble.

SOLDERING CAPACITORS

There are a couple of precautions you should take if you have to replace a capacitor in a circuit.
1. Make sure the new capacitor is soldered in place the same way as the old one when it was installed. If you get too close or too far away from the chassis (the metal box on which electronic components are mounted and wired), it may shift the capacitance. For example, this could result in a hum from a repaired radio.
2. Be sure the new leads are the same length as the old ones. This could affect the resonant frequency (described in Chapter 14).
3. Use a heat sink between the capacitor and the solder

joint, as explained in Chapter 3. Some capacitors have a dielectric that could be melted by the heat from a soldering iron.

TEST YOUR KNOWLEDGE

1. What is the basic unit of measure for capacitance?
2. Change .0001 microfarads into picofarads.
3. List two different types of capacitors.
4. You never touch a capacitor without _____.
5. Because an _____ capacitor is made, in a special way, you must watch its polarity when you put it into a circuit.
6. How long can a capacitor store (hold) its charge?
7. List these readings in order of highest-to-lowest capacitance:

   3 farads.   700 picofarads
   600 microfarads   .03 farads
8. Name two things that affect the amount of capacitance.
9. A potentiometer is another name for a:
   a. Variable capacitor.
   b. Polarized capacitor.
   c. Fixed resistor.
   d. Variable resistor.
10. Copy these colors on a separate paper and, place the number alongside each color that corresponds to the color code found on the resistor.

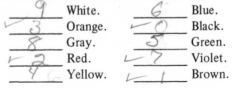

_9_ White.	_6_ Blue.
_3_ Orange.	_0_ Black.
_8_ Gray.	_5_ Green.
_2_ Red.	_7_ Violet.
_4_ Yellow.	_1_ Brown.

SOMETHING TO DO

1. How do you know which tolerance resistor to use in a circuit?
2. Find out how a company actually makes carbon resistors.
3. Visit a store that sells electrical components and see how many different sizes of resistors they sell. Make a list of the sizes.
4. Make a list of the places in your car or home that you use variable resistors.
5. Make up some memory device to help you remember the first letters in the resistor color code.
6. Why is it important to know the failure rate of resistors? Why do some resistors lack a fifth band to show failure rate?
7. Make your own capacitor, utilizing knowledge you

have gained about things used in their construction.

8. See if you can find a capacitor with a rating over one farad. Where are these types of capacitors used?

9. What is the smallest capacitor rating you can find in a book? What is the smallest size capacitor you can find in some circuit?

10. What kind of capacitor holds its charge for the longest period of time? Where is it used?

11. Should capacitors and resistors be soldered into circuits the same way? Should you pay attention to which direction each lead is hooked into the circuit?

12. Find an easy way to store resistors and capacitors when you are laying them out for soldering into a circuit.

To identify a color-coded resistor, turn the varicolored wheels at the top until the color of the wheels matches the color of the bands on the resistor. Then, read the numbers in the windows.

Chapter 5
OHM'S LAW
D.C.

In reading the previous chapters, you came across the words "amps, resistance, power and watts." You were introduced to these words because every one is important to the study of electrical circuits.

This chapter is titled OHM'S LAW because Ohm was the person who discovered how amps, resistance, power and watts are tied together mathematically. George Simon Ohm (1787-1854) was a German scientist who found a very simple method for figuring the third value when the other two values are known.

Take any three numbers, such as 2, 3 and 6. You can arrange these numbers so that if you know any two of them, you can find the third one. How about 2 x 3 = ?

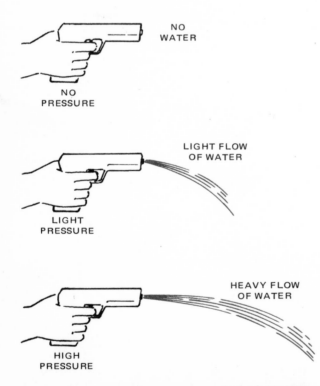

NO WATER

NO PRESSURE

LIGHT FLOW OF WATER

LIGHT PRESSURE

HEAVY FLOW OF WATER

HIGH PRESSURE

Fig. 5-1. Just as increased pressure on a squirt gun forces more water to squirt farther, increased electrical pressure (voltage) forces more electrons to flow.

In this case, you simply *multiply* to find 6. If you have 2 x ? = 6, arrange the numbers you know so you will be able to find the third. That is: *divide* to find the right answer. In this case $\frac{6}{2}$ = ? The result is 3.

Ohm's law is just as easy to use.

THREE VARIABLES OF OHM'S LAW

When using Ohm's Law, you will work with three elements:
1. Voltage.
2. Current.
3. Resistance.

Just as you were able to use a formula for finding 2 x 3 = 6, you can say that voltage in a circuit is equal to current multiplied by resistance. *In a formula, you would have: Voltage = Current x Resistance.*

It has become common practice to substitute letters for words in formulas. Instead of using the word voltage, we use the letter E. Current is replaced with the letter I. Resistance is replaced with the letter R. *This gives us a new formula: E = I x R.* However, it still means that voltage is equal to current times resistance.

ELECTROMOTIVE FORCE

The letter E in the formula stands for voltage. It refers to the pressure which is required to force the electrons through the circuit. Often, therefore, voltage is called "electromotive force." To get a better understanding of this, see Fig. 5-1. The flow of electrons in an electrical circuit is compared with the flow of water from a squirt gun. If no pressure is exerted on the gun, no water will flow. The more pressure exerted on the gun, the greater the amount of water that will flow, and the farther it will travel. This is the same type of pressure that must exist in all electrical circuits.

The pressure or voltage in electrical circuits can come from many different sources. See Figs. 5-2 and

Fig. 5-2. A "cell" is one possible source of electrical pressure in a low voltage circuit. (Burgess Battery Div., Gould, Inc.)

5-3. At this point, we are only discussing direct current (dc) sources, such as batteries. There is another whole group of voltage sources that give alternating current (ac). These sources are described in Chapter 10.

Fig. 5-3. A "battery" is the source of electrical pressure in an automobile electrical system. (ESB Brands, Inc.)

CURRENT

The letter I in the formula is for "Intensity" of current flow. Earlier, we discussed the millions of electrons that pass a given point in the circuit in one second when current is flowing. Numbers this large are hard to work with, so the current flow rate is shortened to read in amperes (amps). *One amp is equal to 6.24 x 10^{18} electrons flowing past a given point in one second.* This given value for one amp is a "standard" issued by the National Bureau of Standards.

The term "ampere" was named in honor of Andre Marie Ampere (1775-1836). As you continue reading, you will find that many laws and even some circuits have names which honor the people who first discovered them.

RESISTANCE

The letter R stands for resistance to, or the opposition to, the flow of electrons. This resistance is measured in ohms. Earlier, we saw from our table of wire sizes that the resistance went up as the length of wire increased. We also saw that as the wire got larger in diameter, we could get a longer piece of wire and still have the same resistance.

All three elements, E, I and R, can be measured with meters similar to the one shown in Fig. 5-4. These are:

Voltage is measured with a voltmeter.

Current is measured with an ammeter.

Resistance is measured with an ohmmeter.

Fig. 5-4. A multimeter is used to measure voltage, amperage and ohms. (Heath Co.)

Many times, meters can be used to help find some missing values. It is also possible to find out the value of E, I or R by using Ohm's Law.

Fig. 5-5. As a visual aid to learning Ohm's Law, try covering the letter for the unknown value. For example, cover I and the formula becomes $I = \dfrac{E}{R}$.

By shifting letters around, it is possible to get different forms of Ohm's Law. A study of Fig. 5-5 should help you follow these changes. Depending on which values you know, you can select a formula for finding the unknown value. The three basic formulas created by Ohm's Law are:

$$E = I \times R \qquad I = \frac{E}{R} \qquad R = \frac{E}{I}$$

CIRCUIT VOLTAGE

The simple circuit shown in Fig. 5-6 offers a problem that you can solve by applying Ohm's Law. In this circuit, the symbol for a battery is used to show the source of voltage. The wires are shown by lines. The symbol for resistance is the load, or bulb. The circular symbol with a large A indicates a test ammeter. The voltage, in this case, is unknown.

Looking at known values, the amount of current flowing in the circuit is 3 amps. The resistance is 2 ohms. To determine the voltage needed to light the bulb, use Ohm's Law as follows:

E = I x R   Most people do not use the x to show multiplication. Instead, they would simply write out the formula as:

E = I R   Then, by substituting known values for the letters

E = 3 amps x 2 ohms

E = 6 volts

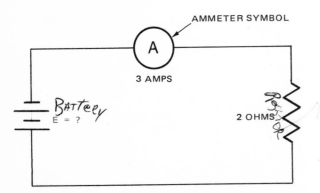

Fig. 5-6. This simple electrical circuit poses a problem that can be solved by applying Ohm's Law, E = I x R.

Therefore, put a 6 volt battery in the circuit. If you mistakenly use a 3 volt battery, Fig. 5-7, the bulb will not light properly. A 3 volt battery does not produce enough pressure (voltage) to push the right number of electrons through the resistance of the bulb.

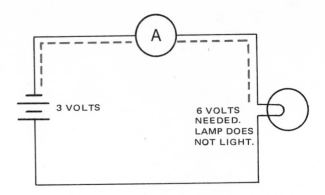

Fig. 5-7. The voltage is too low in this circuit to force electrons through the resistance of the lamp.

If you had selected a 12 volt battery, Fig. 5-8, it would produce *too* much pressure. Battery voltage would push so many electrons through the bulb that the filament would get hot and burn out. Replacing the bulb would not solve the problem. No matter how many new 6 volt bulbs you put in, each new one would burn out. You must reduce the voltage of the battery or add resistance to the circuit to reduce the current flow.

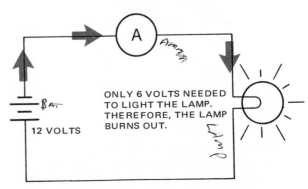

Fig. 5-8. If the voltage is too high for the low resistance of the lamp, you must reduce the voltage or add a resistor to the circuit.

HOTDOG VOLTMETER PROJECT

You can make a hotdog voltmeter, Fig. 5-9, for use in checking whether an electrical circuit is 110 volts or 220 volts. Obtain the following materials and assemble them according to the construction procedure given.

No.	Item
1	R_1—220K resistor.
1	R_2—470K resistor.
1	R_3—180K resistor.
2	NE 2 neon bulbs.
2	Probes.

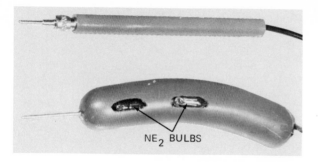

Fig. 5-9. You can test circuit voltage with this hotdog volt-meter. If one lamp lights, the circuit is 110 volts. If both lamps light, the circuit is 220 volts.   (Project by Donald Winchell)

CONSTRUCTION PROCEDURE

Build the project by following the schematic shown in Fig. 5-10. If you wish, build and substitute any other insulated probe instead of the hotdog. Be sure, however, that it *is* insulated.

1. Cut two openings in the side of the hotdog for the neon lamps. See Fig. 5-9.
2. Make a hole lengthwise through the hotdog for the wire and nail.
3. Grind down end of nail to form a point for probe.
4. Wire the circuit as shown in Fig. 5-10.
5. Insert one end of the wire in the hotdog, connect both lamps and the resistors. Solder the end of the wire to the nail head.
6. Push the nail through the end of the hotdog from the inside, and insert the bulbs in their openings.
7. Epoxy the nail and wire at the ends of the hotdog so they will not pull out when the probe is in use.
8. Attach a regular probe to the other end of the wire.

To use the hotdog voltmeter, touch the two probes to the closed circuit being tested. If one light comes on, you have 110 volts. If both lights come on, you have 220 volts.

CIRCUIT AMPERAGE

How much current could flow through a 36 ohm resistor in a car radio powered by a 12 volt battery? Using one formula, you would first list what is known:

$$I = \frac{E}{R}$$

$$I = \frac{12 \text{ volts}}{36 \text{ ohms}}$$

$$I = \frac{1}{3} \text{ amp or } .33 \text{ amp}$$

The completed formula tells you that about 1/3 of an amp of current would flow through the resistor.

Study Fig. 5-11. What size resistor would you pick to keep the current to 3 amps or less? Listing the things you know, you would have:

$$R = \frac{E}{I}$$

$$R = \frac{12 \text{ volts}}{3 \text{ amps}}$$

$$R = 4 \text{ ohms}$$

This means that a resistor of at least 4 ohms should be used.

If you picked a 6 ohm resistor, the current would be kept below 3 amps:

$$I = \frac{E}{R}$$

$$I = \frac{12 \text{ volts}}{6 \text{ ohms}}$$

$$I = 2 \text{ amps}$$

This is lower than the high limit you set, but it is acceptable.

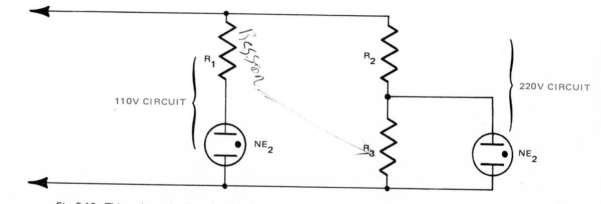

Fig. 5-10. This schematic for the hotdog voltmeter gives the wiring arrangement and basic connections. Probes are shown at the left.

If you picked a resistor of 2 ohms, you would be in trouble:

$$I = \frac{E}{R}$$

$$I = \frac{12 \text{ volts}}{2 \text{ ohms}}$$

$$I = 6 \text{ amps}$$

This setup would result in a blown fuse (if circuit is fused) or an overheated resistor (if circuit is not fused). These examples show the importance of choosing the right resistor in terms of its resistance value.

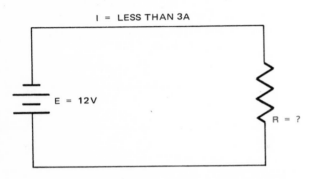

Fig. 5-11. A circuit having an unknown resistance value is shown. Ohm's formula $R = \frac{E}{I}$ will help supply the answer.

WATTAGE

To find the right resistor for a given circuit, you have to pick a resistor of the correct wattage. Earlier, we found that wattage of a resistor tells how much heat it will dissipate. As long as the resistor can transfer heat into the air around it, the resistor will not overheat; if it cannot, the resistor will be ruined.

A formula for "power" has been worked out which uses many of the same terms as Ohm's Law. By putting down known values, the formula will permit you to figure out the correct wattage resistor to use. The formula is:

$$P = E I$$

The unit of electrical power is the watt, named after James Watt (1736-1819). Here, again, the term is named after an important person in the early history of electricity.

A watt is the amount of work done in one second by one volt moving a charge of one coulomb. A coulomb equals 6.24×10^{18} electrons per second. This means we can replace the coulomb with amperes and get the formula $P = E I$. Per second, of course, means "in one second."

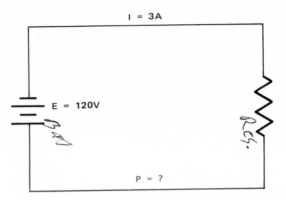

Fig. 5-12. What amount of power (wattage) is used in this circuit? Power formula P = IE will help solve this problem.

KILOWATT HOURS

Fig. 5-12 shows a simple circuit. To find the number of watts being used, work out the formula: P = E I. You know that 3 x 2 = 6 and that 2 x 3 = 6. It does not

APPROXIMATE COST OF OPERATION OF APPLIANCES

APPLIANCE	AVERAGE WATTAGE	EST. COST PER MO. AT $0.04 PER KWH
TOOTH BRUSH	1.1	.003
CLOCK	2	.06
HAIR DRYER	381	.05
SEWING MACHINE	75	.04
SHAVER	15	.04
TOASTER	1146	.13
VACUUM CLEANER	630	.15
RADIO	71	.29
WASHING MACHINE	512	.34
COFFEE MAKER	894	.35
TELEVISION, B & W	45	.33
TELEVISION, COLOR	145	1.06
DISHWASHER	1201	1.21
REFRIGERATOR/ FREEZER MANUAL DEFROST	–	2.33
CLOTHES DRYER	4856	3.31
RANGE W/OVEN	12,200	3.92
REFRIGERATOR/ FREEZER AUTO. DEFROST	–	5.98
WATER HEATER	4474	14.91
CENTRAL A/C	1670/TON	35.00

Fig. 5-13. Wattages and cost of operation are given for many popular appliances. Note that higher cost is in direct proportion to higher wattages. (Edison Electric Institute)

Ohm's Law

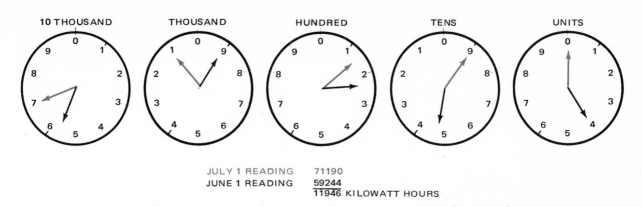

Fig. 5-14. Read kilowatt hour meters from right to left. In this diagram, black hands indicate June reading; blue hands point to July reading. The difference between these readings is the number of kilowatt hours of electricity used.

JULY 1 READING   71190
JUNE 1 READING   59244
11946 KILOWATT HOURS

matter which number comes first in multiplying numbers. Therefore, an easy way to remember the formula for power is to think of PIE (P = IE). In our problem, the circuit has an electromotive force (voltage) of 120 volts and a current of 3 amps. Therefore, substituting values for letters, we would have:

P = I E
P = 3 amps x 120 volts
P = 360 watts

You might ask if 360 watts is a lot of power or a small amount. For a helpful comparison of typical wattage ratings, consider some of the appliances you have around your home. Also, see Fig. 5-13. The cost of operation shown in this table is based on the number of kilowatt hours (KWH) used. The power company charges about four cents for one kilowatt hour. A kilowatt hour is equivalent to one thousand watts of electricity being used in one hour.

To keep a record of how much electrical power you use every month, the power company installs a meter at your home, Fig. 5-14 shows how they keep a record of your power use by means of this meter. By recording meter readings at regular intervals, the company is able to bill you for the exact amount of electricity used.

The power formula (P = E I) is just as flexible as the basic formula (E = I R) established by Ohm's Law. By shifting letters around, as in Fig. 5-15, you can come up with different forms of the power formula for use in finding unknown values in electrical circuits.

MILLI AND MICRO AMPS

In previous examples of electrical circuits, the word "amps" was established as the basic unit for the amount of current flowing through a load. The term "amps" is used when referring to circuits carrying relatively large

VARIATIONS IN OHM'S LAW AND THE POWER FORMULA

E	I	R	P
I R	$\dfrac{E}{R}$	$\dfrac{E}{I}$	I E
$\dfrac{P}{I}$	$\dfrac{P}{E}$	$\dfrac{P}{I^2}$	I^2R
$\sqrt{PR}$	$\sqrt{\dfrac{P}{R}}$	$\dfrac{E^2}{P}$	$\dfrac{E^2}{R}$

Fig. 5-15. All of Ohm's formulas and all power formulas are arranged in chart form. Read down from headings, E, I, R and P to get formula desired.

amounts of current, such as power plants, Fig. 5-16, high tension (voltage) lines, house wiring and in many household appliances.

In many circuits, however, you will find that current flow is less than one amp. In fact, the current may be so low that you will have numbers such as .001 amp. Rather than say "one thousandth or one millionth of an amp," the names "milli" and "micro" are used.

A milliamp (mA) is .001 or one thousandth of an amp. A microamp is .000001 or one millionth of an amp. Very often, you will find the word "micro" replaced with what looks like a smaller letter u (μ), the symbol mu. You saw this symbol in the last chapter when we worked with capacitors.

Consider the circuit illustrated in Fig. 5-17. The voltage is 120 volts and the resistance is 120,000 ohms. What is the current? Laying out the formula, you would have:

$$I = \frac{E}{R} \qquad I = \frac{120 \text{ volts}}{120,000 \text{ ohms}} \qquad I = .001 \text{ amp}$$

77

Fig. 5-16. This highly instrumented control room for a large power plant serves as the brain center for the generation of electricity. (Consumers Power Co.)

The current in this case would be equal to 1 milliamp or 1 mA.

If the resistance in the previous example had been 120 Meg ohms, we would have:

$$I = \frac{120 \text{ volts}}{120{,}000{,}000 \text{ ohms}}$$

$$I = .000001 \text{ amp}$$

The current in this case would be equal to 1 microamp or 1 μ A.

As mentioned, the term "milli" and "micro" are used when referring to circuits carrying less than one amp of current. These terms also find use with reference to low voltage circuits in radios, television sets and many other communications systems, Fig. 5-18. In this case, the proper terms are "millivolts" and "microvolts."

Fig. 5-17. This student is wiring an electrical circuit from a schematic. See the accompanying text for a problem in how to find an unknown current value by applying Ohm's Law.

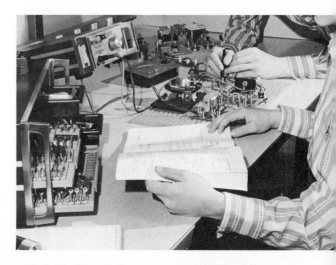

Fig 5-18. These students are testing a completed circuit. (Hickok Teaching Systems, Inc.)

BODY RESISTANCE

Have you ever considered what happens to the human body when it is involved with electricity? Fig. 5-19 shows an electrician doing a typical job of circuit testing. If he were to make a mistake, would he be killed (electrocuted)? The quick reply is "Maybe."

To be able to answer this question properly, several things must be known. First, every person's body is a

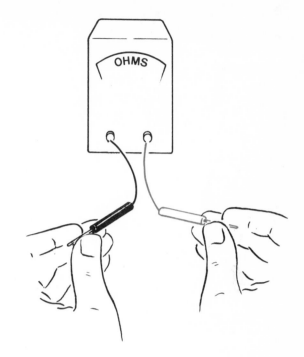

Fig. 5-20. You can measure body resistance by means of an ohmmeter. Note that when you increase pressure on the probes, the resistance reading drops.

Fig. 5-19. Installer/electrician uses an oscilloscope (instrument that shows test pattern on a viewing screen) to test circuits in an electronic telephone exchange. (Western Electric Co.)

giant resistor. To measure your body resistance from hand to hand, hold the ohmmeter probes as shown in Fig. 5-20, and compare the reading to others. You will find that everyone has a different amount of body resistance, and most will fall in a range from 10K to 150K ohms. The reason you differ is because of different weights, heights, size of bones and many other body chemistry factors. In other words, no two people are exactly alike.

If you measure your body resistance on a hot summer day, then measure it again on a cool day, you will get different readings. This is caused by the fact that when you perspire, your body resistance changes. You can prove this by wetting your fingers and measuring your body resistance again. Note that this reading is lower.

PRESSURE VS RESISTANCE

Measure your body resistance by holding each probe loosely between your thumb and forefinger. Take a second reading while squeezing down as hard as you can. Is there any difference in the two readings? The reason for the different results is because "squeezing" places a larger amount of surface area of skin in contact with the probes. The larger the area in contact, the less the resistance.

What does all this have to do with the danger of being shocked or even killed while working with electricity? These examples show that every person has a different body resistance. No two people will be affected the same way by coming into contact with electricity. One person might be shocked while another may receive serious injury from the same source. The result also will depend on weather conditions and whether or not the victim was perspiring. (Remember, when you perspire, your body resistance goes down.)

Never touch electrical parts when working in a damp basement or in an area with water on the floor. Fig. 5-21. Never touch anything electrical soon after coming out of the tub or shower. Even more important never touch anything electrical while taking a bath, Fig. 5-22. If you do, your chances of being seriously hurt or killed by electrical shock have really gone up.

DON'T TOUCH UNGROUNDED APPLI-
ANCES WHILE STANDING IN WATER.

Fig. 5-21. Standing or walking in water on the floor will
reduce body resistance, making the careless person more suscep-
tible to severe electrical shock.   (Consumers Power Co.)

DON'T WEAR
A HAIR DRYER
IN THE TUB.

Fig. 5-22. Learn to respect electricity. Never wear or touch
anything electrical when taking a bath.   (Consumers Power Co.)

CURRENT PATH

Fig. 5-23 shows the path of electricity going through the heart. When this happens the heart seizes and stops beating. Some people think they can release their grip on a wire when they discover that it is carrying current. Electricity travels at the speed of light (186,000 miles

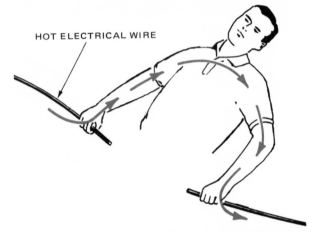

HOT ELECTRICAL WIRE

Fig. 5-23. Picking up a hot electrical wire can cause electrocution because of the direct path of current through the heart.

per second), and no human can react that fast. When you touch a current carrying wire, it is too late to change your mind.

When you touch a "hot" wire, the muscles in your hand and arm contract as shown in Fig. 5-24. This

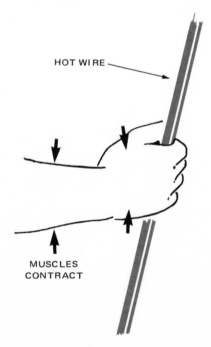

HOT WIRE

MUSCLES
CONTRACT

Fig. 5-24. Holding a hot electrical wire will cause muscles to
contract, making it impossible to let go of the wire.

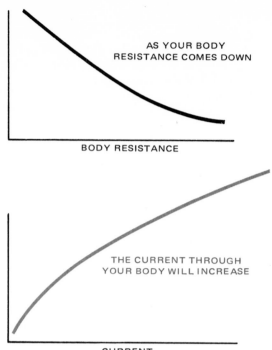

Fig. 5-25. Body resistance is reduced when the grip tightens on a hot wire. When this happens, current goes up and servere electrical shock may result.

means that instead of letting go of the wire, you are forced to grip it even harder. When this happens, the surface area in contact increases and your body resistance goes down. If the current has a path through your heart, you run the risk of serious injury. If there is no path through the heart, your chances of survival are greatly increased. Therefore, keep one hand behind your back or in your pocket (if possible) when working on a current-carrying wire.

Obviously, your best chance of survival would be to turn off the current by removing a fuse or opening a circuit breaker. This eliminates the source of power when working on a circuit.

If the conditions are right, it takes less than one amp to kill you!!

Body resistance has an important effect on your chances of survival when working with electricity. This is easy to understand when you know Ohm's Law.

Imagine that you are working on a 120 volt circuit in your home. Since $E = IR$, you know that $I \times R$ must be equal to 120 volts. Therefore:

$$I = \frac{E}{R}$$

$$I = \frac{120 \text{ volts}}{\text{body resistance}}$$

Body resistance is estimated at 50,000 ohms

$$I = \frac{120 \text{ volts}}{50,000 \text{ ohms}}$$

$$I = .0024 \text{ amp or } 2.4 \text{ mA}$$

As the body resistance falls, the current goes up. See Fig. 5-25.

SOLDERING IRON/ETCHING TOOL PROJECT

Here is a device you can build and use for soldering or etching metal, Fig. 5-26. The carbon tip is used for soldering or it can be replaced with a copper tip for etching (writing on) metal.

Obtain the following materials and assemble them according to construction procedures given in Fig. 5-27.

No.	Item
3	Battery clips 4 in. or 50 amps, lead-plated steel.
1	Handle, 1 1/4 in. dia., 5 in. long wood dowel.
1	15 ft. lamp cord, No. 18 twin lead copper wire.
1	15 ft. ground wire, No. 10 insulated copper wire.
3	Machine screws, 1/2 in. No. 8-32 RH, brass.
1	Tip extender, 3/4 in. dia., 1 1/2 in. long, brass or thinwall conduit.
1	Bushing, 1/2 in. dia., 3/4 in. long, brass or copper.
1	Tip, 3/8 in. dia., 1 1/2 in. long, carbon.
1	Tip, 3/8 in. dia., 1 1/2 in. copper.

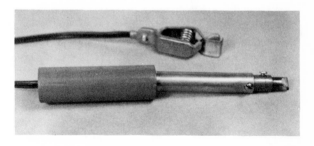

Fig. 5-26. A shop-built soldering iron/etching tool is easy to build and fun to use. (Project by Ken Koenig)

This project is designed to be used with a car battery, Fig. 5-28. To solder with this tool, hold the tip a short distance from the metal workpiece once the arc is started. *It will draw a large amount of current. Therefore, do not solder for long periods of time unless the battery is recharged.* It cannot be used on painted metal. Also avoid metal coated with grease or oil.

When using this tool for soldering, bring the carbon tip into contact with the bare metal. This completes the circuit and creates an arc. The arc is very bright and

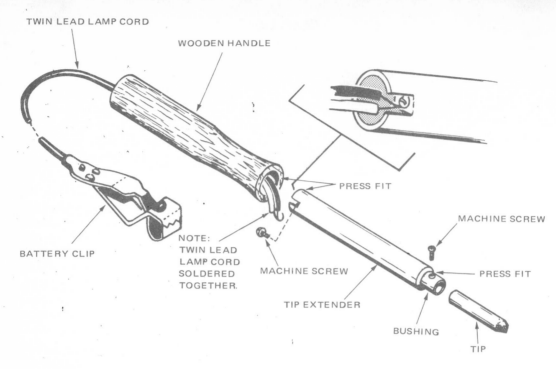

TWIN LEAD LAMP CORD

WOODEN HANDLE

PRESS FIT

MACHINE SCREW

BATTERY CLIP

NOTE: TWIN LEAD LAMP CORD SOLDERED TOGETHER.

MACHINE SCREW

PRESS FIT

TIP EXTENDER

BUSHING

TIP

Fig. 5-27. Construction details are shown for the soldering iron/etching tool pictured in Fig. 5-26.

could harm your eyes. Therefore, wear some type of welding goggles.

If you use the carbon tip for soldering the car body, you do not have to connect the second lead. This is because the car is grounded to the other battery terminal. When you use the tool in other ways, you must connect a heavy wire from the metal being soldered back to the battery. See Fig. 5-28.

If you use a copper tip instead of a carbon tip, the copper will etch the metal. Use the tool this way to mark your property with your name and social security number. Be sure that the wire is large enough to handle the current.

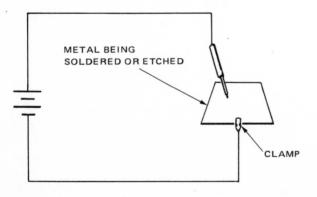

METAL BEING SOLDERED OR ETCHED

CLAMP

Fig. 5-28. This schematic shows the electrical hookup for the soldering iron/etching tool. If the soldering iron is connected to the car's battery and work is done on that car, the second electrical lead is unnecessary.

TEST YOUR KNOWLEDGE

1. Using correct letters, write the Ohm's Law formula in three different ways. $E = I \cdot R$ $I = \frac{E}{R}$ $R = \frac{E}{I}$
2. The unit of measure for resistance is the _____.
3. Explain what is meant by "hot" in an electrical circuit.
4. If we increase the voltage in an electrical circuit containing a light bulb, what might happen? Why?
5. An electrical circuit has a voltage of 10 volts and the current flowing through it is 2 amperes. What is the resistance of the circuit?
6. Voltage means about the same thing as electrical _____.
7. List two devices that can produce a voltage.
8. The term "milli" means _____ of an amp.
9. If a circuit has a current of 2.150 amperes flowing in it, how many milliamperes is that?
10. Give another name for voltage.
11. It takes less than 1 amp to kill you if it travels through your heart. True or False?
12. What is the unit of measure for electron flow in a circuit?
13. The unit of measure for electrical power is the _____.
14. Write the basic formula for calculating electrical power.
15. The unit used to measure the quantity of electricity used in your home is the _____.

82

SOMETHING TO DO

1. How far back can you go in history and find a law named after some scientist or famous person?
2. How did George Ohm discover the law named in his honor?
3. If you wanted to protect a circuit which carried 6 amps, what size fuse should you use? If you wanted to install a fuse in a circuit carrying 1 amp, what size fuse should you use?
4. How much current can flow in a 1/2 watt resistor before it would burn up?
5. Make a reading of the meter that measures the amount of electricity you use in your home. Take another reading 2 weeks later. Find out from the power company what they charge for electricity and use this information to figure out what it cost you for those two weeks.
6. Make readings each morning and evening for 2 weeks on your use of electricity in the home. When do the peak readings take place?
7. Many of your home appliances are rated in watts. How much current flows in each of these appliances? Make a list of the appliances starting with the one which uses the most current and ending with the one which uses the least current.
8. Measure your body resistance on a cool day and again on a hot day. Is there a difference? Why? Do the same thing while lightly squeezing the probes and again by squeezing the probes as hard as you can. Is there any difference? Why?
9. Knowing the speed of electricity, how many times will it travel around the world in one minute? In 30 seconds? In 5 seconds? How long would it take to travel from the east coast to the west coast of this country?
10. Show the number of ways electricity can go through your heart while working with electrical tools or appliances.

An electrical drafter uses a plastic template with various symbols cut into it. The template guides the pencil in laying out the diagram. (Acme Electric Corp.)

Chapter 6
SERIES CIRCUITS

You will find two different types of circuits on most electrical prints. These are known as "series circuits" and "parallel circuits," Fig. 6-1. In this chapter, we will be dealing with series circuits. In the next, we will cover parallel circuits.

There are times when electrical circuits have both series and parallel branches (paths for current to flow) on the same schematic. Note in Fig. 6-1 that there is only one branch in the series circuit, while the parallel circuit has a number of branches.

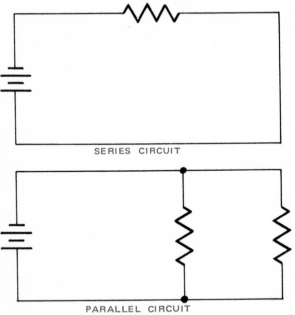

Fig. 6-1. Electrical symbols for the power source, resistors and wires form simple series and parallel circuits.

ELECTRON PATH

Many times, it is necessary to make up a circuit that has its components hooked in series. Fig. 6-2, for example, illustrates a circuit with four lamps connected in series. Electrons flowing in this circuit have only one path to follow. Starting at the source of voltage, the electrons will travel through each lamp and finally return to the source. Therefore, when current is flowing in a series circuit, the same number of electrons will pass through each component.

Basically, then, every part of a series circuit carries the same amount of current flow (amperage). You can prove this by wiring a four-lamp circuit with an

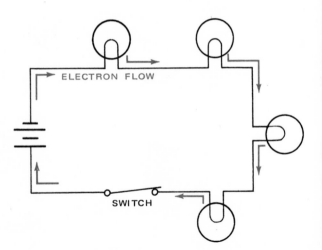

Fig. 6-2. Path of electron flow in a completed series circuit.

ammeter placed in the line behind each resistance (lamp). Close the switch and read the number of amps showing on each meter, Fig. 6-3. Each will read the same. However, if you open the switch or remove one of the wires, Fig. 6-4, all meter needles will drop to zero. This final test proves that the electrons have stopped moving.

NOTE: An ammeter is always connected in series with the components of any electrical circuit. When hooking up an ammeter to an existing circuit, some part of the circuit must be disconnected to permit installation of the tester. Usually, it is placed between a disconnected wire and its terminal.

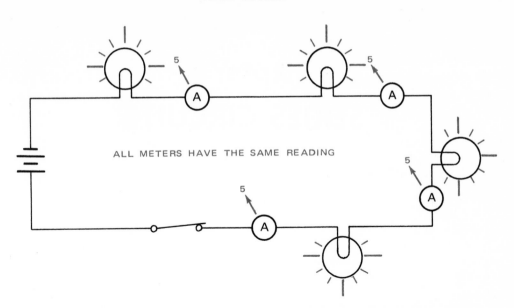

Fig. 6-3. Ammeters placed behind resistances in a completed series circuit will read the same amperage.

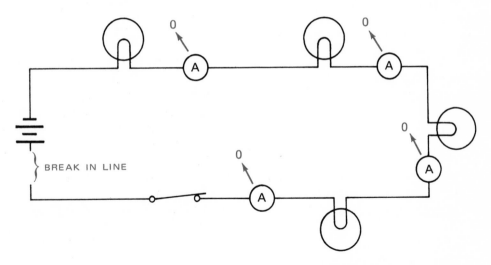

Fig. 6-4. With same circuit shown in Fig. 6-3, a break in the line will cause ammeters to read zero.

A voltmeter, on the other hand, is always connected in parallel to or across the circuit. There is no need to disconnect wires or uncouple electrical parts.

DEAD MAN'S SWITCH PROJECT

A dead man's switch, Figs. 6-5 and 6-6, is a useful project for anyone who owns and/or operates a lathe, drill press or other chuck type machine. Too often, the chuck key is left in the chuck. Then, when the machine is started, the chuck key is thrown out with great force. If the key strikes someone, he or she could be badly injured. A simple way to avoid this problem is to use a dead man's switch inserted in the power line to the machine. The switch will keep the machine from starting to operate while the chuck key is in the chuck.

To build the project, obtain the following materials and assemble them according to Fig. 6-6.

No.	Item
1	Microswitch.
1	Surface mounting outlet box and plain cover.
1	Wood block.
2	Wire nuts.
1	110 volt three wire extension cord.
1	Junction box grounding clip w/wire.
2	1/2 in. connectors.

Fig. 6-5. Dead man's switch. Note that chuck key is inserted in opening to complete the circuit. (Project by John Kleihege)

With the dead man's switch connected to the machine power cord and a 110 volt receptacle, the machine will not operate unless the chuck key is dropped in the hole in the cover of the outlet box, Fig. 6-5. In this position, the key will close the microswitch and allow current to pass to the on-off switch on the machine.

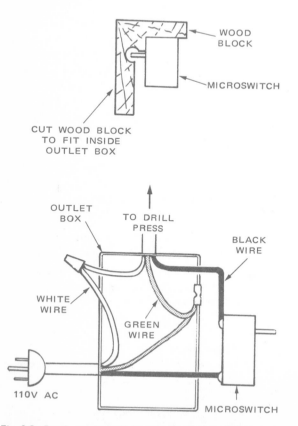

Fig. 6-6. Dead man's switch construction. Switch protects operator from being hit by chuck key left in machine.

AMPS IN A SERIES CIRCUIT

In working with the circuit shown in Fig. 6-3, we proved that every part of a series circuit carries the same amount of current flow. Further proof of this can be had by examining three different test circuits, each having a total resistance of 45 ohms.

First, look at a series circuit with a 90 volt power source, one 45 ohm resistor and a test ammeter, Fig. 6-7. When this simple circuit is completed, the meter will read 2 amps.

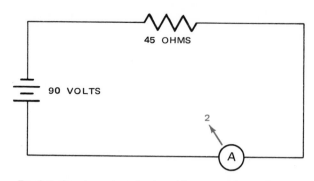

Fig. 6-7. Simple series circuit with one resistor. Electrical values are given.

A second test circuit shown in Fig. 6-8 has a 90 volt power source in series with three 15 ohm resistors and an ammeter. When the last wire is connected, the meter will read 2 amps.

Finally, a third test circuit, using a 90 volt power source has three different resistors (10, 15 and 20 ohms) and an ammeter, Fig. 6-9. Again, 2 amps of current is flowing in the circuit. These three circuits are only used as examples. Do not actually hook them up because the high current will burn up the resistors.

TOTAL RESISTANCE

Each test circuit, Figs. 6-7, 6-8 and 6-9, carries the same number of amps. We proved this by reading the ammeter in each circuit.

In Chapter 5, however, we found that it is possible to figure how many amps are flowing in a given circuit by applying Ohm's Law. We can put this law to use while working with the test circuits just covered. In Fig. 6-7, for example, apply Ohm's Law as follows:

$$I = \frac{E}{R}$$

$$I = \frac{90 \text{ volts}}{45 \text{ ohms}}$$

$$I = 2 \text{ amps}$$

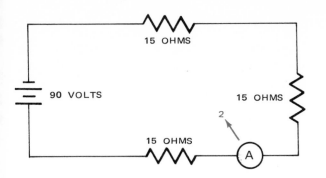

Fig. 6-8. Three resistors of equal value in a series circuit have same total resistance as single resistor in Fig. 6-7.

In order to apply this formula to the other two test circuits, you need to know the total resistance of the circuit. To find the total resistance, simply add the values of the individual resistances:

$$R_{Total} = R_1 + R_2 + R_3 \ldots$$

The ellipsis (...) means that something has been omitted. In this case, it means that you keep adding resistance values until you have accounted for all individual resistors and/or resistances.

In the circuit shown in Fig. 6-8, you have three resistors of equal value. Compute total resistance and how much amperage is flowing as follows:

$$R_T = R_1 + R_2 + R_3$$

$$R_T = 15 + 15 + 15 = 45 \text{ ohms}$$

$$I = \frac{E}{R}$$

$$I = \frac{90 \text{ volts}}{45 \text{ ohms}}$$

$$I = 2 \text{ amps}$$

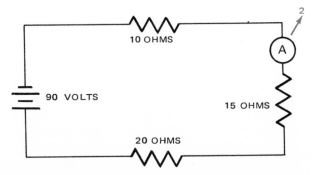

Fig. 6-9. Three resistors of unequal values in a series circuit also have same total resistance as single resistor in Fig. 6-7.

In Fig. 6-9, you have resistors of three different values wired into the series circuit. Compute total resistance and amperage as follows:

$$R_T = R_1 + R_2 + R_3$$

$$R_T = 10 + 15 + 20$$

$$R_T = 45 \text{ ohms}$$

$$I = \frac{E}{R}$$

$$I = \frac{90 \text{ volts}}{45 \text{ ohms}}$$

$$I = 2 \text{ amps}$$

These three test circuits having different individual resistance setups show that total resistance governs how much current is flowing in a series circuit.

Now try a sample problem. What would the total resistance be in the series circuit shown in Fig. 6-10? There are five components in the circuit: a lamp with 15 ohms resistance; a fixed resistor with 45 ohms resistance; a variable resistor set at 100 ohms resistance; a switch, fuse and wire.

NOTE: Although the switch, fuse and wire do have some resistance, it usually is overlooked in circuit calculations because the amounts are so small. Most circuits, for example, will use wire in the range of 12 to 22 AWG (American Wire Gage). Turn to Chapter 3, Fig. 3-11. According to this table of resistances, you would need 629 feet of No. 12 wire, or 61 feet of No. 22 wire, to get one ohm of resistance. Few circuits, other than house wiring, have enough wire to affect total resistance to a great degree. Most circuits which you will be wiring on a bench in the school shop will involve just a small amount of wire.

In the series circuit in Fig. 6-10, then, compute total resistance and amperage as follows:

$$R_T = 15 + 45 + 100 \text{ ohms}$$

$$R_T = 160 \text{ ohms}$$

$$I = \frac{E}{R}$$

$$I = \frac{80 \text{ volts}}{160 \text{ ohms}}$$

$$I = 0.5 \text{ amps or } 500 \text{ mA}$$

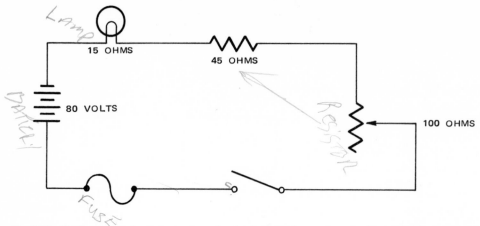

Fig. 6-10. Finding the total resistance of a series circuit is simply a matter of adding individual resistances.

Remember, in a series circuit, you must use the total resistance to find the amperage. To find the amperage in Fig. 6-11, first total the resistances:

$$R_T = R_1 + R_2 + R_3$$

$$R_T = 200 + 1000 + 1200 \text{ ohms}$$

$$R_T = 2400 \text{ ohms}$$

To compute the amperage use Ohm's Law:

$$I = \frac{E}{R}$$

$$I = \frac{24 \text{ volts}}{2400 \text{ ohms}}$$

$$I = .01 \text{ amp or } 10 \text{ mA}$$

We can prove it by reading the milliammeter, which is also in the circuit.

Note that the source of voltage in Fig. 6-11 is a generator. You will find that a generator is used in some circuits instead of (or in conjunction with) a battery. The generator will produce voltage and save the effort of having to change batteries every time they lose their charge.

POLARITY

There are a lot of electrical items that cannot be plugged into a wall outlet for a source of voltage. These items include flashlights, children's toys and portable tape recorders, all of which operate off dry cells. Most use more than one cell because of the need for higher voltage.

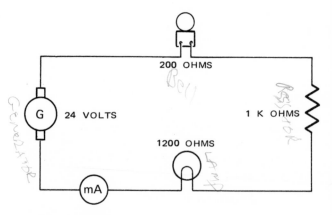

Fig. 6-11. Series circuit with a generator as the power source.

When you prepare to put cells in these devices, note that there are markings on each cell, and on the device, to show you the right way to make the installation. Each cell has a positive (+) pole and a negative (−) pole. To get proper voltage needed to operate a given electrical device, it is necessary to face these poles in the proper direction. This is what is known as "correct polarity." If you face the cells in the wrong direction, you will have "incorrect polarity."

Fig. 6-12 illustrates why polarity is important. Note that when all cells face in the same direction, their individual voltages *add* (1 1/2 + 1 1/2 + 1 1/2 + 1 1/2 = 6). If one cell faces the opposite way, it will reduce total voltage (1 1/2 + 1 1/2 − 1 1/2 + 1 1/2 = 3).

Polarity, then, is the proper identification of positive and negative terminals of a battery, cell or component in an electrical circuit. The polarity of electrical devices will affect: the direction in which a meter needle moves; how bright a flashlight will light; which direction a motor will turn; whether or not some devices will operate.

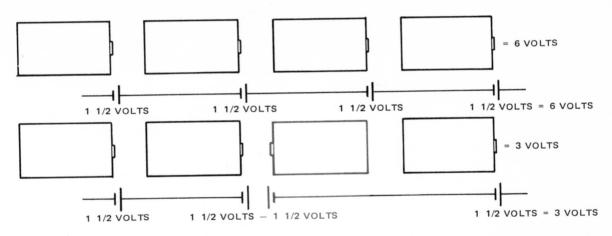

Fig. 6-12. Polarity of cells connected in series. Top. Correct polarity. Bottom. Incorrect polarity, with resulting loss of voltage.

All of which goes to prove that the sum of voltages in a series circuit depends on their polarity. If the polarity is correct, the voltages will add. If not, circuit

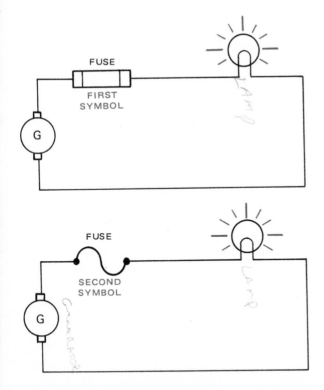

Fig. 6-13. Fuse connected in series with a lamp. Either fuse symbol is acceptable.

voltage will be reduced. One exception is a resistor installed in a series circuit. Resistors do not have polarity problems because they do not have positive and negative poles.

CONTINUITY LIGHT

There are times when you must install wire or components in a circuit so that they are in series with one another. In Fig. 6-13, the fuse (both symbols are shown) is in series with the lamp. After the circuit has been in operation for awhile, something could happen that would create an overload and blow the fuse.

If the lamp in this circuit does not light because of a blown fuse, there are several ways of checking it. The simplest way would be to see if the element is burned through (appears to be broken or completely missing).

Another way of checking for a blown fuse is to use a continuity light, Fig. 6-14. When the element is broken the light will not come on. If the element is good

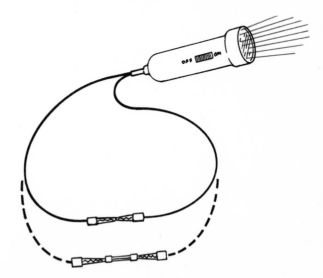

Fig. 6-14. Continuity light will come on if alligator clips touch each other or if they are placed in contact with both ends of a good fuse.

(solid), the continuity light will come on. This tells you that the circuit is continuous, which is why we call the tester a "continuity light."

A continuity light also can be used to check wires to find out whether or not a "break" in a wire exists. If there is a break in the copper under the insulation, you cannot see it. A continuity light will help pinpoint the location of the break.

In another application, a continuity light will reveal both ends of a given wire from among a group of wires bundled together, Fig. 6-15. When we get to the chapter on meters, you will be shown how to make a continuity check with an ohmmeter.

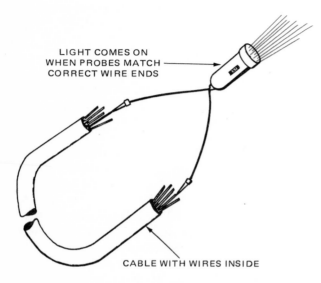

LIGHT COMES ON
WHEN PROBES MATCH
CORRECT WIRE ENDS

CABLE WITH WIRES INSIDE

Fig. 6-15. A continuity light can be used to find two ends of the same wire in a cable or wiring harness.

MULTIPLE OUTLET EXTENSION CORD PROJECT

How many times have you found yourself in need of another electrical outlet? After you hook up an extension cord, you run into the problem of trying to run two power tools. Fig. 6-16 shows details of a little device which will solve that problem. This extension cord outlet box has four receptacles that can be turned on and off with a switch. It also has indicator lamps to tell you whether or not the power is on. If you should overload either side, it is protected by a fuse.

The mutliple outlet extension cord is shown in Fig. 6-16. One view pictures the completed outlet box; another shows details of its internal construction; the schematic illustrates circuit hookups.

To build the project, obtain the following materials:

No.	Item
1	Outlet box.
2	NE 2 lamps with dropping resistors.
2	Fuse assemblies (w/3 to 4 amp fuses).
2	DPST switches.
2	110 volt duplex receptacle w/ground.
1	Male plug (cap) w/ground.
1	18/3 extension cord.
1	Grommet or cord strain relief.

CONSTRUCTION PROCEDURE

Check all parts and decide on size of box. You can either build one or buy it from a local electrical supply store. Proceed as follows:

1. Lay out location of fuses, lamp, resistors, switches and receptacles. Allow enough room, so it will be easy to wire and build.
2. Cut holes in cover for various parts.
3. Mount receptacles in cover and wire them. Use black covered wire to attach to brass electrical fitting on receptacle; attach white wire to silver fitting. Use green wire for grounds.
4. Mount resistors (100,000 ohms or more) and indicator lights, splicing them into line. Keep all connections well insulated, especially if you use a metal box.
5. Mount fuses and switches, then wire them in position.
6. When wiring is completed, close either switch. The indicator lamp should come on. If not, a fuse may be blown.
7. Wire an extension cord of desired length to box as shown in Fig. 6-16. Use a wireman's knot inside box, so wire cannot be pulled out during use.
8. Mount the male plug on the other end of the extension cord. Be sure to use a grounded 110V receptacle to give you the protection you need when using portable power tools.

PUTTING THE PROJECT TO WORK

Plug the extension cord into the 110V receptacle and unwind as much of the cord as you need to bring the multiple outlet box to the job.

You may wish to make a stand on wheels for the multiple outlet box, so you can roll the outlet box and stand to each job. In addition, you can drill a 1/2 in. hole at each corner of the stand and install short wooden dowels. Then, wrap the cord around the dowels.

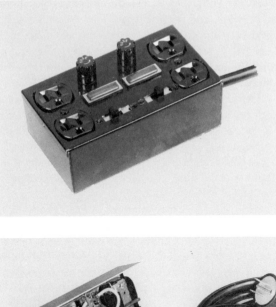

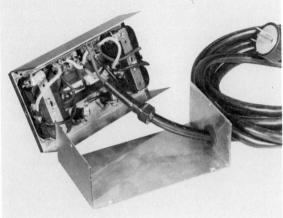

VOLTAGE

In order to study voltages, look back at some of the circuits made up earlier in this chapter. Take the one shown in Fig. 6-7, having a 90 volt power source and a 45 ohm resistor. First, measure the voltage at the source, using a voltmeter as shown in Fig. 6-17. It will read 90 volts. Next, connect the meter across the resistor as indicated in this illustration. It, too, will read 90 volts. Remember, we mentioned that the wiring in these test circuits has almost no resistance.

However, if we add another resistor to the circuit, Fig. 6-18, we start to see some changes. Note the three voltmeters connected across the circuit at various points. All indicate voltage, and the one across the

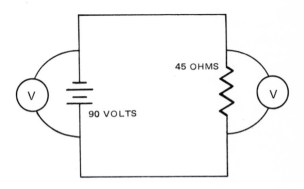

Fig. 6-17. A voltmeter is always connected in parallel with an electrical circuit.

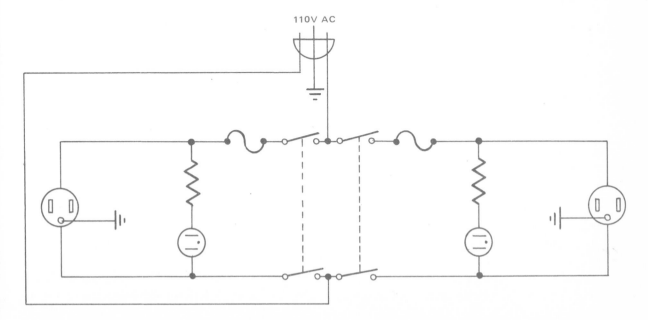

Fig. 6-16. Multiple outlet extension cord. Top. Completed project. Center. Internal view of outlet box. Bottom. Schematic for use in assembling this project. (Project by Mike O'Berski)

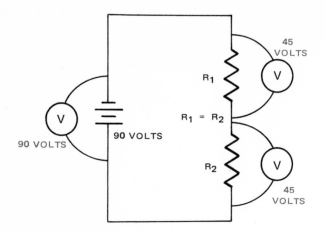

Fig. 6-18. Voltmeter readings taken across two resistors in a series circuit.

power source reads 90 volts. The two voltmeters connected across the resistors are equal, and the sum of their voltages (45 + 45 = 90) is equal to the reading on the voltmeter connected across the power source.

VOLTAGE DROP

Voltage drop is voltage measured across each resistor. Take the circuit shown in Fig. 6-8, having a 90 volt power source and three 15 ohm resistors. Make voltmeter tests at the points shown in Fig. 6-19. Note that the voltmeter across the power source reads 90 volts,

while the voltmeters across the resistors read 30 volts each. If you add these voltages, you will find that they again total the voltage of the power source. No matter how many resistors are placed in a series circuit, the total applied voltage (amount put out by the power source) will be divided among them.

Therefore, we have established that voltage across each resistor is called voltage drop; also that the sum of all voltage drops will be equal to the applied voltage. This discovery was named in honor of the discoverer. It is one of "Kirchhoff's Laws," and it states simply that "Total voltage drop in a series circuit will be equal to the applied voltage."

With this in mind, consider voltage drops in the circuit shown in Fig. 6-20, having four 10 ohm resistors and a generator that produces 40 volts. What would you expect each voltmeter to read? Voltmeters 1 through 4 would read 10 volts each, while voltmeter number 5 would read 40 volts. Each resistor, however, may or may not drop voltage the same amount. The amount of voltage drop depends on the value of each resistor.

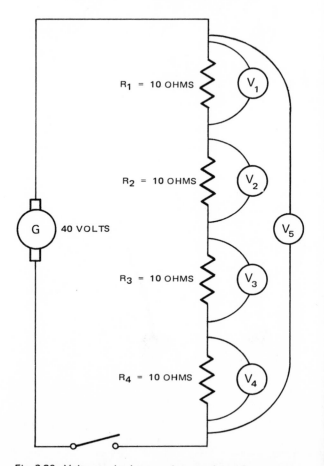

Fig. 6-19. Voltmeter readings taken across three resistors of like values in a series circuit.

Fig. 6-20. Voltmeter lead connections are shown for measuring voltage drops across each of four resistors of like values, and across all four resistors at once.

OHM'S LAW IN SERIES CIRCUITS

The amount of voltage drop across a given resistor can be determined by either of two methods. We can use a voltmeter to measure its value or we can use Ohm's Law. We worked with a voltmeter in getting voltage drop values in the last four circuits. Now we will use Ohm's Law to figure voltage drops across three different resistors in the series circuit shown in Fig. 6-21, then follow up with voltmeter readings.

First, compute the total resistance using symbols and abbreviations learned in Chapter 1.

$$R_T = R_1 + R_2 + R_3$$

$$R_T = 10\Omega + 15\Omega + 20\Omega$$

$$R_T = 45\Omega$$

Next, divide the source voltage by the total resistance:

$$I = \frac{E}{R_T}$$

$$I = \frac{90V}{45\Omega}$$

$$I = 2A$$

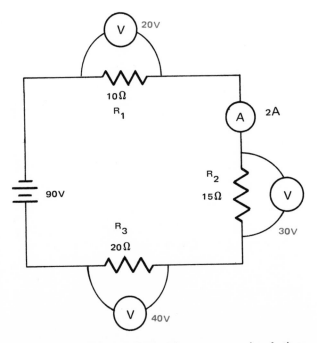

Fig. 6-21. Voltmeter readings taken across each of three resistors of unequal values.

Then, multiply each resistance by the amps running through the circuit in Fig. 6-21. This will give you the voltage drop across each resistor.

The voltage drop across R_1 would be:

$$E_1 = IR_1$$

$$E_1 = 2A \times 10\Omega$$

$$E_1 = 20V$$

You can see that using the E in this formula could be confusing. So instead of using the E, do as many people do: use a V for the voltage drop across each resistor.

$$V_{R_1} = 20V$$

Figuring out the rest of the circuit, you would have:

$$V_{R_2} = IR_2$$

$$V_{R_2} = 2A \times 15\Omega$$

$$V_{R_2} = 30V$$

$$V_{R_3} = IR_3$$

$$V_{R_3} = 2A \times 20\Omega$$

$$V_{R_3} = 40V$$

TOTAL VOLTAGE DROP

Total voltage drop in a series circuit should be equal to voltage at the source. To find total voltage drop in the circuit shown in Fig. 6-21, use the following formula:

$$E_T = V_{R_1} + V_{R_2} + V_{R_3}$$

Next, substitute the individual voltage drop values you worked out in solving the last problem:

$$E_T = V_{R_1} + V_{R_2} + V_{R_3}$$

$$E_T = 20V + 30V + 40V$$

$$E_T = 90V$$

You can see that 90 volts does equal our source voltage.

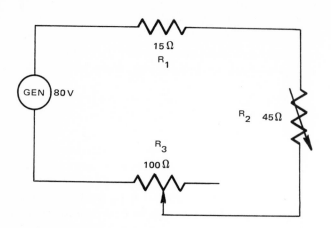

Fig. 6-22. Calculating unknown voltage drops. First determine total resistance, then amperage. Using Ohm's Law, figure individual voltage drop across resistors and come up with total voltage drop for the circuit.

Try another example, using Fig. 6-22. First find the total resistance:

$$R_T = R_1 + R_2 + R_3$$

$$R_T = 15\Omega + 45\Omega + 100\Omega$$

$$R_T = 160\Omega$$

Next, find the amperage:

$$I = \frac{E}{R}$$

$$I = \frac{80V}{160\Omega}$$

$$I = 0.5A \text{ or } 500 \text{ mA}$$

Then, find the voltage drop across each resistor.

$$V_{R_1} = I \times R_1$$

$$V_{R_1} = 0.5A \times 15\Omega$$

$$V_{R_1} = 7.5V$$

$$V_{R_2} = I \times R_2$$

$$V_{R_2} = 0.5A \times 45\Omega$$

$$V_{R_2} = 22.5V$$

R_2 is the symbol for variable resistor.

$$V_{R_3} = I \times R_3$$

$$V_{R_3} = 0.5A \times 100\Omega$$

$$V_{R_3} = 50V$$

Finally, add up the individual voltage drops to see if their sum equals the applied voltage:

$$E_T = V_{R_1} + V_{R_2} + V_{R_3}$$

$$E_T = 7.5V + 22.5V + 50V$$

$$E_T = 80V$$

FOOT SWITCH PROJECT

How many times have you wanted to turn on electrical power, but both hands were full? You could have handled this problem if you had a foot switch, Fig. 6-23, attached to an electrical outlet. By merely pushing down on the switch with your foot, you could have controlled the flow of power to some electrical device.

To build this project, you will need the following items:

No.	Item
1	Foot switch.
1	Double outlet receptacle.
1	Male plug.
2	Black lead wires.
1	White lead wire.

CONSTRUCTION PROCEDURE

Study Fig. 6-23 and proceed as follows:
1. Install white wire and one black wire in male plug.
2. Connect other end of white wire to double outlet receptacle.
3. Connect other end of black wire to foot switch.
4. Use other black wire to connect foot switch to double outlet receptacle.

There are many good uses for a foot switch, including photo enlarging. Operator uses left hand to feed light sensitive paper to an easel under enlarger lens. Foot switch is pressed down to turn on enlarger light. Operator times exposure, releases foot switch and uses right hand to remove exposed paper from easel and drop it in developer.

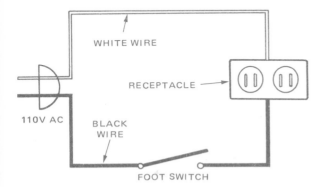

Fig. 6-23. A foot switch connected in series with an electrical device frees both hands. Top. Foot switch and twin receptacle. Bottom. Schematic for use in constructing this project.
(Project by Dave Wightman)

TEST YOUR KNOWLEDGE

1. There is only one _____ in a series circuit, while a parallel circuit has several.
2. Every part of a series circuit carries the same amount of _____.
3. An ammeter is always connected _____ components of an electrical circuit.
4. A voltmeter is always connected _____ components of an electrical circuit.
5. What is polarity?

6. What does R_T stand for?
7. Draw the symbol for ohms.
8. Give the Ohm's Law formula for finding the amount of current flow or amperage in an electrical circuit.
9. If you had three resistors connected in series, each with a value of 30 ohms, how much total resistance would you have?
10. How much current will flow in a circuit having a 24 volt power supply, a 200 ohm resistor and a 2200 ohm resistor in series?
11. To find the voltage drop across a resistor, we can use a voltmeter to measure its value, or we can use

_____.

SOMETHING TO DO

1. Why will several lights in your house go out if you open a circuit breaker?
2. If a continuity light will not light up when checking a coil, what is indicated?
3. Visit a house which is being built. What type and size of wire is being used? Where is the ground wire, and what color is it?
4. List some places or things where you will find three or more sources of resistance connected in series.
5. Explain why a continuity light should never be used on a circuit while it is connected to its source of power?
6. List Kirchhoff's laws and tell what they mean.
7. Make a list of the people who have electrical laws named after them. Show the dates when these people lived. Notice in which years most of their discoveries were made.
8. Where did omega, the name of the symbol used for resistance, first come from? How old is the term?
9. Which electrical devices in your home have the greatest resistance? If they are listed in watts, explain why your answer is correct.

Chapter 7
PARALLEL CIRCUITS

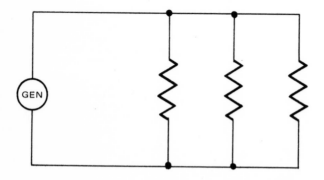

Fig. 7-1. A parallel circuit provides two or more paths for the flow of electricity.

After learning how series circuits work, your next assignment is the study of "parallel circuits," Fig. 7-1. As you work your way through this chapter, you will see why all circuits cannot be wired in series. Parallel circuits have broad application, including radios, television sets and the wiring in your home.

CURRENT IN A PARALLEL CIRCUIT

A parallel circuit is one which has two or more paths (branches) through which electrons flow. For example, electron flow in a parallel circuit may be related to water flowing down a river bed, Fig. 7-2. First, note the full flow down the main branch of the river. Then, when the water reaches the two islands, the river splits and takes three paths. Since each path is the same size, the same amount of water will flow in each.

The same thing is true of electrical circuits. In Fig. 7-3, note that the electrons have only one path to follow when they leave the generator (or battery). Then, when three equal paths are provided, the electrons split up evenly in taking those paths.

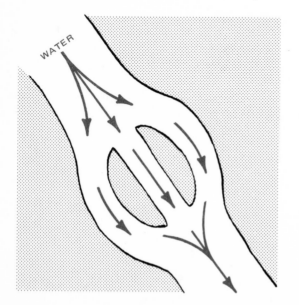

Fig. 7-2. Water flowing downriver will divide evenly when branching into three equal paths.

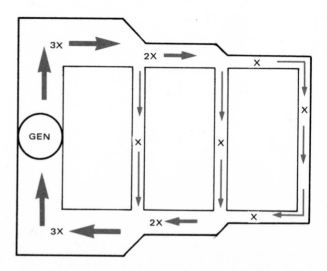

Fig. 7-3. Electron flow in a parallel circuit also divides evenly when three equal electrical paths are provided.

RESISTANCE IN A PARALLEL CIRCUIT

Again relating water flow to electron flow, note that the river shown in Fig. 7-4 also flows down three different paths. However, since these paths are unequal in size, the water will not divide equally. Most of the water will flow through the largest path, which is the

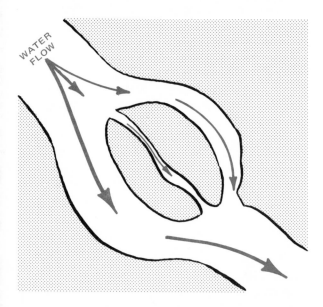

Fig. 7-4. Water flowing downriver will divide unevenly when branching into several unequal paths.

one that offers the least resistance. Lesser amounts of water will flow through the smaller paths, or those offering the most resistance.

The same thing is true of electron flow in an

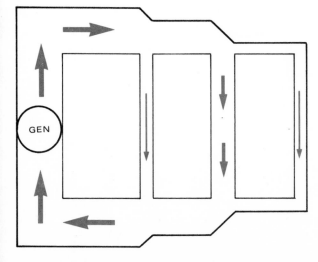

Fig. 7-5. Electron flow in a parallel circuit also divides unevenly when the electrical paths are unequal.

electrical circuit of the parallel type. If the circuit has three unequal paths, the total number of electrons flowing will divide unequally.

Basically, then, the branch of a parallel circuit that offers the smallest amount of resistance will provide the easiest path for electron flow, Fig. 7-5. Ohm's Law can be used to determine the number of electrons flowing but, first, it is necessary to find the total resistance of the circuit.

Total resistance in a parallel circuit can be computed in various ways. One convenient method is by means of the "conductance" formula. Conductance is the reciprocal (opposite) of resistance. It is the current carrying ability of any wire or electrical component. Resistance, on the other hand, is the ability to oppose the flow of current.

Use this conductance formula, then invert the answer to find total resistance:

$$\frac{1}{R_T} = \frac{1}{R_1} + \frac{1}{R_2} + \frac{1}{R_3} \cdots$$

In the portion of a parallel circuit shown in Fig. 7-6, the two resistors connected in parallel have values of 20 ohms each. Substitute these values for factors (numbers or symbols in mathematics) in the formula as follows:

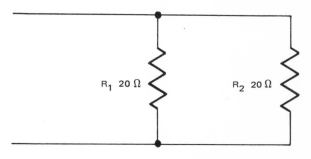

R_1 20 Ω R_2 20 Ω

Fig. 7-6. Parallel paths of equal resistance permit the same amount of current to flow in each electrical path. Total resistance in this circuit is 10 ohms (see text).

$$\frac{1}{R_T} = \frac{1}{R_1} + \frac{1}{R_2}$$

$$\frac{1}{R_T} = \frac{1}{20} + \frac{1}{20}$$

$$\frac{1}{R_T} = \frac{2}{20} = \frac{1}{10}$$

Invert both sides of the equation (equal factors):

$$\frac{R_T}{1} = \frac{10}{1} = 10 \text{ ohms}$$

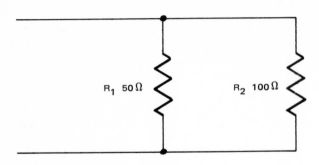

Fig. 7-7. Parallel paths of unequal resistance vary the amount of current flowing in each electrical path. Total resistance of this circuit is 33.3 ohms (see text).

Try another total resistance example. In Fig. 7-7, R_1 is rated at 50 ohms, R_2 is 100 ohms. Again use the formula:

$$\frac{1}{R_T} = \frac{1}{R_1} + \frac{1}{R_2}$$

$$\frac{1}{R_T} = \frac{1}{50} + \frac{1}{100}$$

$$\frac{1}{R_T} = \frac{2}{100} + \frac{1}{100} = \frac{3}{100}$$

Invert both sides of the equation:

$$R_T = \frac{100}{3} = 33.3\,\Omega$$

I.C. SAYS:

"For a quick check to see if your work is okay, look at your answer. The R Total (R_T) must be smaller than either of the resistors used in the circuit."

If your R Total comes out larger than either of the resistors, the answer is wrong. Recheck your work. Study Fig. 7-8 to see why this is so.

COMBINING PATHS
FOR EQUIVALENT SIZE

Fig. 7-8. Two electrical paths combined are equivalent to one path of larger size.

TOTAL RESISTANCE

By using the conductance formula for finding the total resistance in a parallel circuit, you are trying to show an "equivalent circuit," Fig. 7-8. An equivalent circuit is one that takes out all resistors and replaces them with just one resistor. In this way, you can find the amount of current that the power supply is being asked to deliver. It will tell you what effect the load has had on the power supply. Remember, you do not want to overload the circuit or blow a fuse.

Anytime you have resistors in parallel you are making a larger path for the electrons to flow. Anytime you combine two or more paths into one path for electron flow, that path will get larger. *Therefore, when you find the R Total in a parallel circuit, it must offer less resistance to the flow of electrons; and it must be smaller than any one of the resistors in the parallel circuit.*

Fig. 7-9 shows a circuit with three resistors. Compute R_T as follows:

$$\frac{1}{R_T} = \frac{1}{R_1} + \frac{1}{R_2} + \frac{1}{R_3}$$

$$\frac{1}{R_T} = \frac{1}{500} + \frac{1}{500} + \frac{1}{1000}$$

$$\frac{1}{R_T} = \frac{2}{1000} + \frac{2}{1000} + \frac{1}{1000} = \frac{5}{1000}$$

$$\frac{1}{R_T} = \frac{1}{200}$$

Invert both sides of the equation:

$$R_T = 200\,\Omega$$

Note, again, that R Total is smaller than any of the resistors in the circuit.

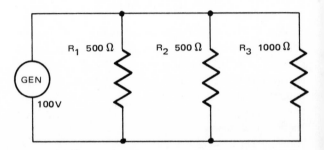

Fig. 7-9. This parallel circuit with three resistors has a total resistance of 200 ohms (see text).

CURRENT

Now that you have found R Total, the next thing to do is find the amount of current flowing in the circuit. Again, we can use Ohm's Law and the values from Fig. 7-9:

$$I = \frac{E}{R}$$

$$I = \frac{100V}{200\,\Omega}$$

$$I = .5A$$

Because R Total was used, the value .5A is the amount of current flowing in the total circuit.

To find the amount of current flowing in each resistor, again use Ohm's Law. In R_1 (and R_2), the amount of current flow would be:

$$I = \frac{E}{R}$$

$$I = \frac{100V}{500\,\Omega}$$

$$I = .2A$$

In R_3 the current equals:

$$I = \frac{E}{R}$$

$$I = \frac{100V}{1000\,\Omega}$$

$$I = .1A$$

Adding these three current values gives .5 amp (.2 + .2 + .1), which is equal to the total current flowing in the circuit.

ERRORS IN CURRENT VALUES

Fig. 7-10 shows the same circuit as Fig. 7-9, with five ammeters installed in the lines. Ammeters 1 and 5 both show .5 amp. Each reading indicates total current flow in the circuit. Ammeters 2 and 3 read .2 amp, and ammeter 4 reads .1 amp. These values represent current flow through each resistor in the circuit. Again, adding these three current values gives .5 amp, which is the total current flowing in the circuit.

These examples serve to illustrate that total current flow in a parallel circuit is equal to the sum of the currents flowing through the branches of the circuit.

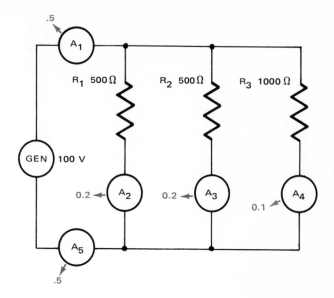

Fig. 7-10. This parallel circuit with ammeters in the lines show effect resistors of different values have on current flow.

Therefore, in a circuit using three resistors, the formula would be:

$$I_{Total} = I_1 + I_2 + I_3$$

I_1, I_2 and I_3 represent the amount of current flowing through each resistor.

I.C. SAYS:

"If you actually hook up this circuit, you may get different results. This difference is because of the tolerance of various resistors. Your results may vary from plus to minus 5 percent if the resistors have a gold fourth band; plus or minus 10 percent if they have a silver fourth band; plus or minus 20 percent if they have no fourth band."

BREADBOARDING

Fig. 7-11 shows a student who has set up a parallel circuit based on an activity in an electricity workbook. The student has wired the circuit, using parts from a kit. When working with circuits, kits are an easy way to get started. However, before you solder any circuit together, it would be wise to "breadboard" it first.

Breadboarding, Fig. 7-12, is a way of laying out the circuit to see if it works. The various components of a given project are mounted on a perforated board and connected temporarily for test purposes. By bread-

Fig. 7-11. Student traces a circuit on an electrical project built from a kit in the school shop. (Hickok Teaching Systems, Inc.)

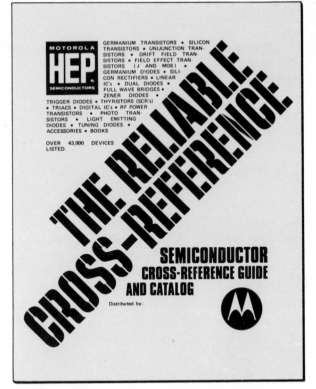

Fig. 7-13. Replacement parts catalogs usually include a cross-reference section for use when substituting replacement parts for original equipment.

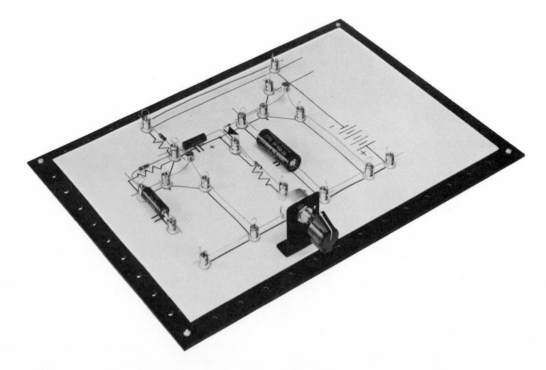

Fig. 7-12. "Breadboarding" a circuit is a means of testing circuit operation before soldering parts and connections. (Cosmic Voice, Inc.)

HEP Semiconductor Replacement Cross-References

DEVICE	HEP	DEVICE	HEP	DEVICE	HEP	DEVICE	HEP	DEVICE	HEP	DEVICE	HEP
399C	R0092	429-0958-41	R0050	576-0003-011	728	617-67	730	690V043H63	254	690V109H44	R0050
399D	R0092	429-0958-42	53	576-0003-012 (NPN)	56	617-68	729	690V047H56	253	690V109H46	53
404-2 (Syl)	729	429-0958-43	Z0408	576-0003-012 (PNP)	717	617-69	238	690V047H57	253	690V109H72	R2502
404A	R0160	429-0986-12	56	576-0003-017	57	617-70	632	690V047H60	632	690V110H30	55
404B	R0162	429-0989-68	R0050	576-0003-018	56	617-71	243	690V047H61	134	690V110H31	54
4C	R0162	430 (Zen)	730	576-0003-019	57	617-117	243	690V049H81	56	690V110H32	50
	R0162	431-26551 A	C6056P	576-0003-020	56	617-87	245	690V052H23	250	690V110H33	50
	R0137	43...(Syl)	243	576-0003-021	56	617-156	134	690V052H24	254	690V110H34	243
				576-0003-022	736	617-161	733	690V052H50	134	690V110H36	243
				...0003-023	709	617-162	709	690V056H31	R0054	690V110H55	242
					3	617-163	R0054	690V056H32	639	690V116H20	53
							245	690V056H33	633	693G	55
							946	690V056H34	632	699 (GE)	736
								690V057H24	254	700-04	801
								...7H25	637	702-810	R9001
								703-1			

Fig. 7-14. Cross-reference listings in parts catalogs give original equipment numbers alongside of replacement parts numbers.

boarding first, you can be sure that:
1. Wiring of components has been done properly.
2. Components of correct value have been used.
3. Circuit will work.

In this manner, you can prove to yourself that the circuit works before you solder it together. "After soldering" is the wrong time to find out the circuit does not work. To remove solder from connections and check out components takes a lot longer than if it were done on a breadboard first.

You will find that some circuits in magazines and books do not work very well. Some do not work at all, or you may have to substitute one manufacturer's components for another's. When this happens, use a substitution manual or look up the cross-reference list in a replacement parts catalog, Figs. 7-13 and 7-14.

Occasionally, a circuit fitted with substitute parts will work poorly or not at all. This occurs because some electrical parts have important and critical (vital) values. If substituted for, circuit design may be changed. It is better to find out about circuit problems on a breadboard first.

USES FOR PARALLEL CIRCUITS

Next, let us see how parallel circuits can be of value to you. Suppose you have a radio and want to attach two speakers to it. You look through a pile of speakers and find some marked 4 ohms, 8 ohms and 16 ohms. However, the back of the radio is marked 4 ohms, meaning it calls for one 4 ohm speaker.

Since you wish to install more than one speaker, you do have a problem. One answer to selecting the right speakers for the project is to make up a parallel circuit, using two speakers marked 8 ohms each, Fig. 7-15. Working out total resistance for this parallel circuit, you will find that two 8 ohm speakers connected in parallel equal 4 ohms of total resistance.

$$\frac{1}{R_T} = \frac{1}{R_1} + \frac{1}{R_2}$$

$$\frac{1}{R_T} = \frac{1}{8} + \frac{1}{8}$$

$$\frac{1}{R_T} = \frac{2}{8} = \frac{1}{4}$$

Invert both sides of the equation:

$$R_T = 4\,\Omega$$

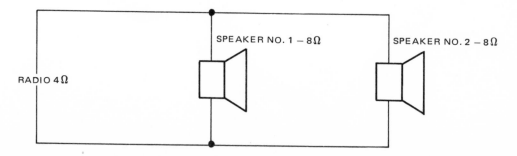

RADIO 4Ω   SPEAKER NO. 1 – 8Ω   SPEAKER NO. 2 – 8Ω

Fig. 7-15. In this circuit, two speakers are connected in parallel, each having 8 ohms resistance to match radio's requirement for 4 ohms total resistance (see text).

The problem is solved. The total resistance of the two speakers now matches the 4 ohms marked on the back of the radio.

ELECTRIC PUZZLE PROJECT

Here is a puzzle that you can build with four switches, a lamp and two dry cells. The whole idea is to wire the project so that the lamp will come on if a mistake is made while working the puzzle.

Each switch, in turn, represents a child, a chicken, corn and a fox. The trick is to operate the four switches so that the child can get all three possessions safely across the river. On the shore is a rowboat which is used to carry the child and only one possession at a time. The "catch" is: the fox will eat the chicken or the chicken will eat the corn if left together on one side of the river.

If you wire the switches as shown in Fig. 7-16, there is a way for the child to get all three possessions safely across the river. Try your skill to see if you can figure out how to do it. Remember, the child can move only one possession per trip across the river.

When working the puzzle, imagine that the switches span the river. S_1 in Fig. 7-16 and 7-17 is for controlling the child's crossings. S_2 is for the chicken. S_3 is for the corn. S_4 is for the fox.

Fig. 7-16. Electric puzzle. Right. Finished project. Below. Drawing shows switches, lamp and wiring hookups. (Project by Dick Butcke)

The answer follows the construction procedure.

No.	Item
2	1.5V dry cells (B_1).
1	3V dc lamp (L_1).
3	SPDT knife switches $(S_1, S_3 \text{ and } S_4)$.
1	DPDT knife switch (S_2).
	Lead wire, wooden box, baseboard, lamp socket and holder for cells.

CONSTRUCTION PROCEDURE

Assemble the materials for this project as follows:
1. Lay out the location of the switches and lamp on the baseboard.
2. Drill all necessary holes in baseboard.
3. Wire and install all switches and the lamp according to the diagram in Fig. 7-16.
4. Assemble the wooden box and install the switchboard on it, as pictured in Fig. 7-16.

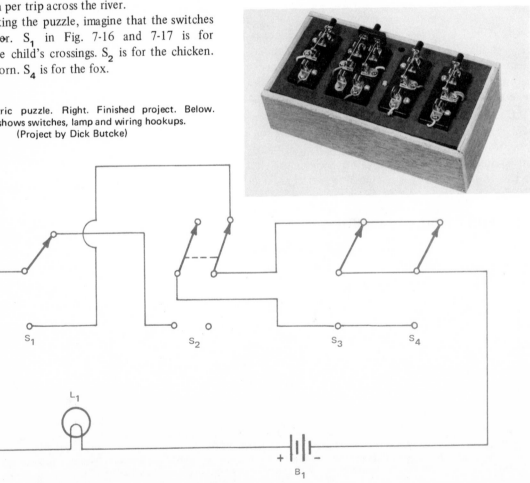

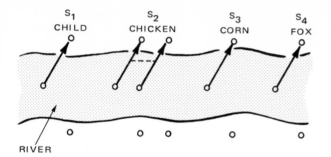

Fig. 7-17. When working the electric puzzle shown in Fig. 7-16, think of the switches as the child, chicken, corn and fox moving back and forth across the river.

Here is the correct switching sequence, Fig. 7-17, that will get the child and possessions to the opposite side of the river:

1st Trip	Child (S_1) and chicken (S_2).
Return Trip	Child (S_1) alone.
2nd Trip	Child (S_1) and corn (S_3).
Return Trip	Child (S_1) and chicken (S_2).
3rd Trip	Child (S_1) and fox (S_4).
Return Trip	Child (S_1) alone.
4th Trip	Child (S_1) and chicken (S_2).

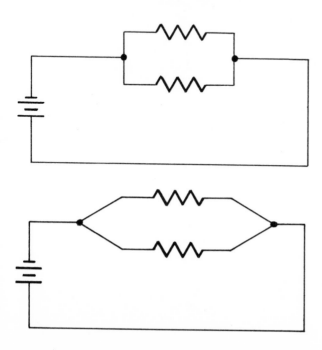

Fig. 7-18. Parallel circuits can be drawn in many different ways. Drawings shown are typical arrangements of power source and resistors.

SCHEMATIC DRAWINGS

You will find parallel circuits drawn in many different ways. So far in this chapter, we have shown simple rectangular circuits with resistors placed parallel to each other in add-on fashion. Fig. 7-18 shows two other ways of drawing parallel circuits.

Note in Fig. 7-19 that the two resistors are connected to separate branches of the circuit, putting them in parallel electrically, but not in parallel position on the drawing. When resistors are *connected in parallel*, it is not necessary to *draw them parallel* to each other.

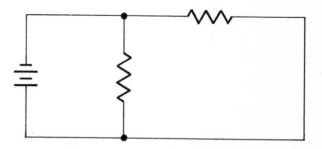

Fig. 7-19. In this parallel circuit, one resistor is shown on a cross wire. The second resistor is placed at right angles rather than parallel to the first resistor.

PARALLEL POWER SOURCES

Power sources also can be connected in parallel. In Fig. 7-20, note that three batteries are connected in parallel across the load. This arrangement is set up when one power source simply cannot supply enough current for an extended time.

In Fig. 7-21, three batteries are connected in parallel to the load (a lamp). This is done for a good reason. There are times when a load will draw so much current, it will shorten the life of the power source. When this happens, it is necessary to install several power sources (batteries, in this case) in parallel. This enables them to

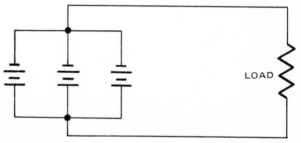

Fig. 7-20. Multiple power sources connected in parallel share the electrical load and last longer.

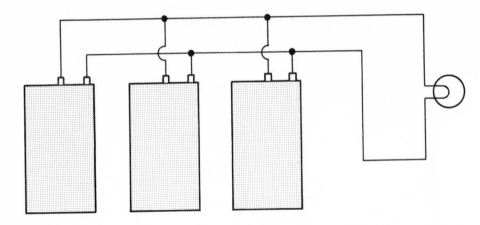

Fig. 7-21. This circuit drawing is another way of showing three power supplies connected to the circuit in parallel.

share the current draw from the load and extend the life of the total power supply. In later chapters, you will see that most people will replace batteries with a power source which converts the ac into dc. Only batteries or power sources that have the same voltage should be connected in parallel.

EQUAL RESISTORS IN PARALLEL

In Fig. 7-22, you see two resistors connected in parallel, each having a value of 100 ohms. When a parallel circuit has two resistors of the same value, another formula can be used to find total resistance. R_T will be equal to R divided by the number of resistors.

In Fig. 7-22, for example, R equals 100 ohms and N is the number of resistors:

$$R_T = \frac{R}{N}$$

$$R_T = \frac{100}{2}$$

$$R_T = 50\,\Omega$$

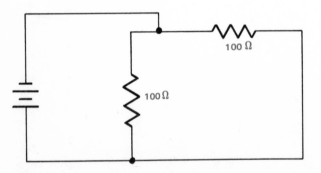

Fig. 7-22. Two 100 ohm resistors connected in parallel offer 50 ohms total resistance (see formula in text).

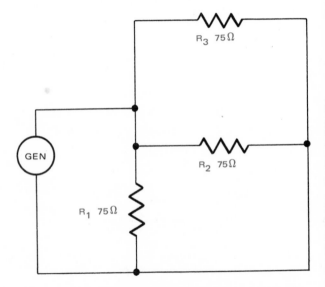

Fig. 7-23. Three 75 ohm resistors in parallel offer 25 ohms total resistance (see text).

In Fig. 7-23, three resistors are connected in parallel. Use the same formula to find total resistance:

$$R_T = \frac{R}{N}$$

$$R_T = \frac{75}{3}$$

$$R_T = 25\,\Omega$$

This means that if you want to substitute one resistor for the three in this parallel circuit, replace them with one 25 ohm resistor.

Fig. 7-24 shows a typical power amplifier chassis with all components in place. A complex electrical

Fig. 7-24. This power amplifier is made up of many parallel circuits. (Fisher Radio)

device such as this has many parallel circuits. It also has series circuits, which were described in the last chapter, and series parallel circuits, which will be explained in Chapter 9.

VOLTAGE

Voltage in a parallel circuit can be found by using a voltmeter and some common sense about hooking up the leads. Using the meter means reading it directly as you did before. However, the voltmeter leads must be connected across the circuit. Look at Fig. 7-25. Note that by connecting a voltmeter across each resistor and the source, all leads are connected to the same two lines. Therefore, all meters are reading the same *source voltage.*

Consider another resistance experiment. If you installed a voltmeter across the line as in Fig. 7-26, you would get a reading of zero. There is no reading because voltage drop across a short piece of wire is very small. It is so small, in fact, that you would need a highly sensitive meter to record the drop.

Take the No. 14 wire in Fig. 7-26, for example. It is carrying 2 amps of current. Going back to the table in Fig. 3-11 on page 39, you will find that No. 14 wire has 1 ohm resistance for 396 feet of wire. Therefore, one foot of wire between the clips of the voltmeter has a resistance of .00252 ohm.

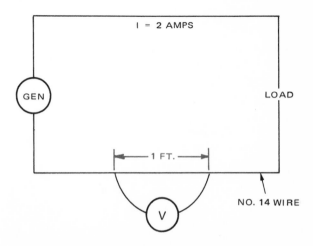

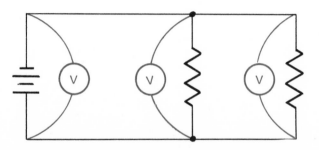

Fig. 7-25. When measuring voltage in this parallel circuit, connect voltmeter across power source or load.

Fig. 7-26. There is no voltage drop when measuring across a wire, because its resistance is so small.

$$R = \frac{1 \text{ ft.}}{396 \text{ ft./ohm}}$$

$$R = .00252 \text{ ohm}$$

Using Ohm's Law to find the voltage drop:

$$E = IR$$

$$E = 2A \times .00252\,\Omega$$

$$E = .00504V$$

You can see that a very sensitive voltmeter would be needed to make this voltage measurement. Since there is practically no voltage drop, we can forget it when using meters. See Fig. 7-26. *Therefore, the voltage in a parallel circuit is the same across each resistor.*

SUMMARY

In this chapter, you have learned that:

1. Total current flow in a parallel circuit is equal to the sum of the currents flowing in the branches.

$$I_{\text{Total}} = I_1 + I_2 + I_3$$

2. Voltage is the same across each path of a parallel circuit.

3. Total resistance in a parallel circuit is the reciprocal (opposite) of conductance. Use this conductance formula, then invert the answer:

$$\frac{1}{R_{\text{Total}}} = \frac{1}{R_1} + \frac{1}{R_2} + \frac{1}{R_3}$$

TEST YOUR KNOWLEDGE

1. What is the total resistance in a circuit having four 200 ohm resistors connected in parallel. Use either formula to solve this problem.
2. Show the formula you used to get the answer to No. 1.
3. How many branches are there in a parallel circuit?
4. What is a quick check to see if the answer to a parallel circuit problem is right?
   a. The answer must be smaller in value than any of the resistors in the circuit.
   b. The answer must be larger in value than any of the resistors in the circuit.
5. In a parallel circuit, how much of the current will flow in the path of least resistance?
   a. A small amount.
   b. Most.
   c. All.
6. Three resistors are connected in parallel. If they are all of equal value, how would you find the equivalent (one resistor) circuit?
7. When you measure the current in a circuit, you may get different results than if you solved the problem using the formula. Why?
8. "Breadboarding" is a method of wiring up a circuit to check its operation before soldering. True or False?
9. Before soldering a circuit together, it would be wise to check its operation. Why?
10. The voltage in a parallel circuit is _____ across all branches.
11. When two resistors are connected in parallel, they must be placed parallel to each other on the circuit drawing. Yes or No?

SOMETHING TO THINK ABOUT

1. Why do speakers have ratings that are different from one another?
2. Why does a voltmeter read zero when it is connected across a piece of wire?
3. Why do you get a voltage drop across a piece of resistance wire and not across copper wire or aluminum wire?
4. What are some of the ways companies use to remove solder when they must repair circuits which have been soldered?
5. Why do some circuits found in books and magazines fail to work?
6. Make a list of some of the places where you can find resistors connected in parallel.
7. Are houses wired in series or parallel?
8. Why are only certain areas affected by power outages when the utility company has trouble with one of its substations?
9. Why can a bird land on electric wires between telephone poles without getting shocked or killed?
10. If you wear rubber gloves, can you handle downed electric lines safely?
11. Why do some houses on the outskirts of towns have only one wire running from the pole to the house?
12. Can power companies use only one wire to transport electricity?

Chapter 8

METERS

In this chapter, we will study two types of meters. First, we will discuss pointer type meters (also called analog meters), Fig. 8-1, then cover digital readout meters, Fig. 8-2. *These meters are sensitive instruments used to measure and indicate electrical values.*

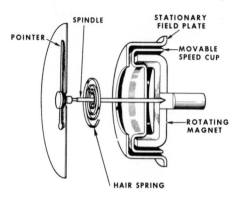

Fig. 8-1. Cross section of a typical pointer type meter. (AC Spark Plug, Div., General Motors Corp.)

POINTER TYPE METERS

A pointer type meter has a calibrated dial (face with marked values) and a needle that "points" to the value of the electrical unit under test. The meter shown in

Fig. 8-2. Compact digital multimeter displays an exact reading of the test result.

Fig. 8-3 will measure current and resistance.

Pointer type meters are used in many fields. Auto mechanics use them to make tests and measurements on automobile electrical systems. Most cars are powered by a 12V battery and an electronic device called an

Fig. 8-3. This pointer type meter used by auto mechanics will measure current (amps) and resistance (ohms). (Sun Electric Corp.)

"alternator." All circuits are dc, so meters are very necessary to help pinpoint the source of electrical problems. Without meters, mechanics would lose valuable time when troubleshooting or making adjustments on customers' cars.

Pointer type meters are also used by electricians in industrial plants, by television repair technicians and by anyone who fixes house wiring, dishwashers, ranges, clothes dryers, furnaces and air conditioning units. Even when products are manufactured, meters are called on to make sure the products pass inspection before they are sold to the customers.

DIGITAL READOUT METERS

Digital readout meters "spell out" the specific value of the electrical unit under test, Fig. 8-2. This type of meter has some special uses, such as com-

Fig. 8-4. Digital readout meter on the instrument panel of an experimental car. (AC Spark Plug Div., General Motors Corp.)

puterized automobile engine analyzers, exhaust emission testers and digital readouts on the instrument panels of cars, Fig. 8-4.

One advantage digital readout meters have over pointer type meters is that they are accurate within 0.1 percent. Most people cannot read a pointer type meter with anywhere near that degree of accuracy. Both meters, however, hook up with circuits much the same way.

Fig. 8-5. Portable digital multimeter has seven segments per digit. Illuminated segments form numbers.
(Weston Instruments, Inc.)

SEGMENTED DISPLAYS

One type of digital readout has groups of seven small segments that form a number on the meter face according to which segments light, Fig. 8-5. In this way, all numbers from zero through nine can be shown by lighting up the proper segments. For example, numbers two and six would light up as shown in Fig. 8-6. Number eight lights up the entire seven segments. Can you figure out the rest of the numbers?

AMMETER

In earlier chapters, we studied series circuits and parallel circuits. We also introduced you to the use of some meters. Now, the first meter we will study further is the ammeter.

Note in Fig. 8-7, that the ammeter is always connected in series with the power source and the load in order to measure load current. You also must pay attention to polarity (positive and negative terminals) when using an ammeter in a dc circuit (see CURRENT READINGS and VOLTAGE READINGS later in this chapter).

With an ammeter connected in series, it is possible to measure the amount of current flowing in the circuit. You know that automotive circuits are dc, because they

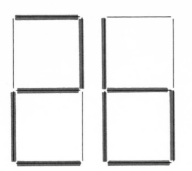

Fig. 8-6. Lighted segments on digital readout meter face form numbers 2 and 6.

are powered by a battery. Our next step in studying pointer type meters is to find out why they work in dc circuits.

HOW AMMETERS WORK

Most dc meters are constructed so that current flows through in one direction only. This is why they are called "dc" or "direct current" meters.

When direct current flows in a wire, it only goes in one direction. We also know that when current flows, a magnetic field is created. You can prove this by setting up a metal filings experiment similar to the one performed earlier. In preparing for this experiment, however, we will put a current-carrying wire through a piece of paper. As a result, the magnetic field will look

like the one you see in Fig. 8-8. Typically, it will be made up of circles around the wire. The only problem with this magnetic field is that it is too weak to do much good. Therefore, we must find a way to make the field stronger. One of the easiest ways to do this is to wrap the wire into the shape of a coil.

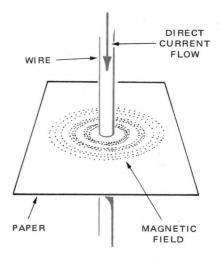

Fig. 8-8. Metal filings show the magnetic field around a wire.

To make the coil, wind the wire around an insulating material called a coil form, Fig. 8-9. Remember, insulation cannot stop the magnetic lines of force; they will cut right through it. This coil will make the magnetic field stronger, and form a field with a north pole and south pole.

If you took an ordinary magnet and placed it in the field, you would notice that the two poles always line up in a certain way. Then, if you pass a direct current through the wire in one direction, the north pole will be established at one particular end of the magnet. However, reverse the direction of current and the north pole will shift to the other end.

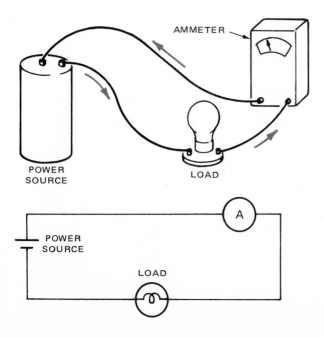

Fig. 8-7. An ammeter is always connected in series with the power source and load.

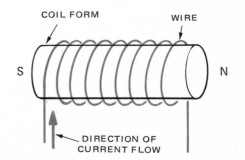

Fig. 8-9. Wire wrapped on a coil form provides the means for making a magnetic field experiment.

LEFT HAND RULE

There is a way that you can tell in advance which end of the coil will have the north pole. If you point the fingers of your left hand in the direction which the current flows, your thumb will point to the north pole. See Fig. 8-10. As a result, this has come to be known as the Left Hand Rule. "Wrap a coil" and see for yourself.

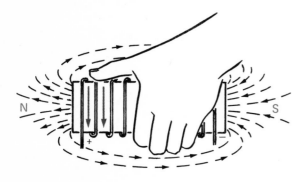

Fig. 8-10. The left hand rule for coil polarity determines which end of the coil is the north pole.

GALVANOMETER

When we have a magnetic field, we will find attraction and replusion. You learned this when working with magnets. Fig. 8-11 shows a galvanometer, which will detect and indicate small amounts of

current. The place where the needle pivots for its swing back and forth is known as a jeweled movement. This movement is made a special way to help reduce friction in the meter. By doing this, the pointer on the meter will move very easily. The more friction in the movement, the less accuracy you get in your meter readings. Therefore, you must keep friction down to the smallest amount possible.

The galvanometer contains a permanent magnet with a north pole and a south pole. Note also, in Fig. 8-11, that there is a coil of wire between the poles. As current passes through the meter a magnetic field is created in the coil. This magnetic field will be affected by the north and south poles of the magnet. As soon as there is magnetic attraction or repulsion, the pointer on the

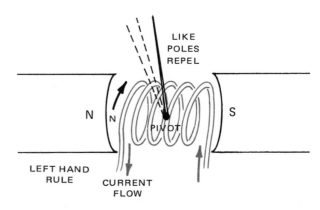
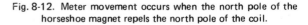

Fig. 8-12. Meter movement occurs when the north pole of the horseshoe magnet repels the north pole of the coil.

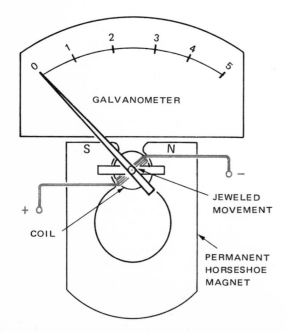

Fig. 8-11. Major parts of a galvanometer. Note the coil and permanent horseshoe magnet.

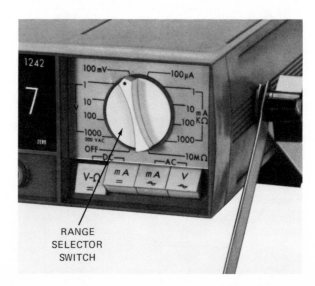

Fig. 8-13. Meters with more than one scale of values generally have a range selector switch. Note the many positions on this digital readout meter switch. (Weston Instruments, Inc.)

meter will begin to move. The greater the flow of current through the coil, the more the pointer will swing. Fig. 8-12 will show you why this happens.

RANGE SELECTOR SWITCH

When making metered tests, it is possible to pass too much current through the meter. A surge of high current will drive the pointer right off the end of the scale and damage the meter movement. To prevent this from happening, many meters have a range selector switch. See Fig. 8-13. When working with unknown meter readings, you should start by adjusting the range selector switch to its highest value, Fig. 8-14. Then, reduce the switch setting until the meter is able to read the current.

This same operating technique is used for all electrical meters with multiple ranges. It will help prevent the problem known as "pegging the meter," which refers to a test that buries the pointer of the meter off the end of the scale, Fig. 8-15. This

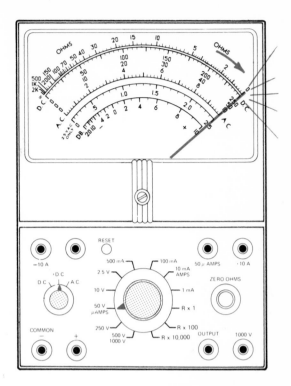

Fig. 8-15. "Pegging the meter" occurs when the range selector switch is set too low and the pointer swings hard against right end of scale.

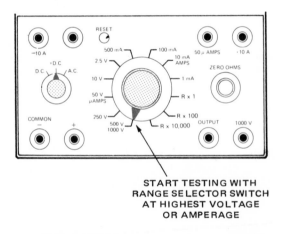

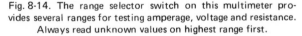

START TESTING WITH RANGE SELECTOR SWITCH AT HIGHEST VOLTAGE OR AMPERAGE

Fig. 8-14. The range selector switch on this multimeter provides several ranges for testing amperage, voltage and resistance. Always read unknown values on highest range first.

mistreatment could ruin a meter or bend the pointer.

Some meters have a peg (small pin) at the end of the scale. When the pointer hits it, the meter is "pegged." The usual cause of pegging is failure to check the setting of the range selector switch.

I.C. SAYS:

"Remember, when reading unknown currents or voltages, always start with the range selector switch in the highest position. This can prevent damage to your meter."

HOW A RANGE SELECTOR WORKS

To get a better understanding of how a range selector works, see Fig. 8-16. Note that when you turn the range selector switch, you add another resistor to the circuit. This resistor is commonly called a "shunt." It provides a parallel path for the current to flow and automatically positions the pointer in a different range of values.

Take a closer look at a shunt resistor in Fig. 8-17. Consider that the moving coil on a meter has a resistance of about 50 ohms, depending on the type of meter used. Since the coil can be rated at 50 ohms, a shunt can be placed in parallel to it. If the shunt is a resistor which has a value less than 50 ohms, and it is connected in parallel to the moving coil, more current will flow through the shunt than goes through the coil.

By using this knowledge, it is possible to put a shunt in parallel to the coil that will allow us to select the range of current which we wish to measure. To make it easier for the user, the manufacturers mark the range on the front of the meter. In effect, when you turn the range selector switch, you are adding or subtracting a resistance in parallel with the meter movement. This also keeps the coil of the meter from being burned up by high current passing through it.

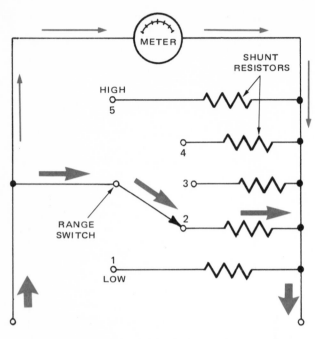

Fig. 8-16. A range selector switch operates on the principle of adding resistors when the switch is turned from range to range.

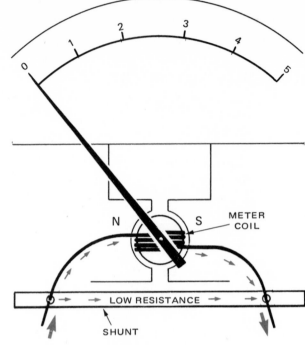

Fig. 8-17. A shunt resistor in a range selector switch provides a parallel path for current flow in a different range of values.

D'ARSONVAL MOVEMENT

The meter we have been looking at has a moving coil movement inside a horseshoe magnet, Fig. 8-18. This is known as a D'Arsonval movement. A D'Arsonval meter has the type of movement found in ammeters, voltmeters, ohmmeters and wattmeters. A D'Arsonval meter has a jeweled movement to cut down the amount of friction and to insure the accuracy of its readings.

While a D'Arsonval meter is reliably accurate, one problem is that the movement is very sensitive. If you drop the meter and the movement receives a shock, it is possible that the jeweled movement or the whole meter will be permanently damaged. In either case, the meter is ruined.

DAMPING SPRING

When using a meter to measure electrical current, it is possible that the sudden surge of current will swing the pointer right off the scale. To prevent this, a damping (shock absorbing) spring is attached to the pointer of the meter. See Fig. 8-18.

When current passes through the meter, the damping spring prevents the pointer from going full scale so fast that it will ruin the meter. Then, when current is removed from the meter, the damping spring keeps the pointer from dropping too quickly.

Damping action, then, is helpful in most electrical test procedures. However, it does create a problem if the flow of electrons lasts for only a few milliseconds. In this situation, a meter equipped with a damping spring will not react quickly enough to give accurate readings.

USING AMMETERS AND MULTIMETERS

Now that you have seen how meters are constructed, we will study their use. In order to take current

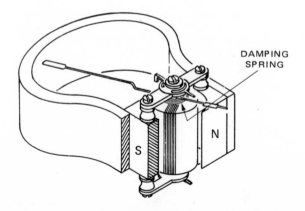

Fig. 8-18. A D'Arsonval meter movement features a moving coil inside a horseshoe magnet. The damping spring keeps the pointer from reacting too quickly to surges of current flow.

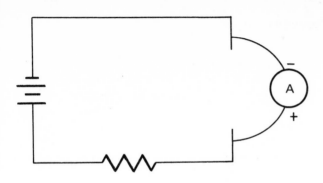

Fig. 8-19. Schematic shows hookup of ammeter in series with circuit: open circuit; connect meter test leads; read meter.

readings with an ammeter or multimeter, you need to know: how to "hook up" to the electrical circuit; how to set the switches and connect the probes (if the meter is so equipped); and how to read the meter.

To hook up an ammeter to take current readings:

1. Open the circuit at some convenient point. The only way we can get readings is by making sure that current is flowing through our ammeter.
2. Connect the negative test lead of the meter (usually black) to the disconnected part of the circuit on the negative side of the power source. See Fig. 8-19. Connect the positive test lead (usually red) to the disconnected part of the positive side of the power source.
3. Read the meter.

I.C. WARNS:

"Remember that current will be flowing when this circuit is under test. Make sure that you do not touch any of the metal on the probes or you, too, will be in the circuit. Shock may result."

Fig. 8-21. Schematic shows correct hookup of multimeter test leads for measuring current flow (in amps or milliamps).

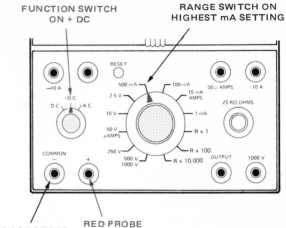

BLACK PROBE   RED PROBE

Fig. 8-20. To read voltage on a multimeter: connect probes to proper jacks; set function switch; set range selector switch; connect meter leads across circuit; read correct scale on meter.

To hook up a multimeter to take current readings:

1. Open the circuit.
2. Set the function switch to the + dc position, as shown in Fig. 8-20.
3. Plug the two probes into their proper jacks: black to the negative terminal on the meter; red to the positive terminal on the meter.
4. Set the range selector switch to the correct position for the circuit under test. If you are working with unknown currents, place the switch in the highest

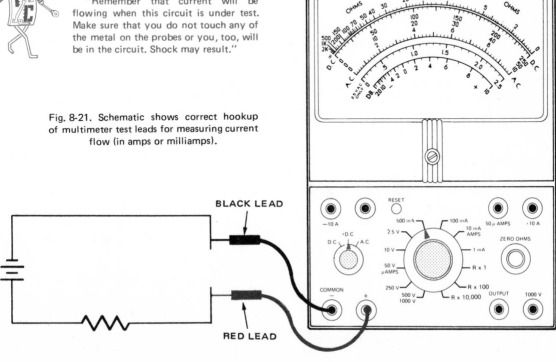

BLACK LEAD

RED LEAD

position, Fig. 8-20. This will prevent you from "pegging the meter."

5. Touch the probes to the disconnected parts of the circuit, Fig. 8-21.

6. Read the meter. If necessary, lower the position of the range selector switch until you are able to get a reading.

READING THE METER SCALES

Reading the current value scales on a multimeter calls for special instructions. If you use a multimeter to measure the number of milliamperes flowing in a given circuit, the pointer will move along the direct current (dc) scales shown in Fig. 8-22. To make it easier to read

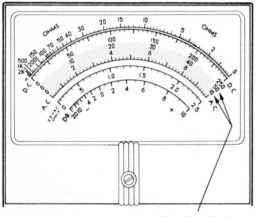

SCALES USED TO READ
DIRECT CURRENT FLOW

Fig. 8-22. Arrows point to scales used to read direct current test results in amps or milliamps.

these current scales, we will look at that part of the meter face in Fig. 8-23.

Our multimeter has four possible ranges of current readings.

1. 500 mA. Set the range selector switch to 500 mA.

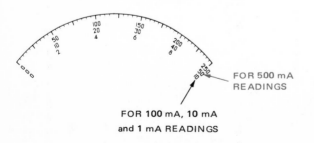

FOR 500 mA
READINGS

FOR 100 mA, 10 mA
and 1 mA READINGS

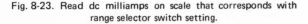

Fig. 8-23. Read dc milliamps on scale that corresponds with range selector switch setting.

Take a reading on the scale ending with 50, then multiply this meter reading by 10. In the example shown in Fig. 8-24, the actual current value would be 23 x 10 = 230 mA.

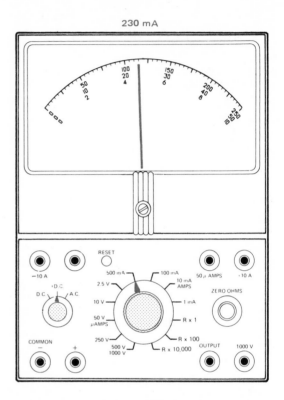

Fig. 8-24. Controls are set for 500 mA range as indicated on the scale ending with 50. Reading is 23 x 10 = 230 mA.

2. 100 mA. Set the range selector switch to 100 mA. Take a reading on the scale ending with 10, then multiply this meter reading by 10. In the example shown at Part A in Fig. 8-25, the actual current value would be 4.6 x 10 = 46 mA.

3. 10 mA. Set the range selector switch to 10 mA. Take a direct reading on the scale ending with 10. In the example shown at Part B in Fig. 8-25, the current value is 4.6 mA.

4. 1 mA. Set the range selector switch to 1 mA. Take a reading on the scale ending with 10, then multiply this meter reading by .10. In the example shown at Part C in Fig. 8-25, the actual current value would be 4.6 x .10 = .46 mA.

METER READING TIPS

In addition to reading the correct scale on a meter, there are some other reading tips worth noting. In Fig. 8-26, for example, a small mirror is located behind the

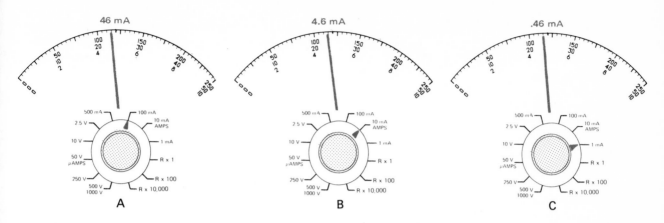

Fig. 8-25. The range selector switch settings for the scale ending with 10 are: A—100 mA. B—10 mA. C—1 mA. The individual readings are: A—46 mA. B—4.6 mA. C— .46 mA.

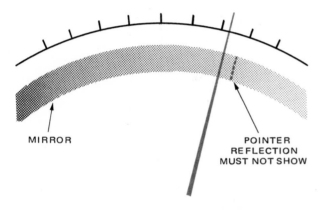

Fig. 8-26. Pointer on the meter must be sighted in a direct line and at right angles to the meter face.

pointer of the meter. *To read this meter accurately, you must position yourself so that the pointer and its reflection in the mirror line up.* If you are able to see the reflection, you are not in the right position to take an accurate reading. When this happens, move your head in either direction until you cannot see the meter's reflection. Then, take the reading.

If the meter does not have a mirror behind the pointer, look straight down at the pointer and meter face for best readings.

When reading multimeters, the most accurate readings can be obtained when the pointer is near the center of the scale. Fig. 8-27 shows the range where most meters have their best accuracy. Try to avoid reading the meter at either end of the scale. The reason why can be found by studying the meter face in Fig. 8-28. Note that the high numbers at the left end of the top scale are very close together. This is the upper range of the ohms scale. If you tried to take a reading in this area, you could make a mistake of several hundred ohms and not even know it.

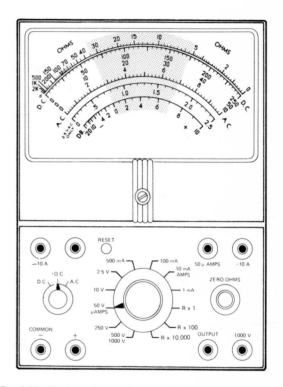

Fig. 8-27. Portion of meter face in color indicates area in which the most accurate readings may be obtained.

MICROAMPERES

There are times when you will have to read currents in the microampere range (μ A). Again, you can use the multimeter to get readings of these small values. Note in Fig. 8-29, however, that the position of one of our probes has been changed. The black test lead stays in the negative jack on the meter; the red test lead must be plugged into the 50 μA jack.

Continue making the settings by placing the range

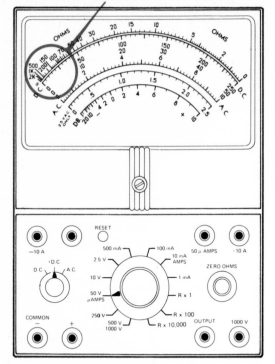

Fig. 8-28. Circled area shows why you should try to avoid taking readings at the extreme ends of a meter scale.

selector switch in the 50 μA position; leave the function switch in the + dc position. Then open the circuit and contact the disconnected parts with the probes while observing the correct polarity (positive probe to the positive side of the power source; negative probe to the negative side).

Take a reading on the meter scale ending with 50. This will be a direct reading in microamperes, as shown in Fig. 8-29. The reading on this test circuit is 34 μA.

AMPERES

It also is possible to use the multimeter for measuring current values up to 10 amps. To do this: connect the black test lead to the − 10 amp jack; connect the red test lead to the + 10 amp jack, Fig. 8-30. Move the range selector switch to AMPS, which also is the 10 mA position. Leave the function switch in the + dc position. Open the circuit and contact the disconnected parts with the probes while observing the correct polarity. Take a direct reading on the meter scale ending with 10. The reading on the test circuit shown in Fig. 8-30 is 6A.

I.C. SAYS:

"Be sure to turn off circuit power before you remove your meter leads. You do not want your body to become a resistor in parallel with your meter."

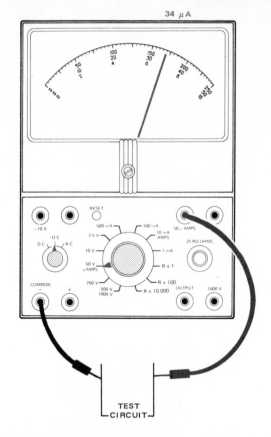

Fig. 8-29. Range selector switch setting for a maximum of 50 microamps. Direct reading on the scale ending with 50 is 34 μA.

Fig. 8-30. Range selector switch setting for a maximum of 10 amps. Reading on scale ending with 10 is 6A.

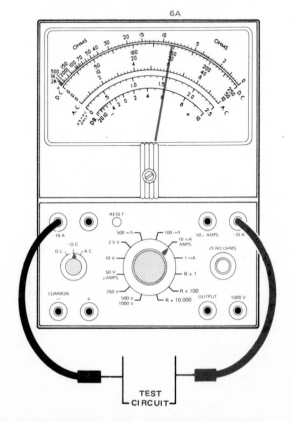

MULTIMETER APPLICATIONS

The multimeter has many uses or applications. You will find that a number of special dc ammeters are available for making specific tests. The ammeter shown in Fig. 8-31, for example, is used by auto mechanics in taking battery cell tests. In this chapter, the multimeter has been used to measure current in amperes, milliamperes and microamperes. It also can be used to measure voltage (volts) and resistance (ohms).

Fig. 8-31. Special ammeters are available with either high or low range scales or with multiple scales. (Sun Electric Corp.)

LIE DETECTOR PROJECT

How would you like to build a "poor man's" lie detector? One particular design is shown in Fig. 8-32. After you have created your own design, mount the parts in compact fashion, keeping the connecting wires as short as possible. Procure the following materials:

No.	Item
1	15 volt battery (B_1).
1	Milliammeter (M_1).
1	12K resistor (R_1).
1	1.5K resistor (R_2).
1	SK3005 transistor (Q_1).
1	SPST switch (S_1).
2	Copper plates, about 4 sq. in. each.
1	Box of your design and wiring.
1	Battery holder.

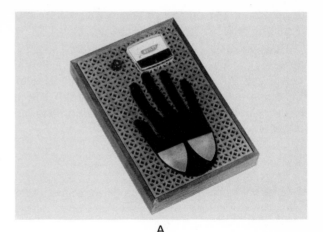

A

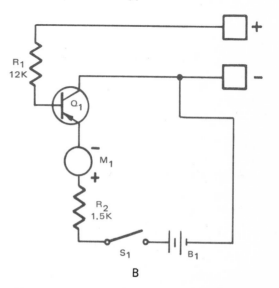

B

Fig. 8-32. Build this lie detector in your school shop. A—Completed project. B—Schematic shows hookup and arrangement of parts. B_1—Battery. Q_1—Transistor. $R_{1,2}$—Resistors. S_1—Switch. (Project by Mike O'Berski)

CONSTRUCTION PROCEDURE

Assemble the materials as follows:
1. Mount the resistors and transistor in the box.
2. Attach the switch and milliammeter to the cover. Use epoxy to attach the copper plates.
3. Wire the lie detector circuit as shown in the schematic in Fig. 8-32.
4. After all other connections are made, install the battery between the switch and transistor.
5. Install the cover on the box.
6. Turn off the switch and zero the meter with the mechanical adjustment screw.

Turn on the switch and use a jumper wire to complete the circuit across the copper plates. You should get a reading of at least 0.8mA. If not, check the condition of the dry cell. In effect, you now have a simple ohmmeter to check body resistance.

Once the project passes this test, have someone put a hand on the copper plates. Turn on the switch and ask a question. If the person tells a lie, moisture on the palm will reduce resistance and the meter needle will deflect to the right. When this happens, you will know that this person is telling a lie.

VOLTAGE READINGS

A voltmeter or multimeter is used to measure voltage values in an electrical circuit. When measuring voltage of unknown value, take the same approach you took

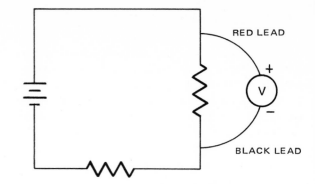

Fig. 8-34. Schematic shows a typical voltmeter hookup across the circuit. Always connect test leads so that voltmeter is in parallel to circuit.

when measuring current with an ammeter. Set the range selector switch to the highest position to avoid "pegging the meter."

When using a multimeter with controls similar to those shown in Fig. 8-33, start testing by setting the range selector switch at 500V. Set the function switch in the + dc position. Remember, too, that meters used in taking dc measurements are affected by polarity. Connect the meter in parallel to (across) the circuit, Fig. 8-34. Touch the positive probe of the meter to a connection on the positive side of the power source; touch the negative probe to a connection on the negative side.

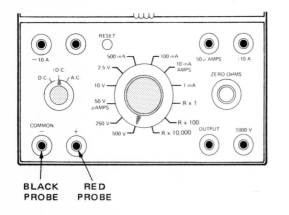

Fig. 8-33. Always start testing for unknown voltage values by placing range selector switch in its highest position.

Fig. 8-35. Proper hookup is shown for connecting a multimeter into a dc circuit to measure voltage drop across a resistor.

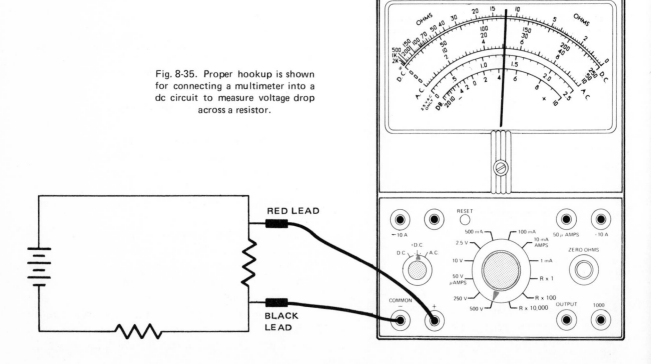

VOLTMETER RANGES

Most voltmeters have one or two ranges or scales. Multimeters have many ranges. Fig. 8-35 shows a multimeter being used to measure the voltage drop across a resistor. Fig. 8-36 provides a closeup of the controls and simplified meter scales used to show voltage values.

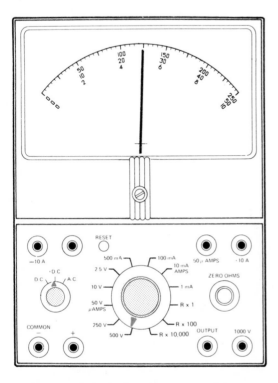

Fig. 8-36. Scale used for multimeter reading shown in Fig. 8-35. Actual voltage value is 25 x 10 = 250V.

Our multimeter has five possible ranges of voltage readings.
1. 500V. Set the range selector switch to 500V. Take a reading on the scale ending with 50, then multiply the meter reading by 10. In the example shown in Fig. 8-36, the actual voltage would be 25 x 10 = 250V.
2. 250V. Set the range selector switch to 250V. Take a direct reading on the scale ending with 250.
3. 50V. Set the range selector switch to 50V. Take a direct reading on the scale ending with 50.
4. 10V. Set the range selector switch to 10V. Take a direct reading on the scale ending with 10.
5. 2.5V. Set the range selector switch to 2.5V. Take a reading on the scale ending with 250, then multiply the meter reading by .01, or move the decimal two places to the left.

Example: Set the selector switch in 2.5V position.
Reading on 250V scale is 235.0
235.0 x .01 = 2.350 or
2.35 Volts

I.C. WARNS:

"Do not touch any meter leads or any part of an electrical circuit unless you are absolutely sure the power is off. It only takes about 0.1 amp to kill you. To get that amperage through your arm, across your heart and out your other arm takes about 30 volts. So before you touch anything electrical, make sure that the power supply is turned off. Otherwise, you run the risk of electrocution."

ALTERNATING CURRENT READINGS

The multimeter you have been using has two sets of scales marked dc and ac. The dc scale, of course, stands for direct current electricity. This is the scale you have referred to in all of your amperage and voltage tests. The ac scale stands for alternating current electricity.

In Chapter 10, we will fully discuss alternating current (ac). For now, a couple of explanatory notes are in order. Possibly you noticed that the function switch on the multimeter has an ac position. To take both current and voltage readings in ac circuits, the function switch should be moved to the ac position. Later, you will learn that polarity is not important when hooking up your meter to an ac circuit.

VOLTAGE-POLARITY CHECKER PROJECT

There will be times when you want to check whether a circuit has ac or dc voltage and a meter is not handy. With the voltage-polarity checker shown in Fig. 8-37, you can check for ac, dc and polarity (if circuit is dc). You can check circuits in the range of 110 to 480 volts. Higher voltages will cause arching in the neon lamp.
Procure the following parts:

No.	Item
1	NE 51 neon lamp.
1	220K, 1 watt resistor.
2	Probes.
3	1 ft. lengths, AWG 18 insulated wire.

Attach a probe to each of two lengths of wire. Connect the other end of one wire to the resistor; connect the end of the other wire to the neon lamp. Connect a third length of wire to the resistor and lamp.

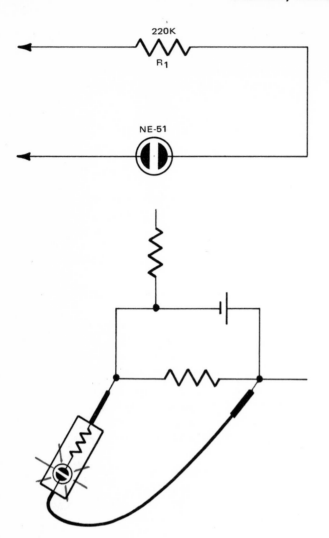

Fig. 8-37. Voltage-polarity checker project. Top. Arrangement of parts. Bottom. Test of dc circuit will light lamp electrode touching negative point in circuit.

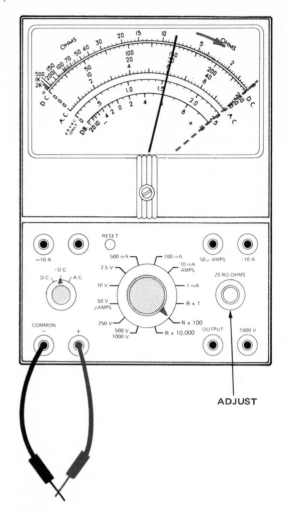

Fig. 8-38. To zero your ohmmeter, touch one probe with the other and adjust the zero ohms control until the pointer aligns with zero.

TEST PROCEDURE

To check voltage, touch the probes across the circuit to be tested. The lamp will light when voltage is present. If both electrodes in the lamp glow, you have an ac power source. If only one electrode glows, the voltage is dc. When only one lamp electrode lights, it will be on the side of the lamp connected to the probe touching a negative point in the system. Therefore, you also have a polarity check on dc circuits.

OHM READINGS

As you have seen, a multimeter can be used to measure amps and volts. It also can be used to measure ohms (resistance) in an electrical circuit. For this reason, a multimeter usually is called a VOM (volt-ohm-milliammeter).

Note that the range selector switch on the multimeter shown in Fig. 8-38 has positions marked R x 1, R x 100 and R x 10,000. These settings tell you that the resistance (R) value on the top scale of the VOM must be multiplied by either 1, 100 or 10,000, depending upon the position of the range selector switch.

ZERO THE METER

When a multimeter is going to be used for measuring resistance, the first thing to do is "zero the meter," Fig. 8-38. To zero the pointer on the ohms scale, simply take the red probe (in the positive jack) and touch it to the black probe (in the negative jack). Note that when the probes touch, the pointer will swing to the right, toward the zero on the top scale of the meter.

If the pointer does not stop exactly on zero, you will have to adjust the "zero ohms pot" (potentiometer) until the pointer does align with the zero mark. See Fig. 8-38. The zero ohms adjustment must be made each time you move the range selector switch to a new resistance position. This will insure an accurate resistance test reading every time. If you do not zero the ohmmeter, your reading will be off the same amount that your meter is out of adjustment.

READING THE METER

With the pointer set to zero ohms, you can proceed with your resistance test. You have a single top scale to read on the multimeter and the range selector switch has three positions: R x 1, R x 100 and R x 10,000.

Consider these settings and readings:

1. If the range selector switch is in the R x 1 position and the pointer on the meter is in the position shown in Fig. 8-39, what is the resistance value in ohms? Note that the pointer is between 5 and 10 ohms. If you count the major marks between 5 and 10, you will find that the actual meter reading is 8 ohms. Since the selector switch is set on R x 1, then the ohms value is 8 x 1 = 8.

2. If the range selector switch is in the R x 100 position, then the ohms value is 8 x 100 = 800.

3. If the range selector switch is in the R x 10,000 position, then the ohms value is 8 x 10,000 = 80,000.

Take a group of resistors and make your own resistance measurements. The only way you can be sure that you are using the meter correctly is by practicing.

OHMMETER SAFETY PRECAUTIONS

A word of caution is in order at this point. *Whenever you work with an ohmmeter, the power must be off in the circuit under test.* All ohmmeters have dry cells inside that act as the power source. Ohmmeters and multimeters need these dry cells to serve as the voltage supply needed to force current through the resistor being tested.

If you accidentally connect an ohmmeter to a live circuit, you probably will damage the meter. If you happen to leave the range selector switch in the ohms position and attempt to test voltage, the meter may explode. In effect, you are using the voltage from the circuit's power source to charge the dry cells in the VOM at a very high rate. This huge rush of electrons could cause the meter to blow apart.

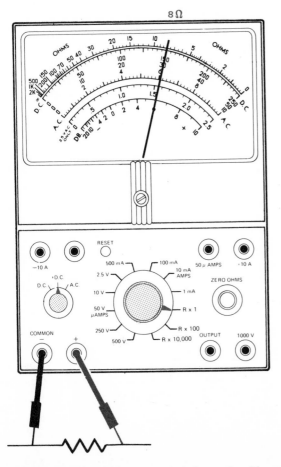

Fig. 8-39. Resistance measurement is being made with range selector switch in the R x 1 position. Reading is 8 x 1 = 8 Ω.

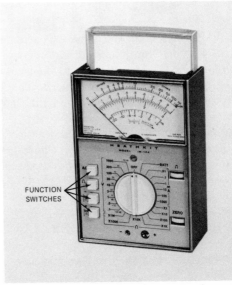

Fig. 8-40. This popular volt-ohm-milliamp (VOM) meter features push button function switches. (Heath Co.)

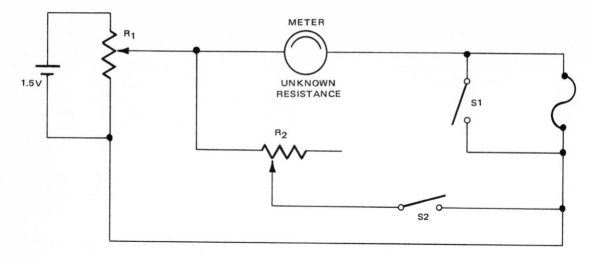

Fig. 8-41. Basic circuit used to determine unknown resistance of meter coil.

Most multimeters have a built-in overload device. If you take a *voltage reading* with the selector switch in the *ohms position,* it will "pop" the reset button. If so, remove the probe from circuit contact, set the range switch to the desired voltage range, and push the reset button. The meter is ready to go again.

In any case, always disconnect the power source from the circuit before connecting an ohmmeter or multimeter. Also, when you are through using a multimeter, put the range selector switch in the highest voltage position, or in the OFF position if the meter has one. This precaution will take the dry cells out of the circuit, and they will last longer.

LIFE OF POWER SOURCE

As the dry cells inside the ohmmeter or multimeter begin to age, the "zero ohms pot" must be adjusted more and more. Finally, it will get to the point where the meter cannot be zeroed. When this happens, replace the dry cells inside the meter. If you do not, all of your readings will be wrong.

Remember, not all VOMs look alike. Get familiar with different "makes." Although they look different, they still make the same kinds of measurements, Fig. 8-40. Note that this multimeter has a range selector switch, zero adjustment and function switches.

COIL RESISTANCE OF METERS

The various meters we have discussed have coils as part of their construction. Since coils have some resistance too, you want to be sure that coil resistance does not affect your meter readings. Meter movements

with the lowest full-scale current ratings are called "sensitive," because the pointer will move with small amounts of current flowing through the meter.

When a voltmeter is rated for sensitivity, it is said to have so many "ohms-per-volt." This rating applies to both the resistance in the meter movements as well as to the resistors in the range selector switch. Voltmeters having the highest ohms-per-volt rating will have the least effect on the measurements you make. The most common types of voltmeters have ratings of 20,000 to 30,000 ohms-per-volt.

Some old meters may not have their resistance rating marked on them. In this case, set up the test circuit shown in Fig. 8-41, and use the following method of finding the coil resistance of the meter.

First, adjust R_1 until the meter reads full-scale deflection, Fig. 8-42. The 1.5V dry cell acts as a power supply. Note that you have a parallel circuit: a portion of the current will flow through the resistor; the rest will flow through the meter. The resistance value for R_1 will start somewhere between 400 and 600 ohms.

After you have obtained full-scale deflection on the meter, close switch S_2, Fig. 8-43. Adjust the R_2 potentiometer (50,000 ohm) until the meter reads half-scale. Now you have an equal split through R_2 and the meter. Close switch S_1 and fine tune the adjustment on R_2 again. Since the current has split evenly between R_2 and the meter, their resistances should be the same. Measure the resistance of R_2 and you also have the resistance of the meter.

Generally, meters are used to measure voltage, current and resistance. Combination meters are also available. The one shown in Fig. 8-44 measures voltage on one side and temperature on the other side.

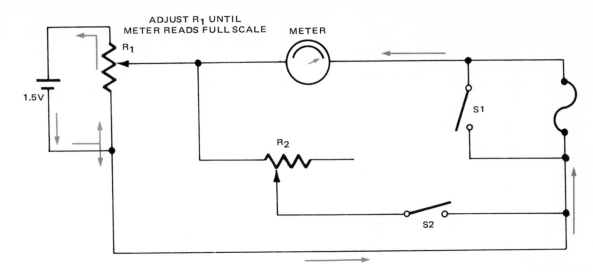

Fig. 8-42. First basic step to find unknown meter resistance is to adjust R_1 to get full scale reading on meter.

VACUUM TUBE VOLTMETER

The last type of meter to be covered in this chapter is known as a "VTVM." The letters stand for Vacuum Tube Voltmeter, Fig. 8-45. A VTVM must be plugged into a 110 volt ac outlet. Before it will work properly, it must warm up for awhile because it has a vacuum tube inside. We will not discuss tubes of this type in this book. However, you know that such a meter exists.

The VTVM has a high input resistance on all ranges. Usually, however, it is read in the megohm region. That means that the VTVM will not affect circuit readings as much as the VOM did. The meter movement in a VTVM is the moving coil, permanent magnet type. Ranges and scales are different on a VTVM and a VOM.

Fig. 8-44. Volts-temperature tester. Note four ranges on the range selector switch and multiple scales on meter face. (Sun Electric Corp.)

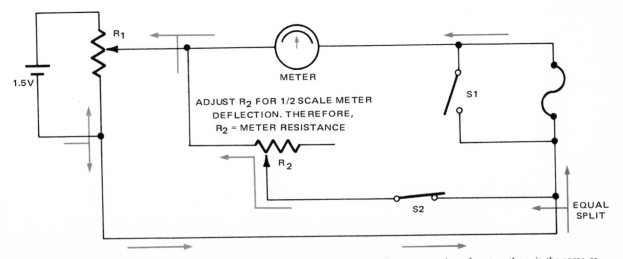

Fig. 8-43. Second step is to adjust R_2 until the meter reads half-scale. Resistance value of meter, then, is the same as resistance value of R_2.

Fig. 8-45. This vacuum tube voltmeter (VTVM) has broad application. Note seven voltage and resistance ranges on the range selector switch. (Heath Co.)

One difference is that the VTVM has a center-zero scale.

VTVMs also use a series of different types of probes, Fig. 8-46. Other differences are beyond the scope of this text. When you get ready to use a VTVM, check with your instructor for assistance.

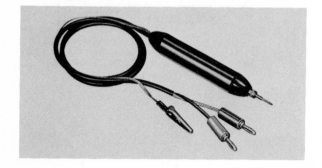

Fig. 8-46. Typical probe assembly furnished with, or as an accessory for, a vacuum tube voltmeter. (Heath Co.)

TEST YOUR KNOWLEDGE

1. Draw a simple circuit showing where a voltmeter should be placed to measure the voltage.
2. Damping a moving coil meter is necessary because _____.
3. Draw a circuit with a dry cell connected to a lamp. Show where an ammeter should be connected to get the proper reading.

4. Name at least three different types of measurements that can be made and the meters used to make them.
5. Give one of the big advantages of a digital readout meter.
6. State the left hand rule.
7. A _____ is a meter used to take measurements of very small electrical currents.
8. When working with unknown voltages, it is best to start taking measurements with the range selector switch set at _____.
9. Describe what is meant by "pegging the meter."
10. A _____ is connected to a meter so that it can be used to measure current values higher than that which the meter originally was manufactured to measure.
11. The _____ probe should be plugged into the positive terminal and the _____ probe should be plugged into the negative terminal.
12. Never touch the metal tips of probes. Why?
13. A microampere is _____ of an ampere.
14. Before you disconnect meter test leads from a circuit, you should _____ to prevent electrical shock.
15. It takes less than 1 amp of electrical current to kill you. True or False?

SOMETHING TO DO

1. Why are there so many different types of meters?
2. Compare the cost of digital readout meters with pointer type meters. How much have they changed in the last five years? Why?
3. How many different types of lighted segments can be found on digital readout type meters? Why do they differ?
4. How is damping done on meters without moving coils?
5. List the costs of different brands of meters which read the same current values. Why do the costs vary?
6. Make a list of voltage readings. Start with the highest voltage you are able to discover anywhere, then go down to the smallest value of voltage you can find.
7. In some situations, electricians working in industry do not turn off the electricity before making measurements. Why not?
8. When you have a downed line caused by a storm, does the utility company turn off the electrical power?
9. Telephone company technicians can tell within a

couple of miles where a broken wire is located. How can they do this without leaving their office?

10. Why does the telephone in your home have such small wire?

11. Some meters made by the same company must have their dry cells replaced more often than others. Why?

12. The dry cells in some modern meters will not run down if you accidentally leave the meter set in the ohms range for a long time. Why not?

13. How does a digital wristwatch get its power to operate?

14. Ask a number of electricians, "Which is the best meter?" Why do they disagree?

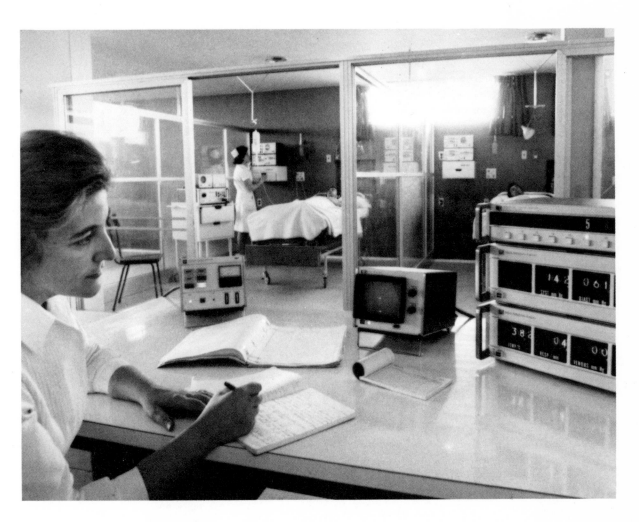

Patient-monitoring systems using digital readout meters save lives in hospitals. Basic unit at bedside of critically ill patient is monitored from a centrally located console, allowing one nurse to check the vital signs of several patients at once.

Chapter 9

SERIES PARALLEL CIRCUITS

When series circuits and parallel circuits are combined, the resulting single circuit is commonly called a "series parallel circuit" or "network." Most series parallel circuits are very complex. However, if you follow some simple guidelines, you will be able to reduce any series parallel circuit to its simplest form to determine current and voltages.

Part A in Fig. 9-1 illustrates a simple series parallel circuit having three 60 ohm resistors. The first step in figuring values in this type of circuit is to determine which resistors are in series with the power supply and which resistors are in parallel. Note that R_1 is in series, while R_2 and R_3 are in parallel.

Use the following formula (from Chapter 7) for finding the equivalent resistance (R_{EQ}) of the two resistors of equal value in the parallel element of the circuit shown at A in Fig. 9-1:

$$R_{EQ} = \frac{R}{N}$$

$$R_{EQ} = \frac{60}{2}$$

$$R_{EQ} = 30\Omega$$

Next, use the following formula for finding total equivalent resistance in the newly formed series circuit. See Part B.

$$R_T = R_1 + R_{EQ}$$

$$R_T = 60 + 30$$

$$R_T = 90\Omega$$

This gives you an equivalent circuit, Part C, having a total resistance of 90 ohms.

LOCATION OF RESISTANCES

You might think that if you placed the series element of a series parallel circuit *ahead of* the parallel element, it would make a difference when figuring total

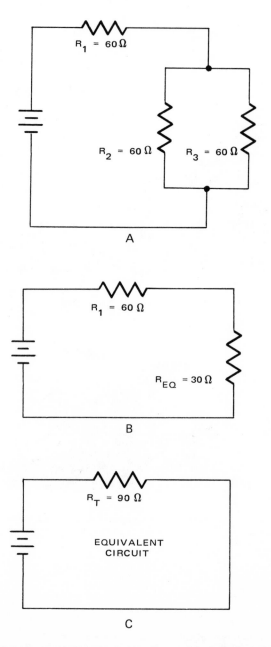

Fig. 9-1. Series parallel circuit has three resistors. A—R_1 is in series, R_2 and R_3 are in parallel. B—Equivalent resistance is 30 ohms. C—Total resistance is 90 ohms.

resistance. This would be true for figuring the amount of current flow in various parts of the circuit, but not for total resistance.

For example, Part A in Fig. 9-2 shows a circuit with the series element ahead of the parallel element. Now if you go through the same steps in figuring total resistance that you did in Fig. 9-1, you *will* get the same answer (90 ohms): In Part B, the equivalent resistance (R_{EQ}) of R_1 and R_2 is 30 ohms. When this is added to the 60 ohms of R_3 Part C, you would again have an equivalent circuit with a total resistance of 90 ohms.

CIRCUIT LAYOUT

To make circuits a little easier to read, most people who draw them follow certain basic rules of construction. First, they try to lay out the circuit so that the top part is positive and the bottom part is negative. See Fig. 9-3. A second general rule is to lay out the circuit so that the input is on the left and the output is on the right, Fig. 9-4. This means that the power supply would be drawn on the left and the load (such as lamps) on the right.

Knowing about these two circuit drawing practices will save you a lot of time when you look at schematics. In the past, these "rules of construction" were not observed. Today, it is easy to see how a simple idea like this can be very valuable in troubleshooting electrical circuits.

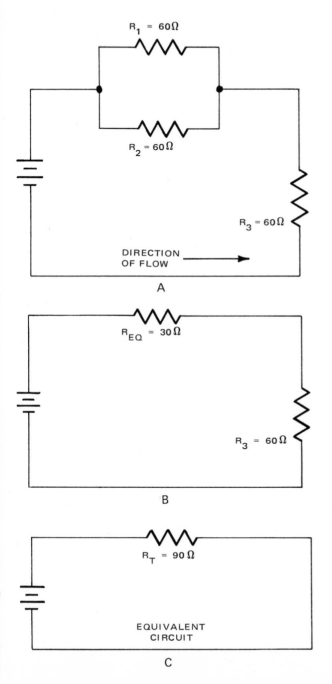

Fig. 9-2. This circuit is similar to Fig. 9-1, but with elements reversed. A—Series element R_3 is ahead of parallel element, R_1 and R_2. B—Equivalent resistance is 30 ohms. C—Total resistance is 90 ohms.

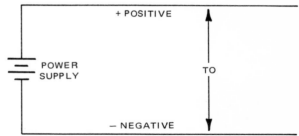

Fig. 9-3. When doing circuit layout work, draw positive side of circuit at top, negative side at bottom.

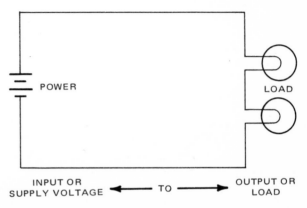

Fig. 9-4. Good circuit layout practice calls for power supply to be drawn at left and load at right.

ORDER FOR SOLVING
TOTAL RESISTANCE

You might think that all series parallel circuits are solved for total resistance by first looking for a parallel circuit. This is not true, although the previous examples make it appear that way. The schematic at Part A in Fig. 9-5, for example, has resistors of 10 ohms, 20 ohms and 30 ohms. The first step in solving this circuit is to take the two resistors in series (R_2 and R_3) and find their equivalent:

$$R_{EQ} = R_2 + R_3$$

$$R_{EQ} = 10 + 20$$

$$R_{EQ} = 30\Omega$$

This gives us the equivalent of two 30 ohm resistances in parallel, shown at Part B in Fig. 9-5.

In order to solve this equivalent circuit for total resistance, use the formula for parallel branches of equal value:

$$R_T = \frac{R}{N}$$

$$R_T = \frac{30}{2}$$

$$R_T = 15\Omega$$

Part C in Fig. 9-5 shows the equivalent circuit (for the series parallel circuit at Part A) with a total resistance of 15 ohms.

TV ANTENNA COUPLER PROJECT

Three resistors can be wired in a special TV antenna coupler circuit so that you can hook up two television sets to one antenna. To build this project, carefully follow the construction procedure while studying the diagram in Fig. 9-6.

Few parts are needed:

No.	Item
3	910 ohm resistors.
	300 ohm TV feed line.

CONSTRUCTION PROCEDURE

When soldering the connections, keep the resistor leads short and run standard TV feed line right up to the soldered joints. Pay close attention to the circuit

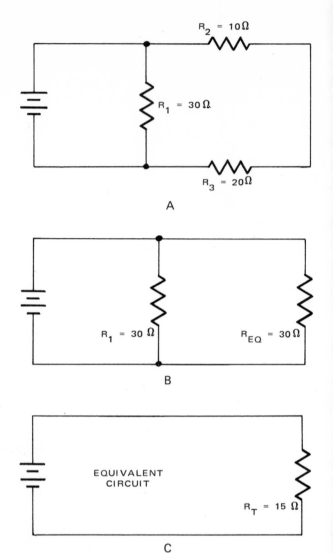

A

B

C

Fig. 9-5. Series parallel circuit has resistors of unequal value. A—First solve the series element, R_2 and R_3. B—Equivalent resistance is 30 ohms. C—Total resistance is 15 ohms.

diagram in Fig. 9-6:

1. Solder R_1 leads to the antenna leads, along with one feed line from each TV set antenna terminal.
2. Solder one lead of R_2 to one lead of R_1. Solder the other lead of R_2 to a TV feed line running between antenna terminals on TV set No. 1 and TV set No. 2.
3. Solder one lead of R_3 to other lead of R_1. Solder other lead of R_3 to TV feed line running between antenna terminals on TV set No. 1 and TV set No. 2.

With this coupler, your need for a second TV antenna is eliminated. It also serves to isolate and cut down on interaction between the two operating TV sets.

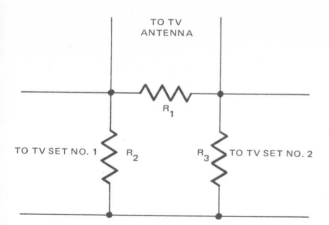

Fig. 9-6. Schematic shows how to hook up two television sets to one antenna. Resistors are 910 ohms each.

POWER SUPPLIES IN SERIES

At this point, we should look at what will happen if our power supplies are hooked up in series. See Part A in Fig. 9-7. Later, we will see how power supplies are used to provide different voltages for different parts of our circuits. This can be done by means of two batteries or by using a "voltage divider" (to be covered later in this chapter).

The series parallel circuit shown in Part A in Fig. 9-7 can be solved for total voltage and total resistance in just a few simple steps. First, note that the power supplies *are* in series. This means that we can add the voltages of the two batteries and come up with the equivalent power supply shown at Part B:

$$E_{EQ} = E_1 + E_2$$

$$E_{EQ} = 12 + 12 = 24V$$

Next, take the two resistors in series and figure their equivalent resistance as shown at Part C:

$$R_{EQ} = R_1 + R_2$$

$$R_{EQ} = 12 + 12 = 24\,\Omega$$

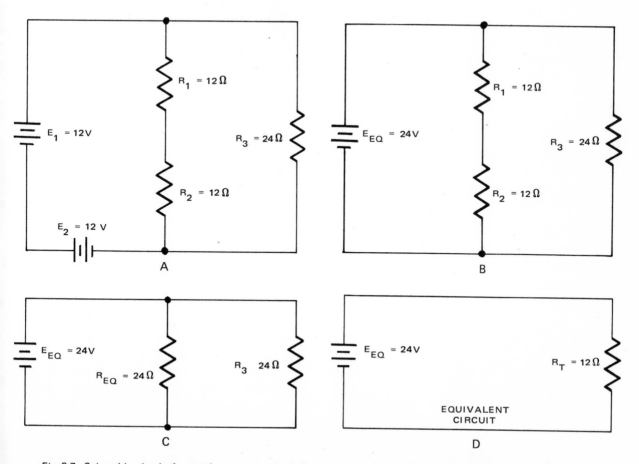

Fig. 9-7. Solve this circuit for total power supply and total resistance. A—Circuit has two power supplies and three resistors. B—Equivalent power supply is 24 volts. C—Equivalent resistance is 24 ohms. D—Total resistance is 12 ohms.

Then, solve for equivalent resistance of two resistors of equal value in parallel. This, in turn, will give us the total resistance of the equivalent circuit shown in Part D in Fig. 9-7:

$$R_T = \frac{R}{N} \qquad R_T = \frac{24}{2} \qquad R_T = 12\Omega$$

The answers to the questions concerning electrical values presented at Part A in Fig. 9-7 are: total voltage = 24V; total resistance = 12Ω.

CURRENT FLOW

You probably are asking yourself, "Why go through all of this work?" The answer is: "We must use these methods to determine the total current flowing in any series parallel circuit."

Circuit analysis is just another way of saying that you must break down the full circuit network to determine the amount of current in each part. The same thing is true for finding the voltage drop across each part of the circuit.

Circuit analysis also reveals open circuits and short circuits. An open circuit occurs when the path for current is broken. This can occur accidentally or through the use of a switch. A short circuit is one which has practically no resistance. Therefore, the current will cause the fuse to blow or the circuit breaker to open.

Looking at Part A in Fig. 9-8, our circuit analysis starts with the power supply:

$$E = 24V$$

Next, we need to figure an equivalent resistance for the two resistors in series:

$$R_{EQ} = R_1 + R_2$$

$$R_{EQ} = 12 + 12 = 24\Omega$$

Then, use the formula for finding total resistance of parallel branches of equal value:

$$R_T = \frac{R}{N}$$

$$R_T = \frac{24}{2} = 12\Omega$$

We now have an equivalent circuit with a power supply of 24 volts and a total resistance of 12 ohms. See Part B. Using these values in a formula based on Ohm's Law, you can solve the circuit for total current (amperage) as shown at Part C:

$$I = \frac{E}{R} \qquad I = \frac{24}{12} = 2A$$

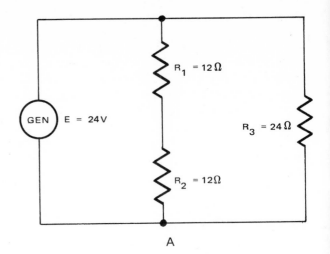

A

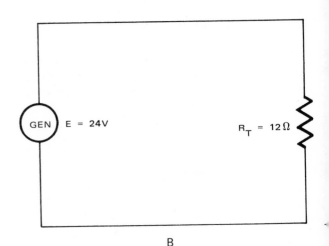

B

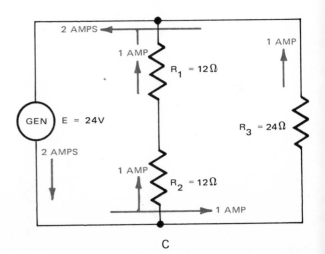

C

Fig. 9-8. Analyze this circuit for current output and amount of current through each element. A—Power supply is 24 volts. B—Total resistance is 12 ohms. C—Current output is 2 amps, with 1 amp flowing through each element of circuit.

Total current, then, is 2 amps. This *does not* mean that 2 amps will be flowing in each branch of our circuit. It *does* mean that the generator must be able to supply 2 amps to the circuit. With this information, you can go back and find the current in each branch:

$$I = \frac{E}{R_{EQ}} \text{ or } I = \frac{E}{R_3}$$

$$I = \frac{24}{24} = 1A$$

CAPACITORS IN PARALLEL

We introduced you to capacitors in Chapter 4. At that time, we covered various types of capacitors and how they are built. You also saw one symbol for a capacitor. Now, you will see how they work in electrical circuits.

Part A in Fig. 9-9 shows two capacitors in parallel, having ratings of 30 microfarads (μF) and 40 microfarads. To get total capacitance, use the following formula for capacitors in parallel:

$$C_T = C_1 + C_2 + C_3 + \dots$$

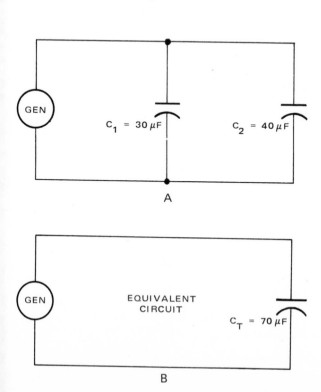

A

B

Fig. 9-9. Find total capacitance of this circuit. A—Two capacitors of unequal value in parallel. B—Total capacitance is 70 μF.

In our example, this would result in the equivalent circuit shown at Part B in Fig. 9-9:

$$C_T = C_1 + C_2$$

$$C_T = 30 \ \mu F + 40 \ \mu F$$

$$C_T = 70 \ \mu F$$

Total capacitance of this equivalent circuit is 70 microfarads.

The formula for finding total capacitance of capacitors in parallel is much different than the formula for finding total resistance of resistors in parallel. It *is* very similar to the formula you used for resistors in series.

The reason why capacitors in parallel differ from resistors in parallel can be seen in Fig. 9-10. Remember,

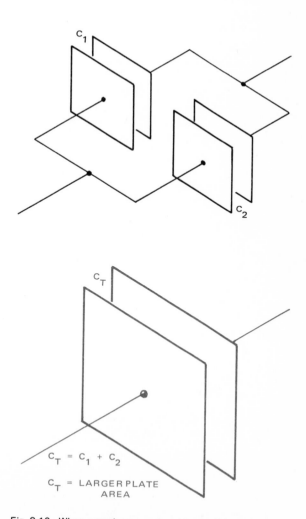

Fig. 9-10. When capacitors are placed in parallel, capacitance in circuit increases because plate area is larger.

one of the things which increases capacitance is a larger plate area. When we have two capacitors in parallel, we have a greater number of plates and more electrons on the plates. Therefore, total capacitance (C_T) really means more plate area.

CAPACITOR SYMBOLS

A popular capacitor symbol is shown at Part A in Fig. 9-11. You can see that this is an easier symbol to draw than the one in Fig. 4-26. For this reason, you will find the simpler one used on many schematics.

To find the total capacitance in the circuit presented at Part A, Fig. 9-11, use this formula:

$$C_T = C_1 + C_2 + C_3$$

$$C_T = 10 + 10 + 10 = 30 \ \mu F$$

Therefore, the equivalent circuit shown at Part B in Fig. 9-11 would have a total capacitance of 30 microfards.

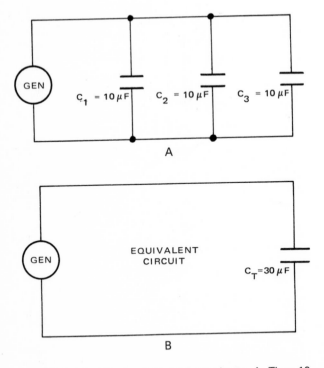

A

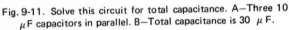

B

Fig. 9-11. Solve this circuit for total capacitance. A—Three 10 μF capacitors in parallel. B—Total capacitance is 30 μF.

CAPACITORS IN SERIES

If we have a circuit with capacitors hooked up in series, how would you find the equivalent circuit? As you might expect, we would use a formula similar to

the one for finding the total resistance of resistors in parallel:

$$\frac{1}{C_T} = \frac{1}{C_1} + \frac{1}{C_2} + \frac{1}{C_3} + \ ...$$

The circuit shown at Part A in Fig. 9-12 has two capacitors in series, each having a rating of 20 microfarads. To find the total capacitance in this series circuit:

$$\frac{1}{C_T} = \frac{1}{C_1} + \frac{1}{C_2}$$

$$\frac{1}{C_T} = \frac{1}{20} + \frac{1}{20}$$

$$\frac{1}{C_T} = \frac{2}{20} = \frac{1}{10}$$

Invert both sides of the equation:

$$C_T = 10 \ \mu F$$

Therefore, total capacitance of this equivalent circuit, Part B in Fig. 9-12, is 10 microfarads.

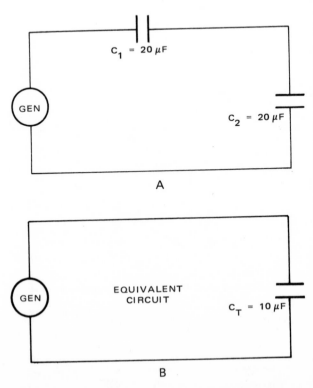

A

B

Fig. 9-12. Analyze this series circuit for total capacitance. A—Two 20 μ F capacitors in series. B—Total capacitance is 10 μF.

DISTANCE BETWEEN PLATES

You may wonder why capacitors placed in series will offer less capacitance than a single capacitor. The reason is quite simple. In Fig. 9-13, you can see that the distance between the outer plates of C_1 and C_2 will increase when placed in series. This, in turn, will lower the capacitance of the circuit.

In Chapter 4, we established that an increase in the distance between the plates of capacitors will lower the capacitance. One way manufacturers do this is by putting a thicker dielectric between the plates. Another way is by making the plates adjustable. In later chapters, we will be working with an adjustable unit known as a "variable capacitor." By adjustment, you can either increase or decrease the distance between plates, and thereby lower or raise the capacitance of the capacitor.

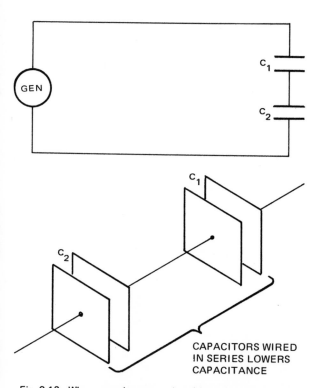

Fig. 9-13. When capacitors are placed in series, capacitance in circuit decreases because distance between plates increases.

SERIES PARALLEL CAPACITORS

If you have a circuit with capacitors in a series parallel network, you can solve it for total capacitance by the same methods used to handle series and parallel circuit problems. Looking at Part A in Fig. 9-14, you have a circuit with three capacitors, each with a rating

of 40 microfarads. To solve this circuit for total capacitance, first work on the parallel element.

Use this formula for capacitors in parallel:

$$C_T = C_2 + C_3$$

$$C_T = 40 \ \mu F + 40 \ \mu F$$

$$C_T = 80 \ \mu F$$

The equivalent capacitance of the two capacitors in parallel is 80 microfarads, as shown at Part B in Fig. 9-14.

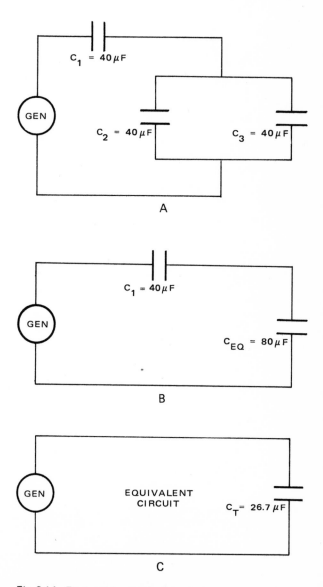

Fig. 9-14. Find total capacitance of this circuit. A—Three capacitors of equal value. B—Equivalent capacitance of capacitors in parallel is 80 μ F. C—Total capacitance is 26.7 μ F.

Then, use this formula for capacitors in series:

$$\frac{1}{C_T} = \frac{1}{C_1} + \frac{1}{C_{EQ}}$$

$$\frac{1}{C_T} = \frac{1}{40} + \frac{1}{80}$$

$$\frac{1}{C_T} = \frac{2}{80} + \frac{1}{80} = \frac{3}{80}$$

Invert both sides of the equation:

$$C_T = 26.7 \ \mu F$$

Total capacitance of this equivalent circuit is 26.7 microfarads, as shown at Part C in Fig. 9-14.

As you can see, total capacitance is smaller than the capacitance of any capacitor in the circuit. If you solve for total capacitance and the answer is larger than the value of the individual capacitors, you have made a mistake and must refigure to find your error. Again, this is just like resistors in series parallel problems which you solved earlier.

CAPACITOR/RESISTOR COMBINATIONS

You know how to find equivalent circuits for both capacitor networks and resistor networks. The next question would seem to be: "What happens when capacitors and resistors are combined in the same circuit?" Circuits of this type cannot be solved by

means of the simple formulas we have used so far. You must cover additional material before you can solve this type of network. When we get to Chapter 13, you will be shown how to do it.

ELECTRIC COMBINATION LOCK PROJECT

A toggle switch can be used in a number of ways. Basically, it is designed to open or close an electrical circuit by a flip of the little switch lever. However, by wiring five toggle switches in series, you can make an electric combination lock.

Make up the circuit by assembling the following parts as pictured in Fig. 9-15:

No.	Item
5	Toggle switches.
1	Right angle plate.
	Insulated primary ignition wiring.

CONSTRUCTION PROCEDURE

The trick to setting up the circuit is to mix up the switch connections; do not have all switch levers "on" in the same position. One good automotive application for this electric combination lock is to wire it into a dash-mounted ignition switch circuit to foil car thieves. Proceed as follows:

1. Wire the switches together, using your own code for "off/on" position of the switch levers. See Fig. 9-15.
2. Disconnect the negative cable from the car battery.

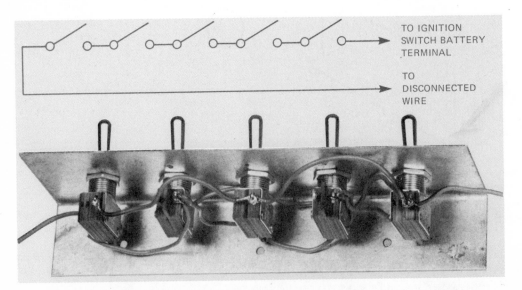

Fig. 9-15. Pictured is an electric combination lock made by wiring five toggle switches and assembling them on a right angle plate. Schematic shows the hookup for an automobile ignition switch. (Project by William Miller)

3. Disconnect the primary wire from the battery terminal of the ignition switch.

4. Run a wire from the disconnected ignition switch wire to the open terminal on the No. 1 toggle switch.

5. Run a wire from the battery terminal of the ignition switch to the open terminal on No. 5 toggle switch.

6. All switches have to be "closed" to activate the circuit.

Only the person knowing the right combination of lever positions can complete the circuit to the ignition switch.

You also can use a toggle switch setup like this to create an interesting game. Install a 6V battery and a lamp or buzzer in the circuit, Fig. 9-16. Then, see who can flip the switches the least number of times to light the lamp or sound the buzzer. Do not solder the connections, because you will want to change the combination and restart the game.

GROUNDS

We have mentioned the term "ground" in previous chapters. *A ground is an electrical connection between a circuit and the earth or a metallic object that takes the place of the earth.*

Fig. 9-17 shows two types of ground symbols that normally appear on drawings. At times, these symbols are used interchangeably. There *is* a difference, but all you need to know at this point is that a ground is a path for current.

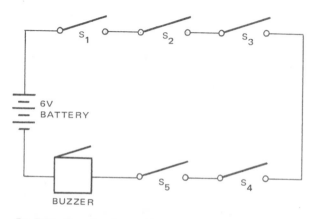

Fig. 9-16. The electric combination lock can be used in this buzzer game. Again, the right combination of switch lever positions will complete the circuit and sound the buzzer.

Fig. 9-17. Two ground symbols used on circuit drawings.

Fig. 9-18 is a schematic of a signaling system that can be used to turn a light on and off. Trace the ground circuit from the power supply through a wire to the load (a lamp). Note, too, that two ground symbols are used, indicating that the ground circuit is to act as a

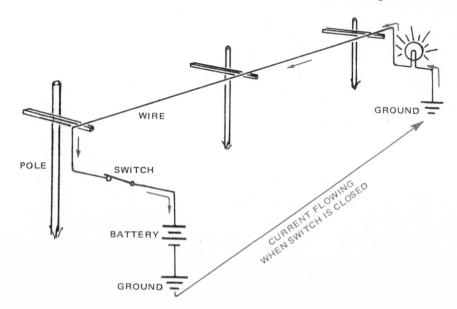

Fig. 9-18. Schematic shows a one wire signaling system, using the ground circuit in place of a second wire in the system.

second wire in the system. When the switch is closed, Fig. 9-18, the circuit is completed and the lamp lights.

Radios and television sets also use this arrangement. In these applications, the chassis serves as the ground circuit.

ONE WIRE SYSTEM

Automobiles use a one wire electrical system. This means that the frame of the vehicle serves as a path for ground, rather than having a return wire to the power supply. The one wire system is easier to trace

and simpler to service. Fig. 9-19 illustrates a series circuit of this type. Fig. 9-20 shows a parallel circuit.

All U.S. car manufacturers ground the negative post of the battery. Some manufacturers of imported cars use the positive ground principle. Regardless of which system is used, all automotive electrical circuits are completed, either directly or indirectly, by using the car frame as the second wire or path.

Fig. 9-21 gives some of the electrical circuits you can find on a car. The ground paths for the headlights and taillights are shown. The ground path for the alternator

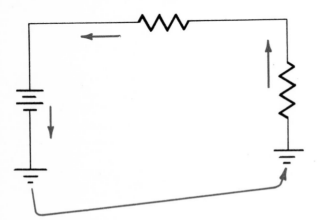

Fig. 9-19. A simple one wire series circuit. Note path from ground connection- to-ground connection.

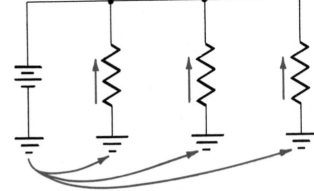

Fig. 9-20. A simple one wire parallel circuit. Note multiple paths to ground in this application.

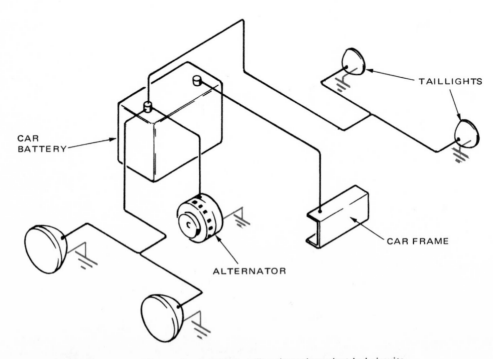

Fig. 9-21. Automobiles use frame for grounding the various electrical circuits.

is through its housing to the engine, then to the engine ground strap attached to the frame or to the negative post of the battery.

There are some real advantages in using the one wire system. It simplifies test procedures. Note in Fig. 9-22 that you can make voltage checks by clamping the negative lead of your voltmeter to a good ground point. Then, use one hand to move the second probe to various points in the circuit to take voltage readings.

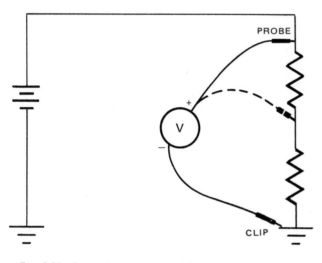

Fig. 9-22. One wire system speeds voltmeter checks for continuity and voltage drop.

The one wire system has other drawing and testing advantages. For example, Fig. 9-23 shows two different ways of drawing the same simple series circuit. By removing the wire from the right side of this circuit and replacing it with ground symbols, you will have the one

wire system shown at Part A in Fig. 9-24. Note that polarity is marked at either end of each resistor. This is helpful when making voltage drop tests.

POLARITY

When the polarity of a resistor is indicated on a drawing, it is done only with respect to the polarity of the power supply. See Fig. 9-24. Part A shows that resistor R_1 has a positive side and a negative side. The side of R_1 nearer the positive post of the battery is marked plus (+), and the side farther away is marked minus (−).

Resistor R_2 is marked in just the opposite way, because the negative post of the battery is closer to it. Therefore, the polarity of this resistor also is marked with respect to the polarity of the power supply.

If you reconnect this circuit without the grounds, you will see the polarities shown at Part B in Fig. 9-24. Again, the two ends of each resistor are labeled with respect to the power supply. Although the lower end of R_1 is marked minus (−), and the upper end of R_2 is marked plus (+), there is no difference between these points.

To illustrate how this form of polarity works, look at the simple series circuit in Fig. 9-25. If you measure

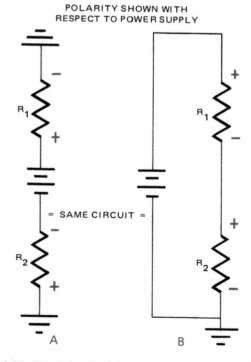

Fig. 9-24. This series circuit has polarity indicated at ends of resistors. A—One wire system grounded at two points. B—Complete circuit with return wire and ground connection.

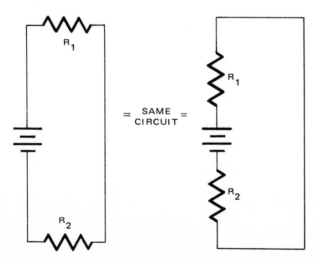

Fig. 9-23. A simple series circuit is shown with resistors placed at different points. Resistance values remain the same.

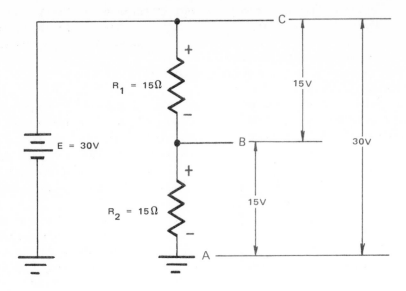

Fig. 9-25. Values are given for voltage drop tests taken across each resistor in turn, then across both resistors. Also note polarity of the resistors.

the voltage between point A and point B, you will get 15 volts. Likewise, the voltage between B and C will be 15 volts. The voltage across both resistors from A to C will read 30 volts (voltage of the power supply).

A resistor always causes a drop in voltage if there is current through it. In Fig. 9-25, you dropped 15 volts across R_2 and its polarity is minus (−) on the bottom and plus (+) on the top. Although the bottom of R_1 is marked minus (−) with respect to the top, it still has a voltage of 15 volts across it.

REFERENCE POINT

In order to have a better understanding of voltage drop, we could pick some reference point and call it zero. Everything above zero would be positive, while everything below zero would be negative.

This reference point can be compared with 0 deg. on a Celsius thermometer. See Fig. 9-26. Again, any reading above zero is positive; below zero is negative. You could get a reading of 20 deg. above the reference point (20 deg. C) or 20 deg. below the reference point (−20 deg. C). In this application, the reference point is where water freezes. With voltages, you also can establish a reference point.

In Fig. 9-27, you have a power supply with a 30 volt potential. By grounding the circuit at the reference point, you can make voltage measurements in relation to that point. *In this situation, the ground is said to be a reference point from which voltage measurements are made.*

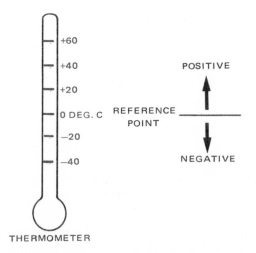

Fig. 9-26. On a Celsius thermometer scale, the zero mark is considered the reference point. All readings above zero are positive, all readings below zero are negative.

If you measure between point B and point A, you will get a reading of −10 volts. If you measure from B to C, you will get a reading of +10 volts. Finally, measuring from B to D, you will get a reading of +20 volts.

Note that from point A to point D, the potential difference is still 30 volts (which matches our power supply). Since you are using the ground as a reference point, you will get both negative and positive voltages from the same power supply. This is the whole idea behind a "voltage divider."

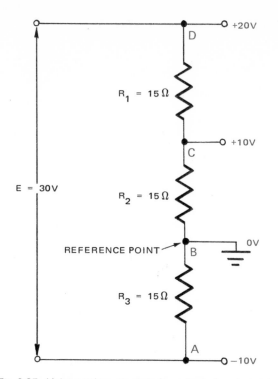

Fig. 9-27. Voltage values are given for voltage drop tests taken across resistors from A to B (−10V), B to C (+10V) and B to D (+20V). Note that the circuit is grounded at reference point B.

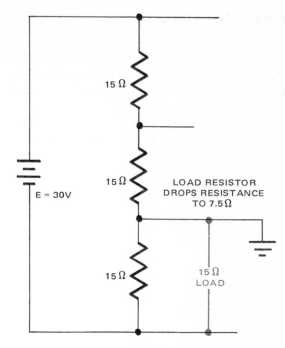

Fig. 9-28. Three 15 ohm resistors and one 15 ohm load resistor serve as a voltage divider that routes different voltages to different parts of the circuit.

VOLTAGE DIVIDER

If you want to have different voltages for different parts of a circuit, you can construct a voltage divider. In effect, a voltage divider is nothing more than a series parallel circuit.

You cannot design a good voltage divider without first looking at the load resistance. Note in Fig. 9-27 that if you make a voltage divider with three 15 ohm resistors, you would get 10 volts drop across each one. However, as soon as you add another resistor (load), Fig. 9-28, there is a further change. The load resistor serves to drop the total resistance of the lower part of the voltage divider. Use this formula for finding the equivalent resistance of resistors of equal value in a parallel circuit:

$$R_{EQ} = \frac{R}{N}$$

$$R_{EQ} = \frac{15}{2} = 7.5\Omega$$

The equivalent resistance of these two 15 ohm resistors in the lower part of the voltage divider is 7.5 ohms. As a result of this resistance change, what will happen to the current in the circuit?

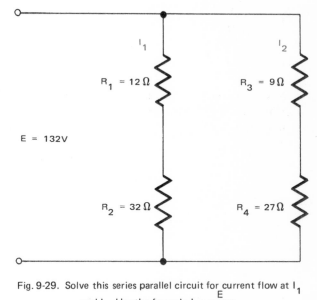

Fig. 9-29. Solve this series parallel circuit for current flow at I_1 and I_2. Use the formula $I = \dfrac{E}{R_{EQ}}$.

Remember that as resistance goes down, current goes up. Therefore, with the addition of the load resistor, the circuit will now carry higher amperage. It is important, then, when constructing a voltage divider circuit, to watch both resistance values and current values. Review Fig. 9-28 to make sure you understand how a voltage divider works.

WHEATSTONE BRIDGE

Of special use in series parallel circuits is a device known as a "wheatstone bridge." It is used to find the value of an unknown resistor in a circuit. Much more accurate than an ohmmeter, a wheatstone bridge will give values within very close tolerances.

Fig. 9-29 shows a simple series parallel circuit with four resistors. Find the current flowing in I_1 and I_2:

$$I_1 = \frac{E}{R_{EQ}}$$

$$I_1 = \frac{132V}{12\Omega + 32\Omega}$$

$$I_1 = \frac{132V}{44\Omega}$$

$$I_1 = 3A$$

$$I_2 = \frac{E}{R_{EQ}}$$

$$I_2 = \frac{132V}{9\Omega + 27\Omega}$$

$$I_2 = \frac{132V}{36\Omega}$$

$$I_2 = 3.66A$$

Now that you have determined the current value in each branch, Fig. 9-30, find the voltage drop across each resistor.

$$V_{R_1} = I \times R_1$$

$$V_{R_1} = 3A \times 12\Omega$$

$$V_{R_1} = 36V$$

$$V_{R_2} = I \times R_2$$

$$V_{R_2} = 3A \times 32\Omega$$

$$V_{R_2} = 96V$$

$$V_{R_3} = I \times R_3$$

$$V_{R_3} = 3.66A \times 9\Omega$$

$$V_{R_3} = 32.94V$$

For our purposes, round off 32.94V to 33V.

$$V_{R_4} = I \times R_4$$

$$V_{R_4} = 3.66A \times 27\Omega$$

$$V_{R_4} = 98.82V$$

Call this 99 volts.

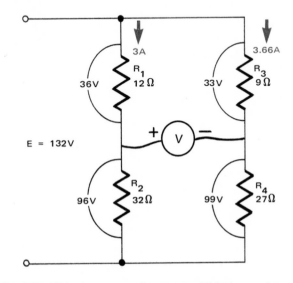

Fig. 9-30. With the amount of current established, next determine voltage drop across each resistor by using the formula $V_R = I \times R$.

VOLTAGE DROP – WHEATSTONE BRIDGE

If you put the individual current values and voltage drops in the circuit shown in Fig. 9-30, what would you get for a voltage reading? You should expect the voltmeter to read 3 volts. Note that there is a 3 volt difference in potential between the 12 ohm resistor and the 9 ohm resistor. This same 3 volt potential exists across the two lower resistors.

The principle behind a wheatstone bridge is to get the voltmeter to read zero. A zero reading would mean there is no difference in potential between the elements of this resistor bridge. If you could redraw the circuit, and make one resistor an unknown value, you would have the setup shown in Fig. 9-31. With a little shifting around of Ohm's Law formulas, you would find that the following ratio exists:

$$\frac{R_1}{R_2} = \frac{R_3}{R_x}$$

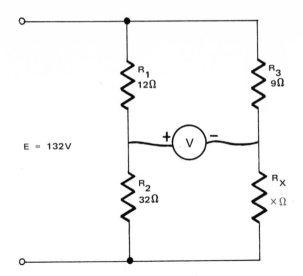

Fig. 9-31. Wheatstone bridge concept: Voltmeter should read zero when voltages are balanced on each side of circuit. Value of resistor R_x should be 24 ohms.

If you substitute known values into this formula, you will get:

$$\frac{R_1}{R_2} = \frac{R_3}{R_x}$$

$$\frac{12\,\Omega}{32\,\Omega} = \frac{9\,\Omega}{x}$$

$$12x = 288$$

$$x = 24\,\Omega$$

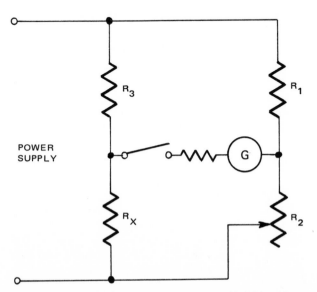

Fig. 9-32. Typical wheatstone bridge setup: Variable resistor R_2 can be adjusted to obtain desired balance of voltages in circuit. This, in turn, pinpoints value of resistor R_x.

This formula shows that if our unknown resistor had a value of 24 ohms, this circuit would be balanced. In both cases, the voltages on each side would be balanced.

In actual practice, three resistors are assembled in one unit, along with an adjustable resistor and a galvanometer. With the sensitive galvanometer, you can get a much more accurate balance in resistor matching process. Fig. 9-32 shows a typical circuit for a finished wheatstone bridge.

CLIMB THE LADDER GAME PROJECT

Here is a game you can make, using an electrical power supply, two pieces of copper tubing and strips cut from old cans. See Fig. 9-33. Wire the metal strips so that it is difficult, yet possible, to move the copper plugs from hole to hole without breaking the circuit to the bulb.

No.	Item
1	1/4 in. plywood base, 5 in. x 10 in.
2	1/4 in. plywood sides, 2 in. x 10 in.
20	Brads.
2	Cans cut into 20 1/4 in. x 1 1/4 in. strips.
2	Lengths of copper tubing, 1/2 in. x 3 in.
1	Flashlight bulb.
1	6 ft. length of 18 gage wire.
1	6V battery.

CONSTRUCTION PROCEDURE

Build the game as follows:

1. Drill ten 1/2 in. holes in plywood base. See Part A in Fig. 9-33.
2. Drill one 1/4 in. hole for bulb.
3. Nail 20 metal strips to underside of base, alongside holes as at Part B.
4. Squeeze one end of each piece of copper tubing to keep it from falling through hole. See Part A.
5. Bend up strips as shown at Part B.
6. Make a small hole in tail end of each strip for wires.
7. Cut wire to lengths shown at Part C and insert wires in holes in strips.
8. Hook up battery and bulb.
9. Test all connections for continuity, then solder.
10. Nail plywood sides to base.
11. Finish to suit.

You can use a step-down transformer, 110V ac to 6V ac, as a power supply for this project. If you make it so that it can be removed, you will be able to use it for other projects. A dropping resistor can be wired in for a 3 volt bulb, if used.

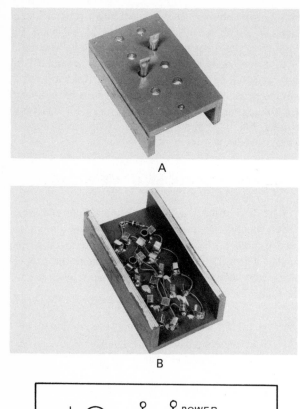

A

B

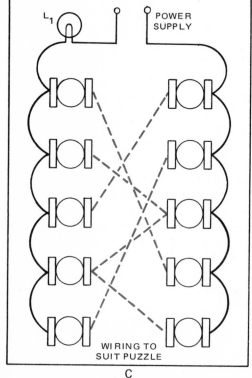

L₁   POWER
      SUPPLY

WIRING TO
SUIT PUZZLE

C

Fig. 9-33. Climb the ladder game. A—Copper plugs must be moved from hole to hole without breaking the circuit to the bulb. B—Underside of game board reveals metal strips and wiring. C—Drawing shows one possible series of hookups for wiring game board. (Project by Randy Schafer)

To play the climb the ladder game, allow the player to move only one hole at a time up to the bulb. By using the right combination of moves, the player should be able to reach the bulb without turning the light off.

TEST YOUR KNOWLEDGE

1. A _____ is another name for a circuit having series and parallel branches.
2. The current in a series parallel circuit is the same in all parts of the circuit. True or False?
3. Tell one of the methods used in drawing circuit layouts to make the circuits easier to read.
4. What is an equivalent circuit for a series parallel circuit?
5. If two 3 volt power supplies are connected in series, the net result will be either _____ or _____.
6. If two capacitors of 4 microfarads are connected in series, what is the total capacitance? Show your work.
7. The larger the plate area on a capacitor, the _____ the capacitance.
8. Draw at least two symbols for a capacitor.
9. If two capacitors of 4 microfarads are connected in parallel, what is the total capacitance? Show your work.
10. An increase in the distance between plates in a capacitor will result in _____ capacitance.
11. What is a short circuit?
12. What is an open circuit?
13. What is a ground?
14. What is a wheatstone bridge used for?

SOMETHING TO THINK ABOUT

1. Why do you see some cars with a black strap hanging down to the ground? Cars are not produced with these straps. Why not?
2. Modern electric drills are made with a 3 prong plug, one of which is a ground. Why do some people cut off the ground prong? Is that really the safe thing to do?
3. Is there a difference between a negative or positive ground found on different cars? Which system is better?
4. What will happen if you are not careful when using jumper cables on cars with weak batteries? Will European and U.S. cars be "jumped" in the same way?
5. Metrics are used in most electrical measurements. Why did it take so long for other trades to convert to

the metric system?

6. Why are there so many ways of drawing the same symbols?

7. Some people claim that they check for voltage by using their fingers. What could happen to those people?

8. Make a list of some of the places where grounds are used.

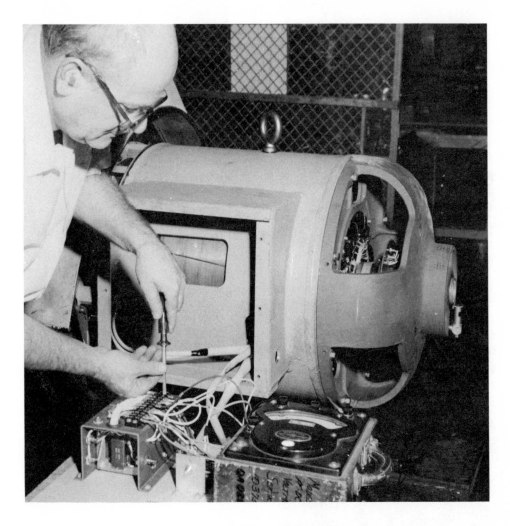

Inspector connects a voltage regulator to an ac generator before making voltmeter tests in an operating final inspection. (Delco Products Div. of General Motors Corp.)

Chapter 10
AC VOLTAGE

When you studied Ohm's Law in Chapter 5, you learned about electrical current that flowed in only one direction. The electrons left the negative post of the battery, flowed through the circuit and returned to the positive post of the battery. This is called "direct current" (dc) electricity.

In this chapter, we are going to deal with "alternating current" (ac) electricity. *Basically, alternating current is any current that reverses its direction of flow at regular intervals.*

In Fig. 10-1, for example, a technician is looking at a "sine wave" on his oscilloscope screen. *A sine wave is a wave form of a single frequency alternating current.* The scope is an instrument that can pick up traces of charged particles moving at a high rate of speed, even though the particles may reverse themselves in an up and down motion.

HOW AC IS PRODUCED

Imagine that you have a pair of magnets like the ones shown in Fig. 10-2. The dotted lines indicate that there is a magnetic field between the poles of the two magnets. (See Chapter 2.) If you pass a piece of wire quickly through the field so that the wire cuts through the lines of force, Fig. 10-3, a current would be "induced" in the wire. At this point, just say that induced means "created in the wire." The induction principle will be explained later in the text.

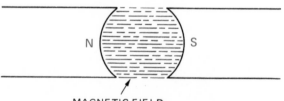

MAGNETIC FIELD

Fig. 10-2. Lines of force between north pole and south pole of magnets show extent of magnetic field.

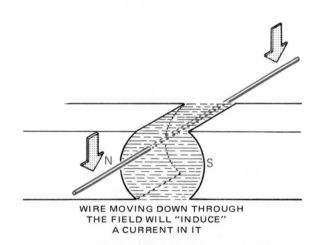

WIRE MOVING DOWN THROUGH
THE FIELD WILL "INDUCE"
A CURRENT IN IT

Fig. 10-3. A wire passing through a magnetic field will cut the lines of force and induce electric current in the wire.

Fig. 10-1. Technician working on a television set uses a sine wave on an oscilloscope to aid in making troubleshooting tests. (Hewlett-Packard)

In order to understand the example shown in Fig. 10-3, there are three important points to remember:

1. A voltage is induced in the wire only when the wire cuts across lines of force.
2. The faster the wire cuts the lines of force, the higher the voltage induced in the wire.
3. If we reverse the motion of the wire, we reverse the polarity of the induced voltage.

Keeping these points in mind, look at the coil of wire lying in a magnetic field in Fig. 10-4. If you cut

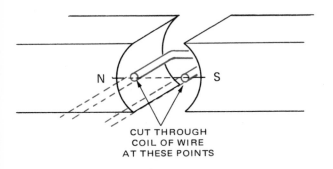

Fig. 10-4. A coil of wire is placed inside the magnetic field. If the coil is cut at the dotted line, cut ends will show what happens when coil rotates.

across the coil at the dotted line, you will have the end view shown in Fig. 10-5. The cut ends of the wire are labeled A and B, so that you can see what happens to each side of the coil as it begins to move, then rotates (spins) inside the magnetic field.

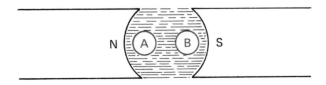

Fig. 10-5. This end view of coil inside magnetic field shows cut ends of coil marked A and B. See Fig. 10-4.

To help you understand how the coil rotates, take a wagon or bicycle and turn it upside down so it cannot move. Mark two spots A and B, Fig. 10-6, and slap the wheel so that it starts to spin. Trace A and B for one complete revolution (turn), and note that they both end up at their original starting point.

The important thing, however, is that both A and B can be traced through one revolution of the wheel. Likewise when the coil of wire cuts through the lines of force of the magnetic field in Fig. 10-7, you can trace A and B throughout the complete revolution. In electrical terms, however, one revolution of the coil is called a "cycle."

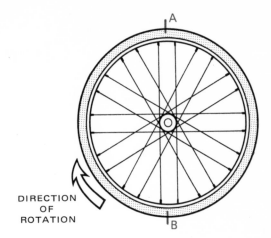

Fig. 10-6. A rotating bicycle wheel illustrates path followed by marks A and B. Marks A and B in Fig. 10-5 follow a similar path.

TRACING POINTS ON THE SINE WAVE

To experiment with the coil and magnet, consider the sine wave shown in Fig. 10-7. Then, start rotating the coil around its axis. At Point 1, the coil of wire is moving parallel to the lines of force in the magnetic field. It is *not* cutting the lines of force. Therefore, no voltage is induced in the coil.

At Point 2, the coil is moving at right angles to the lines of force, and a maximum amount of voltage is induced in the coil. This is the highest point our sine wave will reach.

Again, at Point 3, no voltage is induced. Note that the end of the coil marked A is starting to cut the lines of force in the opposite direction from when it moved past Point 1. This causes the voltage induced in the coil to reverse its polarity. It now starts to generate voltage in the negative direction.

At Point 4, the voltage induced in the coil as it rotates through the magnetic field will reach its lowest point (negative voltage). Here again, the coil is cutting the maximum number of lines of force, but in the opposite direction.

Finally, at Point 5, we have returned to our starting point. No lines of force are being cut, and we are back to the zero point on our reference line.

SINE WAVE FACTORS

The curved line in Fig. 10-7 is known as a "sine wave." When we talk about a sine wave, you will need to know its frequency, period and amplitude.

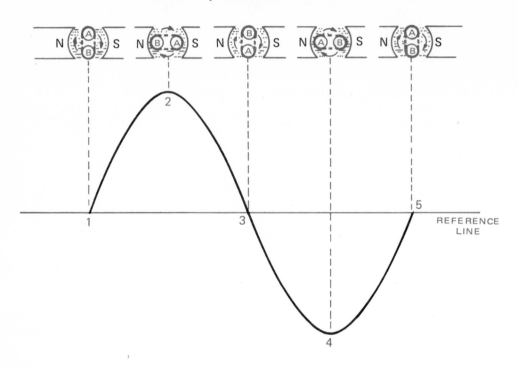

Fig. 10-7. One sine wave (curved line) is generated through one rotation of the coil inside the magnetic field. Coil cuts through maximum number of lines of force in positions 2 and 4.

I.C. SAYS:

"Here are some definitions. Frequency means the number of cycles completed per second. Period is the length of time required to complete one cycle. Amplitude refers to the value of the sine wave."

FREQUENCY

The definition for the electrical term frequency is "number of cycles per second." If the coil pictured in Fig. 10-7 rotated at a speed of five revolutions per second, it would produce five complete sine waves in each second. See Fig. 10-8. *The frequency of this sine wave is five cycles per second, or five Hertz (5 Hz).*

If we rotated the coil faster, the number of cycles would increase and the frequency would be higher. The term "60 cycles per second" is used with reference to house current. This means that your home electrical supply is being generated at a frequency of 60 Hz (Hertz means cycles per second). Some parts of Canada and Australia, and some countries in Europe, generate their electricity at a frequency of 50 Hz.

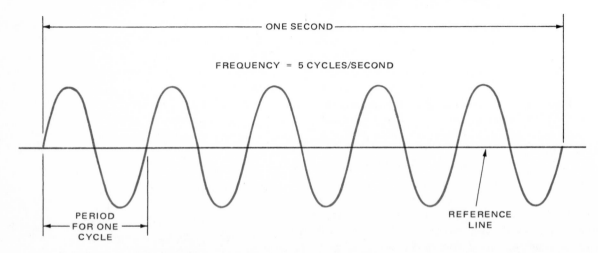

Fig. 10-8. Five sine waves in one second equals a frequency of five cycles per second and a period of one/fifth second.

PERIOD

In Fig. 10-8, the period of the sine wave will be 1/5 of a second. *A period is the time we need to complete one cycle.* Since we generated five complete cycles in one second, the time it takes for one cycle can be found by dividing one by the number of cycles per second (frequency):

$$\text{Period} = \frac{1}{\text{frequency}}$$

$$\text{Period} = \frac{1}{5 \text{ cycles/second}}$$

$$\text{Period} = \frac{1}{5} \text{ second for each cycle}$$

You can see that as frequency increases, the period of time it takes for each cycle gets shorter. For example, if you have 60 cycles per second alternating current in your home, the period will be:

$$\text{Period} = \frac{1}{\text{frequency}}$$

$$\text{Period} = \frac{1}{60 \text{ cycles/second}}$$

$$\text{Period} = \frac{1}{60} \text{ second per cycle}$$

AMPLITUDE

Note that each cycle shown in Fig. 10-8 has a high point and a low point. The highest point is known as the peak value for the sine wave. See Fig. 10-9. *If you*

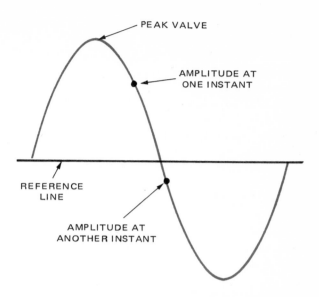

Fig. 10-9. The amplitude of a sine wave is its height above or below reference line at any given time.

wanted to know the height of the wave, you would ask for its amplitude.

However, we are talking about a sine wave that is changing in both directions. At one point in the cycle (peak value), the wave will reach its maximum amplitude in the positive direction. At another point, it will reach its maximum amplitude in a negative direction. This value is taken from a reference line as shown in Fig. 10-10. Therefore, we always have two peaks on a sine wave: a negative peak and a positive peak. The voltage difference between the negative peak and positive peak is commonly called the "peak-to-peak voltage" of a sine wave.

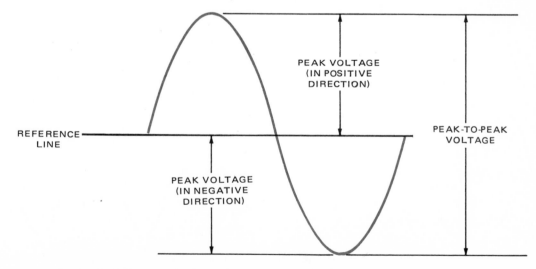

Fig. 10-10. Peak voltage (peak value) is maximum amplitude above or below reference line. Peak-to-peak voltage is sum of the two peak voltages.

PEAK VOLTAGE

Peak and peak-to-peak voltage and current are very important in our study of electronic devices such as diodes, rectifiers and transistors. These devices can be overloaded very easily. If we do not know the value of peak voltage, a diode could be burned up in just a fraction of a second; just about as fast as a fuse rated at 1/100 amp.

In Fig. 10-11, you have an ac voltage with a peak amplitude of 100 volts. Note that the voltage only reaches 100 volts twice during each cycle. Also note that zero volts occurs twice during each cycle.

The symbol most commonly used to indicate an ac power supply is given in Fig. 10-12.

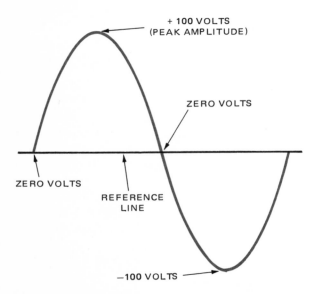

Fig. 10-11. In this specific example of a sine wave, peak amplitude is 100 volts, both above and below the reference line.

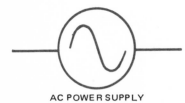

AC POWER SUPPLY

Fig. 10-12. This symbol for the ac power supply is used when drawing schematics.

DEMAGNETIZER PROJECT

Do you want to build a simple project that will demagnetize screwdrivers, drill bits or other metal parts? It can be done easily with a simple coil of wire.

You can salvage one of these coils by disassembling an old motor or you can build one yourself, Fig. 10-13.

To make the demagnetizer, wrap about 2000 turns of No. 28 magnet wire on a 4 in. form. Then, splice and solder the ends of the wire to the two wires of a 110V ac lamp cord. Wrap the entire coil with plastic electrical tape. Then, plug the lamp cord into a 110V receptacle and put the demagnetizer into operation.

To demagnetize an object, simply pass it through the coil. The magnetic field produced by the coil will mix up the pattern of electrons in the metal object, causing it to become demagnetized.

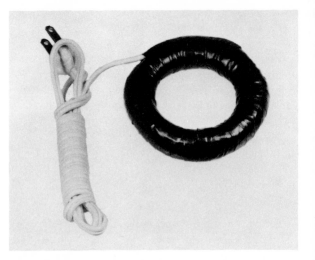

Fig. 10-13. Simple coil "demagnetizer" is wired for 110V ac. Metal parts are demagnetized as they pass through coil. (Project by William Miller)

HEAT FACTOR

In Chapter 5, you read that a certain amount of heat is developed in a resistor when the circuit is completed. For comparison, Fig. 10-14 shows similar ac and dc circuits. One has a 100 volt peak ac power supply hooked up to a 50 ohm resistor. The other circuit has a 100 volt dc power supply hooked up to an identical 50 ohm resistor. Which voltage would you expect to produce more heat? *Remember, ac voltage reaches 100 volts only twice in each cycle, while dc voltage stays at a steady 100 volts.* The answer is: dc voltage will produce more heat in the resistor, Fig. 10-15. (Both a + increase and a − increase cause heating.)

RMS VALUE

After many experiments, scientists have come up with a practical way of comparing the heating values of ac and dc voltages. They found that if you multiplied

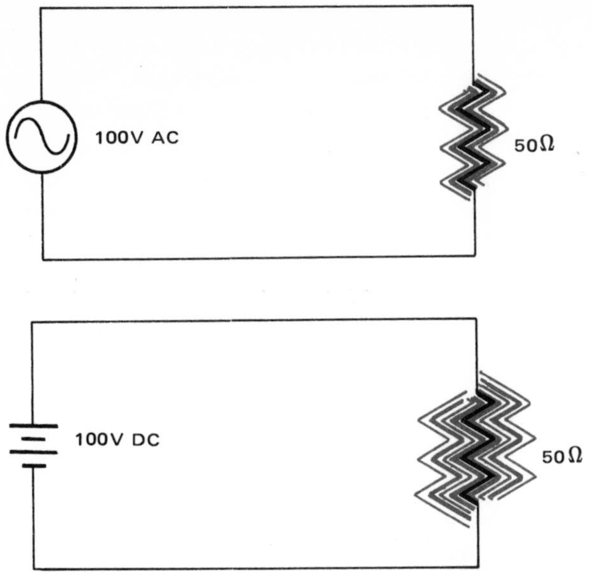

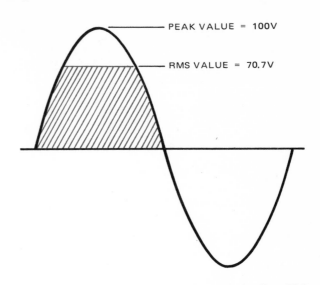

Fig. 10-16. The RMS value of an ac sine wave is, in effect, 70.7 percent of its peak value.

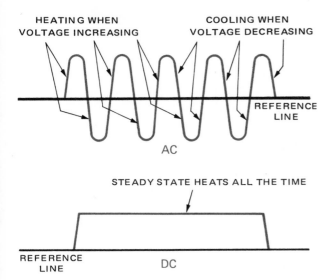

Fig. 10-14. In a test of heat values, identical resistors in similar ac and dc circuits are compared.

HEATING WHEN VOLTAGE INCREASING

COOLING WHEN VOLTAGE DECREASING

REFERENCE LINE

AC

STEADY STATE HEATS ALL THE TIME

REFERENCE LINE

DC

Fig. 10-15. More heat is created by dc voltage. Note that ac voltage heats on rise of sine wave only.

The RMS value of 70.7 volts means that the 100 volt peak ac voltage will have as much heating value as a dc voltage of 70.7 volts. RMS value is so important that whenever people in electrical and electronic work speak of an ac voltage or current, they usually are talking about the RMS value.

A good example of RMS value is the 110 volt power supply found in your home. This is actually the RMS value for the voltage. Peak voltage would be higher. Remember, too, that sine waves can apply to both voltage and current. The formula will work just as well for either of these electrical values.

The letters RMS stand for Root Mean Square. This is just another way of using mathematics to solve problems. You do not have to be concerned with how this factor .707 came about. At this point, we are interested only in knowing how to use the results.

SQUARE WAVE

A "square wave" is another type of wave that is in common use in electricity and electronics. See Fig. 10-17. A square wave is easy to produce by making up a circuit that consists of a battery, switch and resistor.

Consider the circuit shown in Fig. 10-18. When the switch is open, the voltage across the resistor is zero. When the switch is closed at Point A in Fig. 10-19, the voltage across the resistor jumps up to battery voltage. It stays at that voltage until the switch is opened again at Point B. This will happen every time you open and close the switch.

If the switch is closed and opened once every second, we have a square wave with a frequency of one

the amplitude (peak value) of an ac sine wave by the number .707, you would get an RMS value (effective value) for the ac wave. See Fig. 10-16.

The formula is: RMS = peak voltage x .707

Looking back at Fig. 10-14, we can relate the RMS formula to that ac circuit:

RMS = 100V x .707
RMS = 70.7V

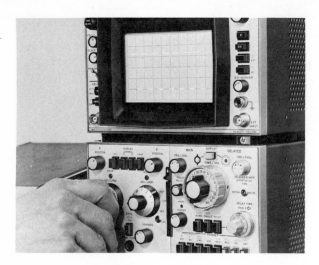

Fig. 10-17. A technician picks out a portion of a square wave for examination, using a delay generator plugged into an oscilloscope. (Hewlett-Packard)

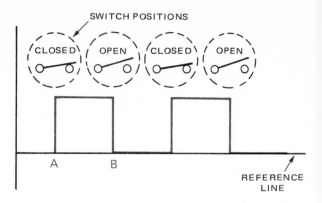

Fig. 10-19. A square wave can be produced by turning the switch on and off at regular intervals.

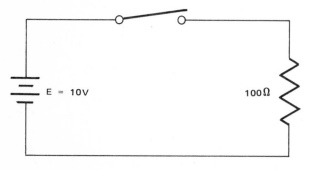

Fig. 10-18. This simple circuit will produce a square wave.

Hertz (Hz). "Hertz" is another way of saying cycles per second (cps) for both square waves and sine waves.

The peak amplitude of the wave shown in Fig. 10-20 is 10 volts. The period of this wave is one second. Note

that even though the reference line is at the "bottom" of the wave form, we still have a square wave whose peak-to-peak amplitude is 10 volts.

As you can see, sine waves and square waves have three things in common: frequency, amplitude and period. These waves are an important area of study in electricity and electronics because they are the two most common types of alternating current.

APPLICATION OF SINE WAVES

You have studied the makeup of sine waves. Now we will look at a practical application. First, take a sine wave with a frequency of 400 Hz and a RMS voltage of five volts. Hook up a loud speaker, as shown in Fig. 10-21, and you will hear a sound. The sound is called a "tone." If you increase the voltage of the wave, the tone will get louder. If you increase the frequency of the wave, the tone will get higher in pitch.

Most people can hear sounds as low as 20 Hz and as high as 20,000 Hz. As they get older, their hearing

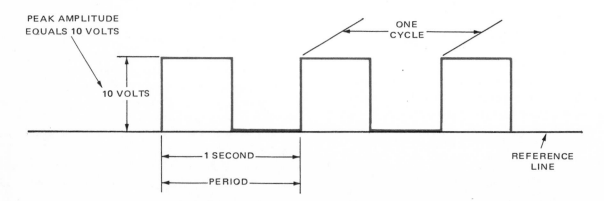

Fig. 10-20. By opening and closing the switch once every second, we can produce a square wave with a frequency of one Hertz.

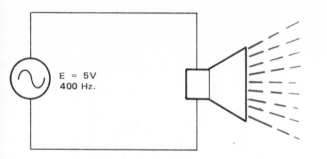

Fig. 10-21. We can produce sound by means of sine waves by putting a loud speaker in the circuit.

range gets smaller. They may only be able to hear sounds between 100 Hz and 15,000 Hz. Usually, frequencies between 20 Hz and 20,000 Hz are called "audio frequency or AF waves."

SIGNAL GENERATORS

Most ac waves you will deal with in electronics can be produced by a machine called a "signal generator." See Fig. 10-22. These machines are used to produce ac

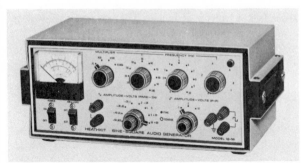

Fig. 10-22. Signal generators produce ac voltages having a broad range of frequencies.

voltages having a frequency as low as 20 Hz or as high as 200 million Hz, Fig. 10-23.

Signal generators have many control knobs, three of which are found on all makes, Fig. 10-24. The frequency selector knob lets the operator choose the frequency. The amplitude control adjusts the amplitude of the wave produced. The function selector permits the operator to choose either a sine wave or a square wave.

There are many different types of signal generators made. It is impossible to describe every type. Because of this, you may have noticed several switches on your generator that we have not mentioned. If you are unsure about operating any of the controls, ask your instructor for an explanation.

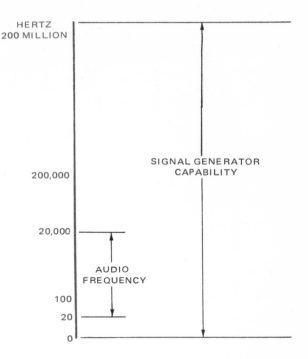

Fig. 10-23. A signal generator can produce sounds far beyond audio frequency (range of human hearing).

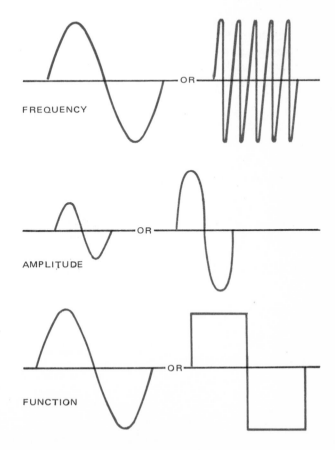

FREQUENCY

AMPLITUDE

FUNCTION

Fig. 10-24. Symbols are given for controls of signal generators.

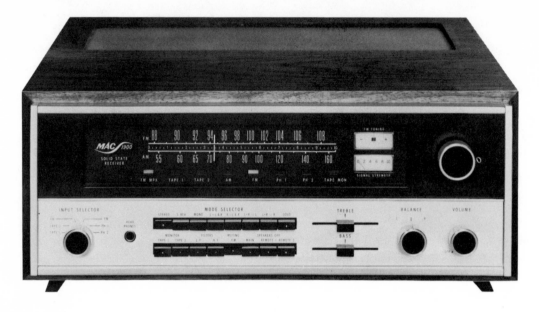

Fig. 10-25. This AM radio dial covers a band from 54 to 160, or 540 kHz to 1600 kHz. (McIntosh Laboratory, Inc.)

Why do signal generators produce frequencies higher than 20,000 Hz if we cannot hear them? In Fig. 10-25, the numbers on the AM dial go from 540 kHz to 1600 kHz (1 kHz = 1000 Hz) even though you cannot hear frequencies this high. However, radio stations *are* able to send both voice and music through the air by means of waves that your radio *can* pick up and change into frequencies that we *can* hear. See Fig. 10-26.

Waves at these high frequencies are called radio frequency waves or RF waves. Radios probably will be covered in your next course, but the most important point to remember is that waves are around us everywhere and are a very important part of electronics.

PHASE RELATIONSHIP

Phase relationship is the last topic in our study of waves. In Fig. 10-27, we have two waves drawn on the same time scale. The black sine wave represents voltage, while the blue wave represents current. Note that each wave reaches its maximum amplitude at the same time, and both reach their zero level at the same time. When this happens, we say that voltage and current are "in phase."

Actually, an in-phase situation occurs when the circuit contains resistance only. In actual practice, most circuits contain resistors, capacitors, inductors and other components. To give you the basics you need to establish a foundation for working with other, more complex circuits, we *must* start at this level. Once you know the basics, it is easier to understand why other combinations of components have particular circuit applications.

In the next chapter, we will introduce you to "inductors." Then, we will combine resistors, capacitors and inductors in various circuits. At that point, you will see that voltage and current may be out-of-phase, Fig.

Fig. 10-26. Radio frequency (RF) waves are transmitted to radio receivers that convert the high frequencies to frequencies you can hear.

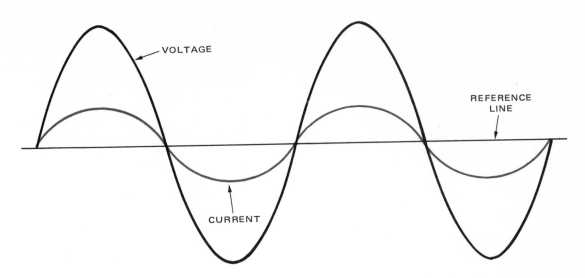

Fig. 10-27. Two sine waves are shown "in phase." Note that they peak at the same time, also reach zero level (reference line) at the same time.

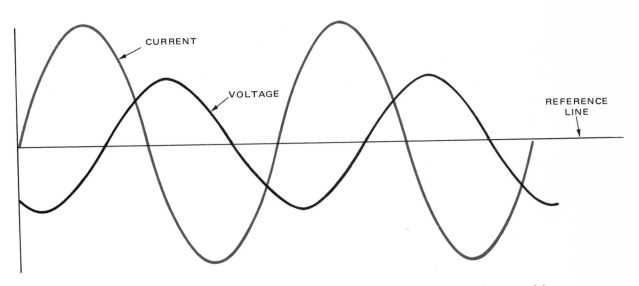

Fig. 10-28. These voltage and current waves are "out-of-phase." Note how the peaks vary, and the waves do not meet at the reference line.

10-28. Then, you will understand how all of these components work together toward accomplishing a desired result.

"FAKEQUAD" PROJECT

You can build a conversion unit that connects to your stereo set to help produce simulated quadrophonic sound. Since the sound *is* faked, this project is called "Fakequad."

Study Fig. 10-29, then procure the following items and assemble them as directed in the construction procedure:

No.	Item
2	20 Ω , 20W resistors.
2	Terminal strips, 2 posts.
2	Terminal strips, 4 posts.
1	Breadboard, 6 in. x 10 in.
1	Plastic box, 6 in. x 10 in. x 2 5/8 in.

A

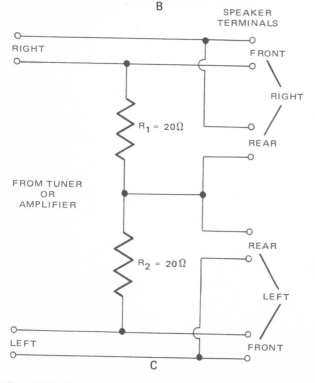

B

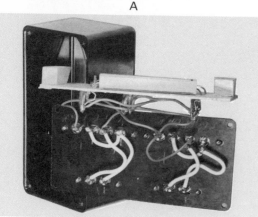

SPEAKER
TERMINALS

RIGHT

FRONT

RIGHT

$R_1 = 20\Omega$

REAR

FROM TUNER
OR
AMPLIFIER

REAR

$R_2 = 20\Omega$

LEFT

LEFT

FRONT

C

Fig. 10-29. "Fakequad" is a project designed to fake quadra-
phonic sound by bringing sound from two tracks through four
speakers. (Project by Ron Simons)

CONSTRUCTION PROCEDURE

Set up and assemble the project materials as follows:
1. Locate and drill holes for terminal posts and mounting screws.
2. Install small wooden blocks on bottom of breadboard to serve as spacers for bottom clearance in plastic box.
3. Install resistors on bottom of breadboard, feeding lead wires through to top of board.
4. Assemble and mount terminal pins on top of breadboard. See B in Fig. 10-29.
5. Using the schematic shown at C, wire terminal pins on breadboard to terminal strips on top of box.
6. Interconnect proper pins to top terminal posts, as shown at B.
7. Use an ohmmeter to check for shorts or opens in wiring. Repair as required.
8. Solder all connections.
9. Install cover on box and label terminal screws as shown at A.
10. Bring in leads from stereo tuner or amplifier to right and left terminals of "Fakequad" unit.
11. Hook up four speakers to right front, right rear, left front and left rear terminals.
12. Position speakers around room.

TEST YOUR KNOWLEDGE

1. Draw a sine wave.
2. What is the voltage of the three wire service which is brought into your home?
3. Frequency is _____.
4. Period is _____.
5. What is the frequency of the current in your home?
6. If you had 100 volts peak, would it give you more or less heat than 100 volts RMS? Show your work.
7. What is the audio range for most people?
8. Say that the resistance of a heating element on an electric range is 20 ohms. How much current does the element use if it is connected to a voltage source of 240 volts? Show your work.
9. Is the voltage in your home measured in RMS or peak voltage?
10. A signal generator is used to produce sine waves and square waves. True or False?
11. Change the following:
   1600 kHz equals _____ Hz.
   200 million Hz equals _____ kHz.
   20,000 Hz equals _____ kHz.

SOMETHING TO THINK ABOUT

1. Why does the power company transport its electricity in ac instead of dc?
2. Recently, a few companies have switched to dc as a means of transporting power for long distances. Where are these companies? Why are they doing it?
3. Make a list of some of the different frequencies used by radio stations in your city.
4. Make a list of the different frequencies used by television stations around your area of the country. Why are these different than radio stations?
5. The National Electric Code governs most electrical construction. What is it? Why do building inspectors follow it?
6. Besides the sine wave, what are some of the other types of waves found in electrical work?
7. What happens to your hearing as you get older? Why does that affect the range of frequencies that you can hear?
8. A dog can hear higher frequencies than humans can. Why?
9. New ovens are being used to cook food in just a few minutes instead of the long periods of time previously needed. How do they work? Would these ovens hurt you if they were turned on without the door being in place?
10. Where are most of the major power companies located in your area of the country?
11. When companies began shifting to atomic energy as a means of generating electricity why were so many people unwilling to have those plants located near their homes?
12. Where is the biggest generating plant in your area located? How much electricity does it generate in a single day? How long would a city of 30,000 people be able to operate on that much energy?
13. How much electricity does your family use in a single year?
14. Make a list of five countries which consume the most electricity.

This 85 ft. high antenna powered by electric motors is one of 14 around the world in NASA's manned space flight network for tracking and acquiring data from manned spacecraft at lunar distances.

Chapter 11
ELECTROMAGNETIC INDUCTION

As you learned in Chapter 2, a current passing through a wire will set up a magnetic field around the wire. In Chapter 10, you found that the reverse effect is also true: that a magnetic field will induce current to flow in a wire. In this chapter, we will discuss *electromagnetic induction, which is a means of generating EMF (electromotive force or voltage) in a conductor by creating a change in the magnetic field. This, in turn, causes electrons to flow in the conductor.* See Figs. 11-1 and 11-2.

HOW VOLTAGE IS INDUCED

To prove to yourself that electromagnetic induction works, perform the following experiment:

Obtain a dc galvanometer, a hollow coil of wire, a bar magnet that is small enough to pass through the coil

and some hookup wire. Assemble the equipment as shown in Fig. 11-1. Thrust the magnet into the coil, then quickly remove it, while watching for movement of the galvanometer pointer.

If you observe carefully, you will detect a slight pointer movement, both when you put the magnet into the coil and when you remove it. This means that current is flowing in the wire *only* when the magnetic field is changing. If you fail to get pointer movement, you probably did not move the magnet quickly enough.

Remember that any magnet has a magnetic field coming from its north pole. Also bear in mind that the lines of force in this field will go right through any insulating material, such as the coating on the wire in the coil. Further proof of this is found in working with a simple electromagnet, Fig. 11-2.

If we pass direct current through the coil of the electromagnet, the iron rod within the coil will attract iron filings. This will happen even though we are using insulated wire. Keep this fact in mind because much of what you learn about the applications of electricity is affected by this source of electricity called "electromagnetic induction."

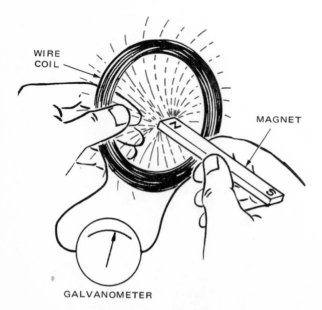

WIRE COIL

MAGNET

GALVANOMETER

Fig. 11-1. In this experiment, sudden movement of a magnet inside a coil of wire will change the magnetic field and induce voltage in the coil.

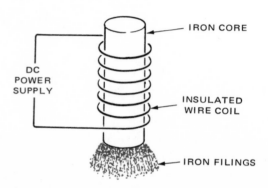

IRON CORE

DC POWER SUPPLY

INSULATED WIRE COIL

IRON FILINGS

Fig. 11-2. In a demonstration of magnetism by induction, iron filings will be attracted to the iron core of an electromagnet when direct current flows through the wire coil.

INDUCED EMF

In Fig. 11-1, the current flows in the wire coil because the changing magnetic field has induced (caused) electromotive force (EMF) or voltage in the coil. This voltage is often called "induced EMF."

Consider, too, that instead of thrusting the magnet into the coil, we could have held the magnet in a fixed spot and moved the coil over it. EMF still would have been induced in the wire. This type of induction will be explained in more detail in Chapter 12.

INDUCTANCE

In Chapter 10 and again earlier in this chapter, you found that a changing magnetic field will induce EMF in a coil of wire. In Fig. 10-7, the magnetic field is, in effect, changing because the coil is rotating in the field. This, you recall, caused major changes at 90 deg. and 270 deg. where the coil cut through the maximum number of lines of force. At 0 deg. and 180 deg., the coil cut through the minimum number of lines of force.

MAGNETIC COUPLING

Now look at the hollow coil of wire shown in Fig. 11-3. If electrical current is passing through the wire, a magnetic field will be created around the coil. In a dc circuit, the magnetic field will expand and remain there.

EXPANDS AND REMAINS

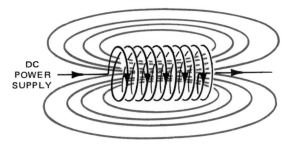

Fig. 11-3. Passing a direct current through a hollow coil of wire will create a magnetic field that will expand, then remain around the wire coil.

However, if current flow in the coil changes, as in the ac circuit shown in Fig. 11-4, a *changing* magnetic field is set up. The magnetic field in an ac circuit will expand and collapse. Note in Fig. 11-5 that the changing field set up by each turn of the coil passes through other turns of the coil. This is known as "magnetic coupling."

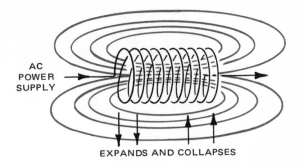

Fig. 11-4. Passing an alternating current through a hollow coil of wire will create a changing magnetic field that will expand and collapse.

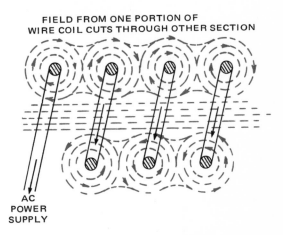

Fig. 11-5. The changing magnetic field created by alternating current in each turn of the coil will cut through the other coils, resulting in "magnetic coupling."

Actually, the changing magnetic field in one turn of the coil will set up an EMF in the other turns of the coil. If the turns are closer together, the lines of force will cut through all the turns of the coil. The further you stretch out the coil, the fewer the number of lines of force that can cut through the wires of the coil.

You might ask, "How does this EMF affect current in a coil of wire?" The answer is: "*Voltage induced tends to oppose or resist the current in the wire, and any part of a circuit that has this property is called an inductor.*"

The two students working on a circuit in Fig. 11-6 are using an oscilloscope. If we used a scope, we could close the switch for the circuit and see that the "wave" on the scope screen does not immediately climb to its peak. Instead, there is a small delay in its rise from the time reference line. Fig. 11-7 shows that the pattern would look like a slanted line on a scope, since the voltage induced tends to oppose the current. This is why the induced voltage is called a "counter EMF."

Fig. 11-6. Students use diagrams and a parts list to aid in assembling a project. The oscilloscope at left serves to verify whether or not the electrical circuit is satisfactory. (Hickok Teaching Systems, Inc.)

CURRENT IN A COIL

CLOSE
SWITCH

→ TIME →

Fig. 11-7. An oscilloscope pattern of a circuit having induced voltage will show a slanted line because counter EMF tends to oppose changing current flow.

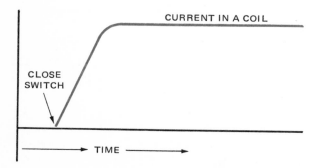

I.C. SAYS:

"Any part of a circuit that opposes a change in current is called an "inductor.""

INDUCTOR SYMBOLS AND USES

The most common form of an inductor is a coil of wire. As might be expected, the more turns of wire you have, the higher the inductance of the coil.

There are several symbols for inductors. The basic symbol, along with related symbols, is shown in Fig. 11-8. Note that they are drawn like a coil of wire, which is just what an inductor is supposed to be.

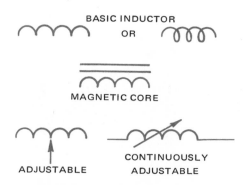

BASIC INDUCTOR
OR

MAGNETIC CORE

ADJUSTABLE

CONTINUOUSLY
ADJUSTABLE

Fig. 11-8. Various symbols may be drawn to indicate the use of inductors in electrical circuits.

Occasionally, you will hear an inductor being called a "choke." A radio frequency (RF) choke would be one which is used where high frequency waves are found. Fig. 11-9 shows one such RF choke in a capacitive discharge ignition (CDI) system for an automobile.

We will not go into CDI circuitry. However, the purpose of the system is to increase the life of spark unit for inductance is the "henry." By definition, if one volt of EMF is induced when the current is changing at the rate of one ampere per second, the coil has an inductance of one henry (H).

INDUCTOR IN A DC CIRCUIT

Consider the circuit shown in Fig. 11-11. When the switch is open, the current through the coil is zero. Then, when the switch is *closed*, the current tries to start flowing through the inductor. Since the current is now changing (increasing in value), the inductor will oppose this change. When the current finally reaches its maximum, there is no further change (increase) in current, so the inductor will no longer oppose the current.

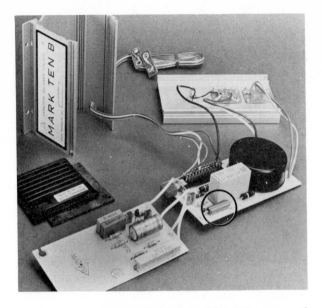

Fig. 11-9. Inductors are called "chokes" (circled component) in high frequency wave applications, such as this capacitive discharge ignition system. (Delta Products, Inc.)

plugs and reduce the number of engine tune-ups needed. The chokes can have either air or an adjustable ferrite (soft iron) core, Fig. 11-10.

HENRY

Just as we needed a unit of resistance for resistors, we also need a unit of inductance for inductors. The

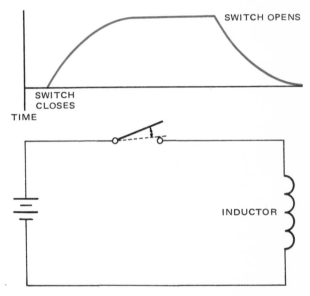

Fig. 11-11. An inductor opposes changes in current flow, particularly when a switch is closed or opened. Note the effect of "switching" on the wave pattern.

From these examples, you can see that an inductor only has an effect on a dc circuit when the current starts to flow. The same thing is true when we *open* the switch. The inductor will oppose the change in current when the voltage falls to zero.

Based on changing magnetic fields, you can see that an inductor will have little or no effect on a dc circuit, but it will have a large effect when used in an ac circuit.

WHAT AFFECTS INDUCTANCE

There are several things that affect the amount of inductance in a coil, Fig. 11-12:

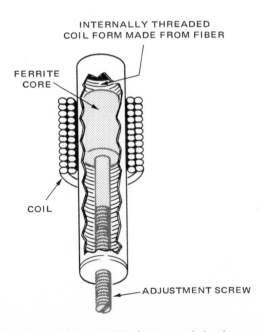

Fig. 11-10. This inductor, or high frequency choke, has an adjustable ferrite core. Many of these variable inductors are used in radios and television sets.

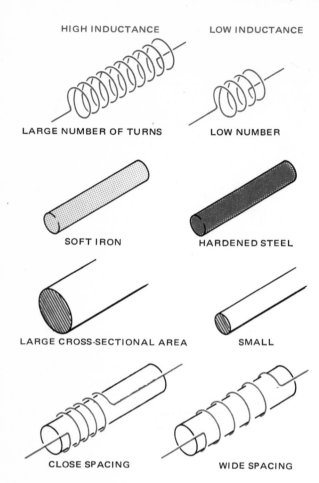

HIGH INDUCTANCE   LOW INDUCTANCE

LARGE NUMBER OF TURNS   LOW NUMBER

SOFT IRON   HARDENED STEEL

LARGE CROSS-SECTIONAL AREA   SMALL

CLOSE SPACING   WIDE SPACING

Fig. 11-12. The amount of inductance in a coil of wire is affected by the physical makeup of the coil.

1. Number of turns of wire in coil.
2. Type of core.
3. Cross-sectional area of core.
4. Length of coil compared to number of turns of wire.

NUMBER OF TURNS

The more turns of wire on a coil, the more lines of flux there will be to act on other turns. If you had only one turn of wire in the coil, the amount of inductance could be forgotten. However, if you multiply that inductance by the number of turns, it will have a large effect on the circuit.

CORE

If you use a core material such as soft iron, you will get more inductance than if you use hardened steel. The flux lines of the magnetic field created in the core actually add to the flux lines from the coil. Together, these two create a greater counter EMF. It should be noted, however, that magnetic cores have a greater inductance than those with air cores or insulating cores.

CROSS-SECTIONAL AREA

If the core of the inductor has a large cross-sectional area, it will create more flux lines than a small core. Lengthening the core is not practical, because then you must space out the wire in the coil and lose the value of using a core.

SPACING

Spacing the turns of wire in the coil does affect the amount of inductance in the circuit. If you space out the turns, the lines of flux from each turn cannot cut through one another.

MAGNETIZER/DEMAGNETIZER PROJECT

Here is a project that will allow you to magnetize or demagnetize small tools. There are times when a magnetized screwdriver can be a big help in tight quarters. There are other times when a magnetized tool will hinder an assembling operation.

See the two views of the magnetizer/demagnetizer pictured at A and B in Fig. 11-13. Then, study the schematic shown at C. Procure the necessary parts and assemble the project according to the construction procedure that follows.

No.	Item
1	Handmade inductor.
1	20 μF, 200V electrolytic capacitor.
1	400 mA (or higher), 300V silicon diode.
1	200 Ω, 5 W wirewound resistor.
1	Momentary contact push button switch.
1	DPDT toggle switch.
1	110V ac lamp cord.
1	6 in. x 3 in. x 2 in. plastic box and cover.

CONSTRUCTION PROCEDURE

Build the inductor first, then install it in the cover of the box. See A and B in Fig. 11-13. Complete the project as shown at B and C.

1. Saw off both ends of ten 20 penny nails, leaving them approximately 2 1/4 in. long.
2. Spray each nail with a thin coating of lacquer.
3. When dry, wrap nails with adhesive tape to form the core of the inductor.
4. Cut two circles of cardboard and put one at each end of the core, leaving about 1 in. of space between the cardboard circles.
5. Wrap layers of No. 36 magnet wire around the core, totaling about 4500 turns.
6. Before cutting the wire, scrape enamel to bare a

A

B

To use the magnetizer/demagnetizer, plug the unit into a 110V ac outlet. Place the toggle switch in the magnetizer position. While depressing the push button, pass the small tool across the core a few times to magnetize the tool. To demagnetize a tool, put the toggle switch in the demagnetize position. While depressing the push button, pass the tool across the core several times.

INDUCTORS IN SERIES

As mentioned earlier, inductance values are given in henrys (H). Lesser values are measured in millihenrys (one thousandth of a henry) or microhenrys (one millionth of a henry). Their abbreviations are: mH for millihenry; μH for microhenry.

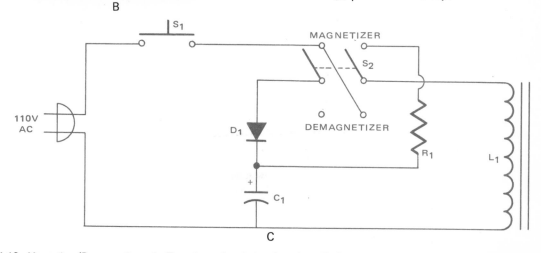

Fig. 11-13. Magnetizer/Demagnetizer. A—Top view of completed project. B—Parts attached to underside of cover. C—Schematic includes: L_1—Inductor. C_1—Capacitor. D_1—Silicon diode. R_1—Wirewound resistor. S_1—Push button switch. S_2—DPDT toggle switch. (Project by Robert Wong)

spot for attaching an ohmmeter lead. Bare starting end of coil wire and connect the other ohmmeter lead to it. Resistance should be between 425 and 450 ohms. If resistance is lower, add turns of wire to coil and recheck resistance value.

7. Cut a hole in the cover just large enough for one end of the core to protrude about 1/4 in. through the cover.

8. Tape the finished coil, leaving both wire ends outside the tape. Scrape enamel from wire ends and solder two hookup wires to them. Tape soldered joints to coil.

9. Epoxy the core to the cover so that it cannot move. Allow 24 hours drying time.

10. Install and wire other parts to underside of cover. See B and C in Fig. 11-13. Observe correct polarity when connecting silicon diode (D_1) and electrolytic capacitor (C_1).

When an inductor is shown in a circuit drawing, it is labeled with a capital L. Inductors will be found in series, parallel and series parallel circuits. In Fig. 11-14, three inductors (L_1, L_2, L_3) are used in series with an ac power supply.

When analyzing a series circuit for total inductance, we use the following formula:

$$L_{Total} = L_1 + L_2 + L_3 \ldots$$

In the circuit shown in Fig. 11-14:

$$L_T = L_1 + L_2 + L_3$$

$$L_T = 2 \text{ mH} + 3 \text{ mH} + 4 \text{ mH}$$

$$L_T = 9 \text{ mH}$$

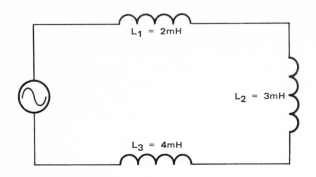

Fig. 11-14. Inductors of three different millihenry values are used in this ac-powered series circuit. Find total inductance.

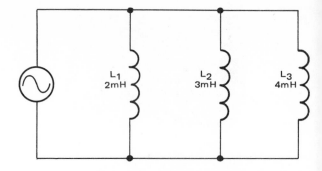

Fig. 11-15. In a setup similar to Fig. 11-14, three inductors are wired in parallel with the ac power supply. Find total inductance.

INDUCTORS IN PARALLEL

If the same three inductors were found in a parallel circuit, Fig. 11-15, we would use this formula:

$$\frac{1}{L_{Total}} = \frac{1}{L_1} + \frac{1}{L_2} + \frac{1}{L_3} + \ldots$$

Solving for total inductance, we would find:

$$\frac{1}{L_T} = \frac{1}{L_1} + \frac{1}{L_2} + \frac{1}{L_3}$$

$$\frac{1}{L_T} = \frac{1}{2\ mH} + \frac{1}{3\ mH} + \frac{1}{4\ mH}$$

$$\frac{1}{L_T} = \frac{6}{12\ mH} + \frac{4}{12\ mH} + \frac{3}{12\ mH}$$

$$\frac{1}{L_T} = \frac{13}{12\ mH}$$

Invert both sides of the equation:

$$\frac{L_T}{1} = \frac{12}{13}\ mH\ or\ 923\ \mu H$$

The same concept used to solve total resistance in series parallel circuits can be used to analyze inductor values. Consider the circuit shown in Fig. 11-16. You recall that an inductor has no effect on a dc circuit, except when the switch is closing or opening. Therefore, we can analyze this circuit as if the inductor is shorted out.

Note in the equivalent circuit shown in Fig. 11-16 that only E (voltage) and R (resistance) are given. Based on these values, you can solve this circuit for I (amperage) by using Ohm's Law:

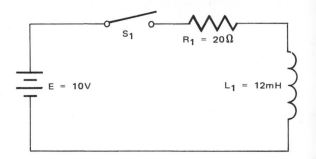

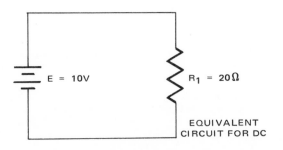

Fig. 11-16. Since an inductor has no effect on the constant current flow of a dc power supply, it can be left out when solving circuit problems.

$$I = \frac{E}{R}$$

$$I = \frac{10V}{20\ \Omega}$$

I = .5A or 500 mA

INDUCTOR/RESISTOR COMBINATIONS

You have been shown how to find total inductance in various types of electrical circuits. Ordinarily, the next step would involve the analysis of a circuit that contains both inductors and resistors. However, just as we found in working with capacitors and resistors in Chapter 9, more information is required before you can analyze these more complicated circuits.

When we get to Chapter 13, you will be shown how to solve circuits having inductor/resistor combinations.

MUTUAL INDUCTANCE

One use of inductance can be demonstrated by placing two coils of wire close to each other, Fig. 11-17. Connect a changing current to coil No. 1 and a changing magnetic field will be produced around it. Some of the lines of force of the magnetic field around coil No. 1 will cut through the windings of coil No. 2. These changing lines of force will induce voltage in the windings of coil No. 2. This induced voltage, in turn, will cause a current to flow in coil No. 2. This is quite a trick, since we caused a current to flow in the second coil without a wire joining the two coils.

When two coils of wire are close enough to be linked by a magnetic field, they are said to have "mutual inductance." This is the basic principle of transformer operation. You can use different kinds of metal in the cores of the coils and see what effect they have on the inductance. Another thing you can do is place pieces of

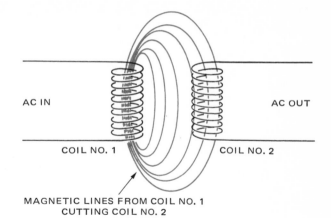

MAGNETIC LINES FROM COIL NO. 1
CUTTING COIL NO. 2

Fig. 11-17. In this demonstration of mutual inductance, ac flow in coil No. 1 will induce voltage across coil No. 2.

metal next to each coil as in Fig. 11-18. The lamp will give you visual proof of the test results.

The amount of mutual inductance in this experiment will depend upon three things:
1. Inductance of coil No. 1.
2. Inductance of coil No. 2.
3. Number of lines of force set up by coil No. 1 that cut across coil No. 2.

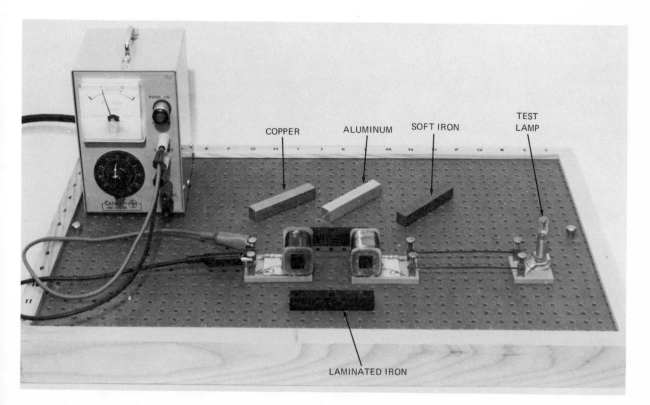

Fig. 11-18. Testing how different core materials affect inductance in an ac circuit.

FORMULA FOR MUTUAL INDUCTANCE

The formula for mutual inductance is:

$$M = K\sqrt{L_1 \, L_2}$$

M = Mutual inductance.

K = Coefficient of coupling (percent of coupling).

L_1 = Inductance of coil No. 1.

L_2 = Inductance of coil No. 2.

Suppose we have two coils placed near each other: one with an inductance of 12.5 henrys; the other with an inductance of 2 henrys. Alternating current is passed through coil No. 1, creating lines of force. When this occurs, it is found that 50 percent of these lines of force cut across coil No. 2. The problem is to find the mutual inductance between the two coils.

First, we know that L_1 = 12.5 H and L_2 = 2 H. We also know that K = .5 (50 percent). Then, substituting these values for factors in the mutual induction formula, we have:

$$M = K\sqrt{L_1 \, L_2}$$

$$M = .5\sqrt{12.5 \times 2} = .5\sqrt{25}$$

$$M = .5 \times 5$$

$$M = 2.5 \text{ H}$$

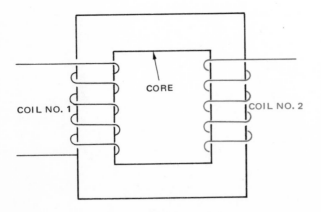

Fig. 11-19. The basic idea for transformer design is a pair of wire coils wound around the same core.

TRANSFORMERS

The main use of mutual inductance is in transformer operation. A transformer usually consists of a pair of coils wound around the same iron core or shell, Fig. 11-19. The coils are wound on the same core to insure that the magnetic coupling between the two coils is as high as possible.

Fig. 11-20 shows a basic type of transformer. This type is known as a "laminated core transformer." The

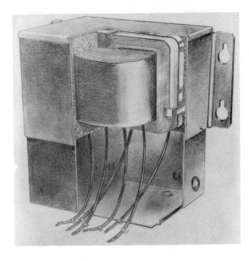

Fig. 11-20. Cutaway view shows the makeup of a laminated core transformer. (Acme Electric Corp.)

core is made from thin strips of metal that look like the letter E. See Fig. 11-21. These pieces are stacked in a special way to produce a closed core assembly, Fig. 11-22. Another design of laminated core transformer uses strips in the shape of the letter U. These strips are stacked in "shell core transformers," Fig. 11-23.

Fig. 11-21. E-strip core material for building transformers is fed into a heat treating furnace. (Acme Electric Corp.)

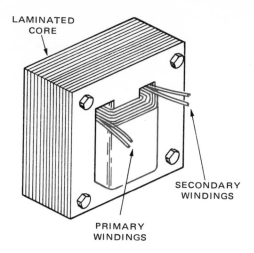

Fig. 11-22. This laminated core transformer is called a "closed core" type.

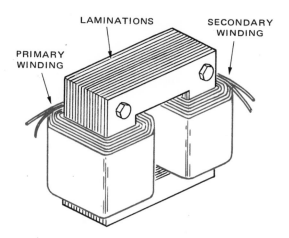

Fig. 11-23. When the laminations are U-shaped, the assembled transformer is called a "shell core" type.

WHY THE CORE IS LAMINATED

If a solid core was used in a transformer, or if the thin strips of metal in a laminated core were stacked one on top of the other, a lot of heat will result. Heat means a loss of energy, so this heat buildup must be avoided.

Heat is created in the transformer when alternating currents (Eddy currents) are forced to flow in the core because of induced voltage. See Fig. 11-24. The larger the voltage created by the fields, the greater the heat generated in the transformer. The greater the heat, the greater the energy loss. In the trade, this loss is known as "Eddy-current loss."

One way transformer manufacturers cut down on transformer heat is by coating each metal E or U strip

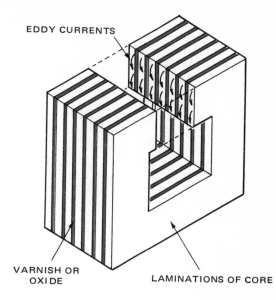

Fig. 11-24. Alternating current flow in the transformer core will break down into tiny loops or Eddy currents.

with varnish or oxide. In effect, then, each piece in the stacked core is separated by an insulator (similar to capacitor construction). This arrangement keeps large voltages from developing in the core, thereby holding down the alternating current flow that builds heat.

The copper for transformers also must be separated by an insulating material, just like insulation on coils of wire. The copper is wound into coils, Fig. 11-25. The

Fig. 11-25. "Winding the coils" is an important step in the early stages of the transformer manufacturing process. (Acme Electric Corp.)

coils are put together with the laminated cores, Fig. 11-26, then they are bolted together and made ready for sale. A view of the finished transformer in Fig. 11-27 will give you an idea of how it all fits together.

A second basic transformer is the ferrite-core type. See Fig. 11-28. It consists of a single, powdered iron bar wound with at least two coils. Ferrite-core transformers are used extensively as power transformers in high frequency applications.

PRIMARY AND SECONDARY

Basically, a transformer's job is to transfer electrical energy from one circuit to another by means of electromagnetic induction. The transformer winding connected to the energy source is called the "primary." The winding from which you receive the energy is called the "secondary."

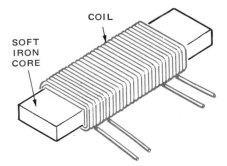

Fig. 11-26. Coils and laminated cores are put together in the start of final assembly of the transformers.

Fig. 11-28. The ferrite core transformer is another popular design. Basically, the coils are wound around a soft iron core. See Fig. 11-18.

STEP-UP TRANSFORMERS

Transformers are widely used in electronics because they can step-up or step-down voltages. The transformer shown in Fig. 11-29 has 100 turns of wire in the primary winding and 600 turns in the secondary winding. If we connect 110V ac to the primary, we will get 660V ac from the secondary. This is an example of a "step-up" transformer.

The relationship between voltage and the number of turns can be expressed by the following equation:

$$\frac{N \text{ primary}}{N \text{ secondary}} = \frac{E \text{ primary}}{E \text{ secondary}}$$

N equals number of turns

E equals voltage

Using the previous example, here is how we get our result:

Fig. 11-27. Laminated core transformers are bolted together and come off the final assembly line for packing and shipment. (Acme Electric Corp.)

$$\frac{N_P}{N_S} = \frac{E_P}{E_S}$$

$$\frac{100}{600} = \frac{110V}{E_S}$$

$$100\,E_S = 66,000V$$

$$E_S = 660V$$

By putting 110 volts in the primary, we are able to get 660 volts out at the secondary. You can see by this increase why they call this a "step-up" transformer.

Fig. 11-30. This step-down transformer carries a nameplate that gives input and output voltage, wattage and cycles per second. Note, too, that the leads are marked primary and secondary. (Acme Electric Corp.)

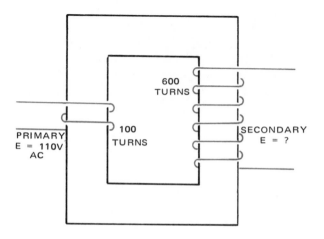

Fig. 11-29. In this simplified version of a step-up transformer, output voltage (E) would be 660 volts. See equation in text.

STEP-DOWN TRANSFORMERS

Next, we will look at a "step-down" transformer, Fig. 11-30. If we apply 1000 volts to a primary winding of 50 turns and the secondary winding has 10 turns, what output would we get?

Use the same formula:

$$\frac{N_P}{N_S} = \frac{E_P}{E_S}$$

$$\frac{50}{10} = \frac{1000}{E_S}$$

$$50\,E_S = 10,000$$

$$E_S = 200 \text{ volts}$$

VOLTAGE-CURRENT RELATIONSHIPS

Transformers are used in electrical circuits because they are very efficient. Almost all the power applied to the primary winding is transferred to the secondary winding, Fig. 11-31. You may recall from Chapter 5, that power is equal to the voltage times current (P = EI). We can take this formula and use it in transformer circuit analysis. If we say that transformers are 100 percent efficient and have no losses we would have:

$$\text{Power}_{\text{Primary}} = \text{Power}_{\text{Secondary}}$$

This equation is equal to:

$$E_P I_P = E_S I_S$$

POWER

Look at the transformer drawing in Fig. 11-31. If the power on the primary winding is 600 watts and the voltage of the secondary is 150 volts, what is the current in the secondary?

Use the following formula:

$$\text{Power}_{\text{In}} = \text{Power}_{\text{Out}}$$

$$\text{Power}_{\text{Primary}} = E_S I_S$$

$$600W = 150V\,I_S$$

$$I_S = \frac{600}{150} = 4A$$

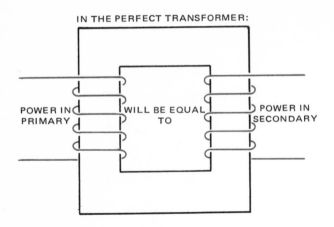

IN THE PERFECT TRANSFORMER:

POWER IN PRIMARY   WILL BE EQUAL TO   POWER IN SECONDARY

Fig. 11-31. Since transformers are very efficient, you can assume that power (E x I) applied to the primary equals power (E x I) produced by the secondary.

HIGH VOLTAGE TRANSFORMERS

In Fig. 11-32, you can see the coils of a high voltage transformer being wound. This transformer looks different because it is not in the same shape as the others we have discussed. Note that these transformers have a shape similar to a can (cylinder). Again, the coils must be separated by insulation. This prevents the coils from touching one another and "shorting out."

High voltage transformers have ratings of over 500 kVA. The abbreviation, kVA stands for thousands of volt amperes of electricity. This type of transformer would not be used in school power supplies or around your home. High voltage transformers are found in industry where they have the need for large amounts of power. Fig. 11-33 shows a finished high voltage transformer.

Fig. 11-33. This high voltage transformer is mounted in its case, ready for stepping up voltages.
(Acme Electric Corp.)

VARIABLE VOLTAGE TRANSFORMERS

A variable voltage type of transformer is shown in Fig. 11-34. This transformer is capable of varying the voltage from a few volts up to 150 volts. Note the selector knob for switching output to the voltage desired.

The first 11 chapters of this book covered most of the components found in electrical circuits. Chapters 16, 17 and 18 will cover some so-called "solid state"

Fig. 11-32. High voltage transformers look different, but follow the same design principle of coil windings separated by insulation and installed on a metal core.

Fig. 11-34. This is a variable voltage type of transformer. Note the selector knob for switching output to the voltage desired.

components. Before that, we will show you how to use resistors, capacitors and inductors in various circuits. We also will show you how you can find out what different combinations of current, voltage, resistance, etc., will do to circuits.

Fig. 11-35 shows a four channel receiver. If you look closely, you can see many of the components we have covered so far. How many can you identify?

Fig. 11-35. This four channel radio receiver is loaded with electrical components. How many can you identify? (Fisher Radio)

TEST YOUR KNOWLEDGE

1. Name the two windings found on a transformer.
2. Is ac or dc produced in the secondary winding of a transformer?
3. A transformer connected to a 120 volt ac source has 400 turns on the primary winding. If we want to obtain 6 volts from the second winding, how many turns of wire do we need for the secondary? Show your work.
4. Lines of flux cannot cut through insulating material. True or False?
5. The part of a circuit that opposes a change in current is called _____.
6. Draw the symbol for an inductor and a capacitor.
7. The unit of inductance is the _____.
8. An inductor will have no effect on _____ circuit.
9. Name three things which affect the inductance of a coil.
10. If we had three inductors connected in series, each having a value of 4 mH, what would the total inductance be? Show your work?
11. Laminations on the core of a transformer help cut down on the _____ loss.
12. The two types of transformers that get their name from how they affect the voltage are: _____ and _____.

SOMETHING TO DO

1. Visit a plant which makes transformers or inductors. How do they wind the coils? Why do these components cost more than resistors?
2. Make a list of the places you find transformers around your house.
3. Some neighborhoods have a device which looks like a large black can attached to its telephone poles. These are transformers. What voltages do they produce?
4. Why do industrial plants operate so much of their equipment at 460 volts instead of the voltage of the appliances in your home?
5. The large electromagnets found in junk yards operate on what voltage?
6. Power companies clear an area under all the high tension transmission lines. Why do they do this?
7. How is it possible to run power lines underground without fear of electrocuting people walking over them?
8. The coil on a car is another type of inductor. What would it look like if you cut it into two pieces?
9. Can you make a radio without an inductor?
10. What kind of wire is needed to make an inductor or transformer? What would happen if the wire was not coated with insulation?
11. How do they encapsulate a transformer?
12. If the core of a transformer came apart, what would it sound like? What would happen to the transformer?

Chapter 12

MOTORS

In the 1830s, Michael Faraday made a discovery that has made work easier for all of us. While experimenting with magnets, he found that moving a wire through a magnetic field would produce an electric current. We have seen how this was done in earlier chapters.

The magnet-and-wire experiment is important because it led to the discovery of the operating principles of electric motors and generators. In this chapter, we will study those principles, along with motor and generator types and applications.

MOTORS AND GENERATORS

A motor is a rotating machine that converts electrical energy into mechanical energy. Motors turn our washing machines, dryers, fans, furnace blowers and much of the machinery found in industry, Fig. 12-1. Today, electric motors provide power for producing most of the mechanical energy used in the U.S.

A generator, on the other hand, is a rotating machine that converts mechanical energy into electrical energy.

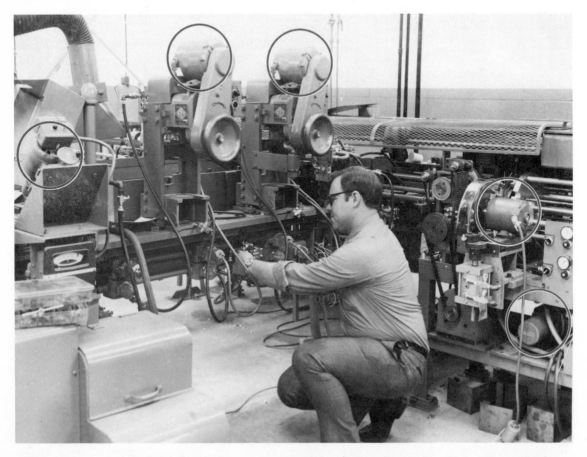

Fig. 12-1. Electric motors (circled) are used extensively in industry. (Packard Electric Div., General Motors Corp.)

This mechanical energy could be supplied by a waterfall, nuclear power or in other ways described in Chapter 1.

In the 18th century, a scientist by the name of Oersted discovered that there was a relationship between magnetism and electricity. Oersted noticed that when he connected a wire to the poles of a battery and held the wire *over* a compass needle, the needle moved in one direction. If he held the wire *under* the compass needle, the needle moved in the opposite direction. This was an early discovery of the magnetic lines of force.

SYMBOLS

You recall that if a coil formed by many turns of wire is moved up and down over a magnet, the current created will move the needle on a galvanometer. This is called "induced current." Electric motors and generators work on principles based on the laws of magnetism and induced current.

Fig. 12-2 gives the popular symbols used to represent motors and generators in electrical drawings.

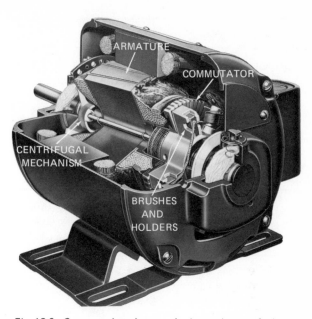

Fig. 12-3. Cutaway view shows major internal parts of a heavy-duty electric motor.

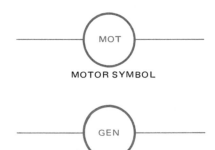

Fig. 12-2. These symbols are commonly used to designate motors and generators in electrical diagrams.

PARTS AND MAKEUP

Both motors and generators are made up of magnets, an armature (moving coils) and a commutator, Fig. 12-3. *A commutator is a device used to reverse electrical connections.* It is required because the armature coils inside motors and generators are in motion, Fig. 12-4. (Shortly, we will see how this works.)

Brushes or terminals are used with the commutator as a means of getting current out of a generator or into a motor armature. When studying motors, you will find a motor that does not need brushes. The most common motor of this type is known as a "squirrel-cage"

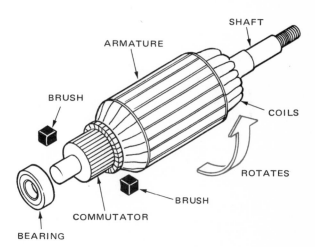

Fig. 12-4. The armature/commutator assembly forms the rotor of a dc motor or generator.

induction motor. Since there are so many different types of electric motors, we will explain some of the methods used to classify them.

CLASSIFYING MOTORS AND GENERATORS

There are several ways of classifying motors and generators. First, we can analyze whether they are single phase or polyphase:

NOTE: Single phase ac power is available from home outlets. Three phase ac power is common in industry. Polyphase means more than one phase. Phasing is covered in Chapters 10 and 13.

Single Phase	Polyphase

A. Split phase
  1. Resistance-start
  2. Split-capacitor
  3. Capacitor-start
  4. Repulsion-start
B. Commutator
  1. Series
  2. Repulsion
	A. Squirrel Cage
	B. Wound Rotor

Another way of classifying motors is by ac or dc:

Alternating current	Direct Current

A. Induction
  1. Phase Wound
  2. Squirrel Cage
    a. Split Phase
    b. Shaded Pole
B. Synchronous
	A. Series
	B. Shunt
	C. Compound

You might ask why we must "classify" motors. Since there are hundreds of different kinds of electric motors, you *must* pick the one way of classifying motors which best suits your needs. Relating this to another situation may help you to understand motor classification a little better. Most people, for example, know that automobiles are classified several ways. Some of the ways include:

1. Manufacturer
2. Model
3. Color
4. Engine size
5. Weight

You can see that some of these choices overlap. If you want a car with a specific horsepower rating, you might find that this engine is available only in certain models. Normally, a small four cylinder engine would not be installed in a car weighing over 4500 lb. Likewise, you would not expect a light-duty electric motor to perform satisfactorily in a heavy-duty application.

SELECTION

To find the right motor for your needs, first look at what it will be required to do. Another way is to decide whether you need an ac or dc motor. You also might want to look at torque (twisting force), and possibly

even speed and load given on the motor identification plate. Then, your next step would be to start narrowing down the available types of motors to one which meets your specific needs.

Although we cannot cover all of the different kinds of motors, we will give you the basic principles of operation for most electric motors.

HOW THEY WORK

Start your study of principles of operation by seeing what makes a motor turn. *Both a dc motor and a dc generator work in a similar manner. The basic difference is that a motor is made to turn by putting current into it. A generator, on the other hand, will produce a current when some outside force causes it to turn.* See Fig. 12-5.

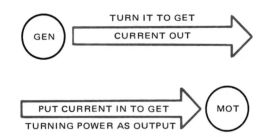

Fig. 12-5. A generator must be turned mechanically to produce current. Current must be applied to a motor to make it turn.

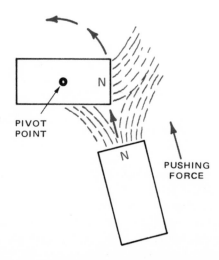

Fig. 12-6. Like poles of two magnets repel each other. If one magnet is pivoted at the center, the pushing force created will cause it to turn.

TURNING MAGNETS

There are two ways you can make a magnet turn. As an experiment, lay down the first magnet so that it is free to move, Fig. 12-6. Then, bring the north pole of the second magnet close to the north pole of the first magnet. Since like poles repel, the first magnet will turn away from the second. The same repelling action (repulsion) holds true if two south poles are used in the experiment.

Another way of making a magnet turn is by setting up as before, but instead of using *like* poles bring the north pole of the second magnet close to the south pole of the first magnet. Since *unlike* poles attract, the first magnet will be attracted to and turn toward the second magnet. See Fig. 12-7.

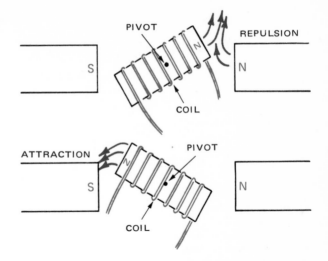

Fig. 12-8. To strengthen the magnetic field and improve the turning action of the pivoting magnet, coils of wire are wrapped around the magnet to form an electromagnet.

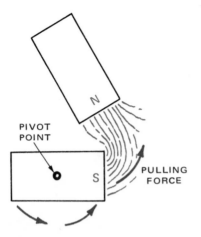

Fig. 12-7. Like poles of two magnets attract each other. If one magnet is pivoted at the center, the pulling force created will cause it to turn.

When this happens, the magnetic field of the coil opposes the magnetic field of our magnet. The repelling action of like poles makes the armature start to turn, Fig. 12-10. As the armature turns, its north pole is attracted to the south pole of the magnet. Just as the two unlike poles are about to line up, the polarity of the armature is changed by the commutator, Fig. 12-11. Again, the two like poles will repel each other, and the armature will continue to spin. Therefore, we get this motion from our motor by means of current flow through the set of brushes to the commutator.

COILS

Instead of making a motor with just one coil, most motors have several sets in the armature, Fig. 12-12.

In a further experiment, substitute a coil for the first magnet and a pair of magnets for the second, as shown in Fig. 12-8. To set it up, wrap a coil of wire around a pivoting metal bar and pass a current through the coil. This will create a magnetic field around the coil.

Then, bring the coil and its magnetic field between the opposite poles of the pair of magnets, Fig. 12-8, and the coil will begin to turn. This turning action is the result of attraction or repulsion, depending on the polarity of the parts.

Therefore, the laws of attraction and repulsion (which you learned earlier) are the basis of electric motor operation.

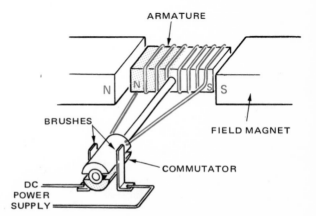

Fig. 12-9. A simple dc electric motor can be formed by connecting the leads of the coil (armature) to the commutator. Fixed brushes carry dc electricity to and from the commutator.

DC MOTORS

To relate our attraction/repulsion experiment directly to electric motor operation, we will start with a simple dc motor, Fig. 12-9. We can pass current through one brush, to the coil and back out the second brush.

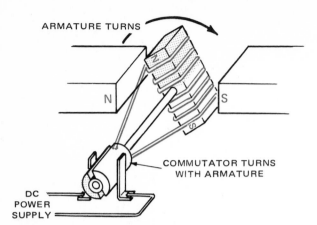

Fig. 12-10. Note that like poles repel each other, causing the armature to start to turn. Once the armature completes a quarter turn, the unlike poles will attract each other and the armature will continue to turn.

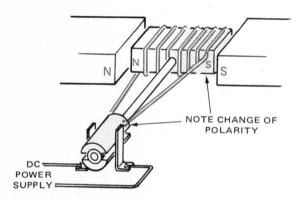

Fig. 12-11. After one-half turn of the armature, its poles are reversed by the commutator and the repelling/attracting process is repeated. Note how the split commutator changes the polarity of the armature.

More coils are necessary because we want to get a large magnetic field in the motor housing. If we used just one coil, we would have a very weak motor and it would not provide smooth operation.

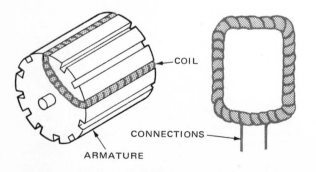

Fig. 12-12. Typical way in which the first coil is wrapped on an armature core.

BRUSHES AND COMMUTATOR

Brushes are one of the big problems with electric motors of this type. Usually, they are made of a carbon material. Because of this, the brushes tend to wear out and must be replaced. Many of the small motors used on sewing machines, power drills and other home appliances are equipped with these brushes. See the power tool in Fig. 12-13.

I.C. WARNS:

"If you replace brushes in an electric motor, be sure to unplug the power cord before you start to work. Brushes carry the current which makes motors spin. Avoid electrical shock."

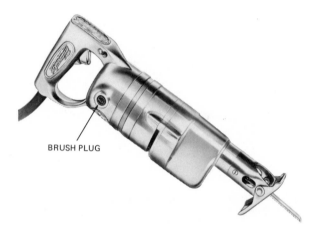

BRUSH PLUG

Fig. 12-13. Some power tools use an ac electric motor with armature and commutator. Plug removal permits access to spring-loaded brushes.

Brushes usually are easy to remove. Many can be taken out by using a screwdriver to loosen the brush holder, Fig. 12-14. Most brushes are spring loaded to keep them in contact with the commutator. Be careful on disassembly. Do not lose the springs when you remove the brushes.

If the brushes are not making proper contact with the commutator, the motor will have a weak starting torque. Bad contact can be caused by worn brushes, weak brush spring or brushes sticking in their holders. A dirty commutator also can cause poor electrical contact.

The commutator is made from copper. It should be cleaned with a very fine sandpaper rather than emery cloth, or you may have "shorting" problems later. When you start to clean the commutator, it probably will have a dull, blackened look. As the copper segments clean up, they will regain the shiny copper look.

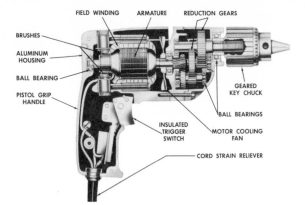

Fig. 12-14. Sectional view shows makeup of an electric drill. Note brush type motor at rear.

BEARINGS

One of the major causes of electric motor failure is bearing wear. Since motors must spin freely, it is important that the bearings have proper lubrication. Pay close attention to the manufacturer's instructions for lubrication. Some bearings need just two or three drops of oil. If you use too much oil, it will attract dirt and dust that will gum up the bearings. Other pre-lubricated bearings do not need any oil at all. These bearings usually have the right amount and kind of lubricant sealed inside.

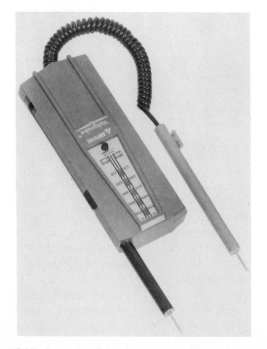

Fig. 12-15. An industrial voltmeter can be used to make voltage tests at various points in the electric motor circuit.

VOLTAGE CHECKS

A number of different troubles can cause a motor to fail to start. First use a voltmeter to see if there is voltage at the motor terminals. Fig. 12-15 shows a voltmeter which many industrial electricians carry with them. Note that it reads both ac and dc voltage. Neon lights inside the meter will light up to tell you the voltage level.

Many times, motors have been removed from the mounting to be checked in a repair shop, only to find nothing is wrong with them. Because of their size, Fig. 12-16, some motors have to be lifted by means of a crane. Therefore, if a check with the motor in place

Fig. 12-16. Certain industrial applications call for electric motors so large that a crane must be used to move them. Applications include machine tools, pumps, pulp and paper machines and conveyors. (Delco Products Div., General Motors Corp.)

proves that there is no voltage at the terminals, a lot of time and work will be saved. No voltage would mean that a part other than the motor is at fault. There are some other voltage checks which can be made, Fig. 12-17, without first taking the motor off the machine.

Fig. 12-17. A service technician uses a voltmeter to check a voltage regulator and an ac generator suitable for use in ships, railroad refrigeration cars, hospitals and factories. (Delco Products Div., General Motors Corp.)

BLOWN FUSE INDICATOR PROJECT

Have you ever spent a lot of time troubleshooting a circuit, only to find a "blown" fuse? To avoid this problem, make up the simple circuit shown in Fig. 12-18. It can be added across the fuse in the original circuit to warn you when the fuse blows. When it does, the lamp will light.

Parts needed for the project include:

No.   Item

1   NE-2 neon lamp.
1   Resistor (value depends on circuit voltage).
         Connecting wire.

In order to assemble this project, you must know the voltage across the fuse. This information will allow you to select the right size resistor to use. The formula is:

$$\text{Resistance of } R_1 = \frac{\text{voltage}}{.0025}$$

For example, if the fuse is across 110 volts:

$$R_1 = \frac{110V}{.0025}$$

$$R_1 = 44,000\,\Omega$$

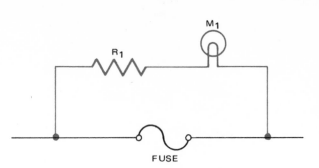

Fig. 12-18. Schematic shows how to connect a "blown fuse indicator" across the fuse in the original circuit. R_1 is a resistor. M_1 is a neon lamp.

Based on this computation, then, the 110 volt circuit will require a 44,000 ohm resistor.

Since an NE-2 neon lamp requires between 60 and 65 volts to ionize, this blown fuse indicator is only good for sources greater than 65 volts.

INDUCTION MOTORS

The second major type of electric motor we will cover is the induction motor found on many types of machinery. It does not use a commutator or brushes. Instead, it is constructed with a rotor that spins inside a stator.

One way to get a better grasp of the construction of induction motors is to think of the stator as stationary and the rotor as rotating. The stator consists of a winding or windings on a laminated steel core. The rotor is made of laminated steel, and it has copper bars that fit into slots on the steel core. Two end shields are fitted with bearings to support the rotor shaft.

If you have ever seen a squirrel cage, or treadmill for small animals, you will know why this rotor is called "squirrel cage." See Fig. 12-19.

OPERATION

A magnetic field is created in the stator of an induction motor by passing current through its windings. The lines of force of this magnetic field revolve around the stator windings, inducing a current in the copper bars of the rotor. The rotor, in turn, becomes an electromagnet, and its magnetic field is attracted to the magnetic field of the stator.

Looking at Fig. 12-20, the north pole created in the stator winding will attract the south pole of the rotor winding at point 1. As the magnetic field revolves around the stator windings, it exerts a magnetic pull on the rotor, causing it to turn to point 2, then point 3 and

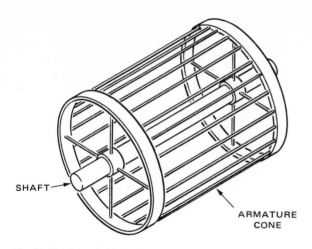

Fig. 12-19. A squirrel cage rotor serves as the core for an induction motor. Coils of wire are wrapped on the core to form the moving part of the motor.

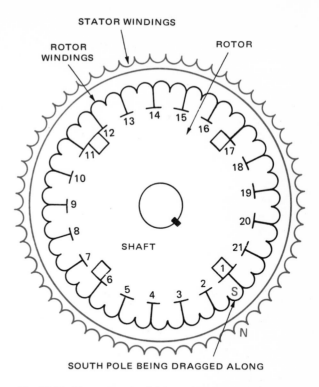

SOUTH POLE BEING DRAGGED ALONG

Fig. 12-21. The magnetic field revolving around the stator windings moves just ahead of the rotating rotor.

so on. Finally, it will get to point 21 and start all over again.

Since the rotor can turn in its bearings, it starts to rotate just behind the magnetic field of the stator, Fig. 12-21. These magnetic fields follow one another around the squirrel cage, with the rotor always behind the magnetic field of the stator. If the rotor did manage to

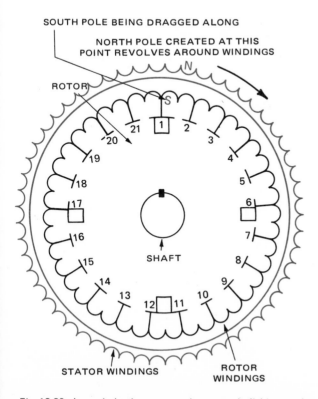

Fig. 12-20. In an induction motor, the magnetic field around the stator windings creates a north pole which, in turn, attracts the south pole of the rotor windings, causing the rotor to turn.

"catch up" with the rotating field of the stator, the conductors in the rotor would be standing still with respect to the rotating field. If this condition occurred, no voltage would be induced in the squirrel cage which, in turn, would not have current in it nor magnetic poles set up. There would be no attraction between the rotor and the rotating field in the stator. The rotor would stand still.

For proper operation of an induction motor, the rotor must revolve just a bit slower than the rotating field in the stator can cut through the rotor conductors. This difference is called "slip." Without slip, the rotor will not turn. If the load on the rotor is increased, such as making it turn a heavy weight, we would get greater slip and the motor would run a little bit slower. However, for most purposes, the squirrel cage induction motor is considered to be a constant speed device.

SYNCHRONOUS MOTORS

A third type of motor we will look at is known as the "synchronous motor." In your study of induction motors, you have found that as you pass a current through the stator, you set up a magnetic field. The rotor, in turn, is pulled along by this magnetic field with a small amount of slip. The synchronous motor operates in a similar manner. See Fig. 12-22.

The synchronous motor gets its name from a term used to describe the natural speed of the rotating magnetic field of the stator. In the U.S., the natural speed is tied to the frequency of applied ac power. Here, this frequency is regulated by the power companies at 60 Hz.

In the case of the synchronous motor, the rotating magnetic field is designed to lock onto the rotor. By locking the rotating magnetic field to the rotor, they will operate at the same speed. Because of this locking idea, synchronous motors can be found in various timing devices such as electric clocks. Since power companies regulate the frequency at 60 Hz, these clocks keep very accurate time.

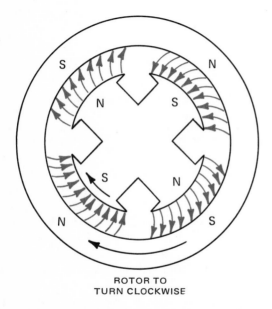

ROTOR TO
TURN CLOCKWISE

Fig. 12-22. Arrows show how a synchronous motor works. The rotating magnetic field locks onto the rotor and turns it at a natural speed.

POLES AND RPM

Other uses can be made of synchronous motors by building them to operate at any one of a number of constant speeds. However, each motor will run at the one speed for which it was built to operate. To get a synchronous motor to operate at another speed, the manufacturer must produce one with a different number of poles.

The formula for figuring the speed of a synchronous motor is:

$$\text{rpm} = \frac{\text{frequency x } 120}{P}$$

P = Number of poles in motor

Since the frequency of applied ac power in the U.S. is 60 Hz, all we have to do to change the rpm of a synchronous motor is change the number of poles. For example, if you have a motor with two poles, what is its speed?

$$\text{rpm} = \frac{\text{frequency x } 120}{P}$$

$$\text{rpm} = \frac{60 \text{ x } 120}{2}$$

$$\text{rpm} = 3600$$

This means that a two pole synchronous motor will run at a constant speed of 3600 rpm. Synchronous motors do not have variable speeds. If you want a motor that will run at 1200 rpm, how many poles would have to be built into the motor?

$$\text{rpm} = \frac{\text{frequency x } 120}{P}$$

$$P = \frac{\text{frequency x } 120}{\text{rpm}}$$

$$P = \frac{60 \text{ x } 120}{1200} = \frac{7200}{1200}$$

$$P = 6 \text{ poles}$$

Therefore, to get 1200 rpm from a synchronous motor, six poles would have to be built into it. Synchronous motors always have an even number of poles, ranging from two up to 90 poles. A 90 pole motor of this type would run at 80 rpm.

GENERATORS

A dc generator is almost the same as a dc motor. Again, the main difference is that the shaft of the generator must be turned by some means of mechanical power. When the armature starts to turn, magnetic fields are cut inside the generator and current begins to flow. By putting brushes on the commutator, we can then pick off this flow of current.

The commutator is not one continuous ring, Fig. 12-23. Instead, it is made of two separate pieces. To make sure they do not touch each other, a piece of insulating material such as mica is placed between them. If they do touch, it would, in effect, give us a short circuit. All the work done by the generator would be wasted. Instead of getting current out, it would circulate inside the coils of the generator and overheat.

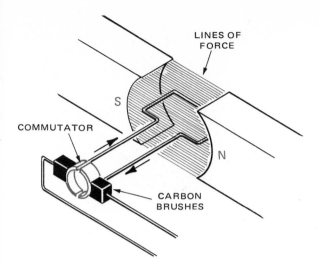

Fig. 12-23. The commutator in this simple generator consists of two metal pieces. Brushes make contact with commutator to pick off flow of current generated in the armature.

DC GENERATOR OUTPUT

The output from a dc generator is not pure dc. It does not give us the same wave we would get out of a battery, Fig. 12-24. Instead, it is a pulsing dc, Fig. 12-25. From Fig. 12-26, you can see that if we could use a four piece commutator, we would get two overlapping sets of pulsing dc. If you look at the peaks

in Fig. 12-27, you can see that we are getting closer to the straight line you think of as being dc.

By using capacitors, Fig. 12-28, we can filter the dc to take out the ripple. The rotating part of the

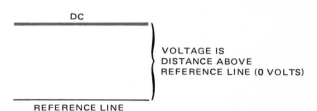

Fig. 12-24. This dc waveform is typical of battery output. The wave shows a steady voltage above the reference line.

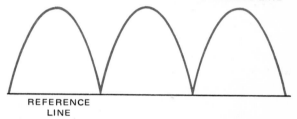

Fig. 12-25. A dc generator produces a pulsing waveform with all waves above the reference line.

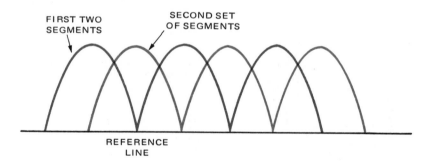

Fig. 12-26. A four-piece commutator will produce overlapping sets of pulsing dc. This is called a "ripple pulse."

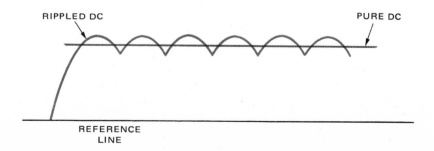

Fig. 12-27. With rippled dc, the resulting waveform more closely follows the pure dc wave.

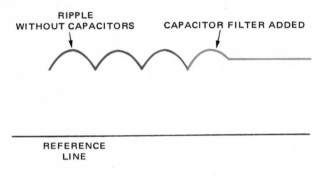

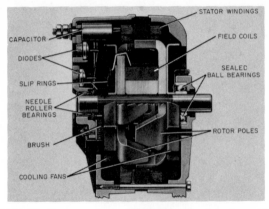

Fig. 12-30. Cutaway view reveals the internal makeup of an automotive ac generator (alternator). (Chrysler Corp.)

Fig. 12-28. Capacitors can be added to a dc generator circuit to smooth the ripples from rippled dc.

generator is known as the rotor; the stationary part is the stator. The armature usually is part of the rotor, while the field coils are part of the stator.

EDDY CURRENTS

We dicussed eddy currents in transformers in Chapter 11. Eddy currents are also found in generators. When the generator armature cuts through the magnetic field, current is induced in the armature windings and it also circulates within the metal. See Fig. 12-29. The armature is made of several parts and is laminated. The armature windings are wound in the slots of the core. This construction is designed to keep eddy currents as small as possible in the metal core of the armature.

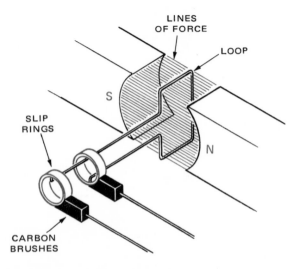

Fig. 12-31. This simple ac generator has two slip rings and two brushes. Note that each end of the coil wire end leads to a separate slip ring.

the coil turn without twisting or breaking any wires. The ends of the coil are soldered to pick off points. As the coil turns, the slip rings, or commutator, slide past the fixed brushes. This lets us obtain electrical output from the wires attached to the brushes.

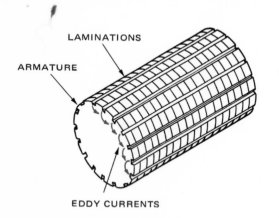

Fig. 12-29. The metal core of a generator armature is laminated to reduce the flow of eddy currents.

AC GENERATOR/ALTERNATOR

The ac generator is called an alternator, especially when it is used in an automotive charging system. See Fig. 12-30. The ac generator uses slip rings, Fig. 12-31, in place of the commutator found on dc generators. The slip rings, or the commutator, are designed to let

THREE-PHASE ALTERNATORS

Many motors and generators use a "three-phase system," Fig. 12-32. Three phase means that the sine wave starts at every 120 deg. instead of the normal 360 deg. In this manner, we can plug the valleys left in a normal sine wave to give us less of a stop-start effect from the dips in the sine waves.

Three-phase alternators are commonly used in automobiles to charge the battery and operate electrical devices. To do this, a rectifier or diode assembly is used,

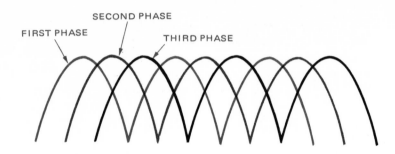

FIRST PHASE SECOND PHASE THIRD PHASE

Fig. 12-32. The three-phase system used in generators starts a new sine wave every 120 deg., instead of every 360 deg., to level off dips in the sine waves.

Fig. 12-30. We will mention rectifiers again when we cover filter circuits. They will be explained in detail in the final few chapters.

For now, just think of a rectifier, or diode, as a one-way street. It limits electrical flow to one direction only. In an automotive alternator, the diodes change the ac to dc.

The alternator is driven by mechanical energy produced by the automobile engine. It will generate current through all three phases of its windings. Because of this, the alternator is able to charge the battery at low engine speed while the dc generator cannot.

TEST YOUR KNOWLEDGE

1. A commutator is a device used to _____.
2. Brushes in a motor are used to _____.
3. Because of their large size, induction motors must have specially made brushes to maintain good contact. True or False?
4. Find the rpm of a motor which has four poles and is operated on 60 Hz. Show your work.
5. Why do motors have a varied number of poles?
6. Before you work on the brushes of a motor, you should turn off the power. Why?
7. Name one of the major causes of motor failure.
8. The alternator on an automobile is a _____ electrical system which is used to _____ the battery on a car.
9. The rotating part of a generator is called the _____ and the stationary part is called the _____.
10. Eddy currents in a generator are reduced by using _____.
11. The big advantage of a dc generator over an ac generator is that its windings produce pure dc which does not have to be filtered. True or False?

THINGS TO DO

1. Get an old motor and take it apart. Label the various parts.
2. Why are brushes for motors made from carbon, which is the same material used for resistors?
3. Make a list of places in your home where you use motors. Be careful. Motors may be hidden in places you might not think about.
4. Research to find the specific sizes of the largest and smallest motors being made. Where are they being used?
5. Generators on cars have been replaced by alternators. What is the big advantage of an alternator?
6. Many motors have sealed bearings. How are they made? What makes them so different from regular bearings?
7. Many electrical devices are laminated. What holds them together? List some of the assemblies that are laminated.
8. The alternator on an automobile engine uses a three-phase system. Get the electrical diagram of such a system and see if you can figure out how it works. Where else do you find three-phase electrical systems?
9. See if you can find some places where motors with more than two poles are used. Make a list of them.
10. Can a motor be used as a generator?
11. Can a generator be used as a motor?
12. How do they "wind" motors?
13. What makes a motor stay together even though it is spinning at high speed? What keeps the windings from coming apart?
14. Start a motor that drives a generator. Is it possible to take the output of the generator, feed it back to the motor so that each one would keep the other going?

Chapter 13
INDUCTIVE REACTANCE AND IMPEDANCE

In this chapter, we will look at the behavior of inductors in an ac circuit. In order to find out whether or not inductors have an effect on an ac current, set up a circuit as shown in Fig. 13-1. You will need a 12 volt battery, a switch, test lamp, inductor and connecting wiring.

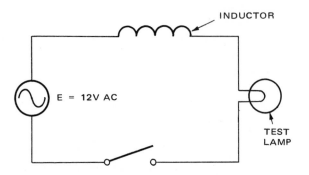

Fig. 13-2. An inductor placed in an alternating current circuit will offer inductive reactance (opposition) to current flow. Adding the inductor will dim the lamp.

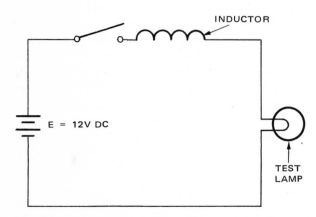

Fig. 13-1. An inductor will oppose a change in direct current flow. When the switch is closed, the lamp will be dimly lit until current flow in the circuit reaches its maximum value.

When you have the circuit assembled, close the switch and watch the test lamp. Note that, after a slight pause, the lamp lights to full brilliance. This is not too surprising. From your studies in Chapter 11, you know that an inductor will oppose a change in current. Once that current reaches its maximum (determined by the resistance of the bulb), the inductor will no longer affect the current. Therefore, the bulb will glow with full brightness.

Now, replace the 12 volt battery with a 12 volt ac source, Fig. 13-2. Since the RMS value of the ac wave is 12 volts, you would expect the bulb to glow as brightly as before. However, when you close the switch, note that the light is less bright. So it seems that an inductor does offer some resistance to the alternating current.

INDUCTIVE REACTANCE

The opposition offered by inductors to any change in current is called "inductive reactance." This is an important part of our study of electronics.

The amount of reactance that an inductor presents to ac depends upon two factors:
1. Inductance of the inductor.
2. Frequency of the varying current or voltage applied to the inductor.

The symbol for inductive reactance is X_L. Its value can be found by working out the following formula:

$$X_L = 2 \pi fL$$

Where

f   is the frequency of the applied voltage.
L   is the inductance of the coil in henrys.
π is the constant 3.14.

Example:
If an ac voltage with a frequency of 318 Hz and an amplitude of 200 volts is applied to an inductor whose inductance is .5 H, find the inductive reactance of the circuit, Fig. 13-3.

$$X_L = 2 \pi fL$$

$$X_L = 2 \times 3.14 \times 318 \text{ Hz} \times .5 \text{ H}$$

$$X_L = 999 \,\Omega$$

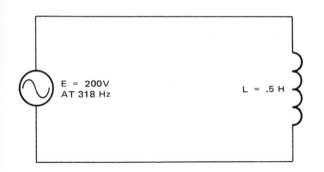

Fig. 13-3. The inductive reactance (X_L) of this ac circuit is 999 ohms, found by using the formula $X_L = 2 \pi fL$.

INDUCTIVE REACTANCE = OHMS

The value of X_L is measured in ohms. This is done because reactance affects ac the same way that resistance affects dc. In fact, if we refer back to our example, we can compute the amount of current that is flowing in the circuit.

Use Ohm's Law, replacing the R (resistance) with X_L (inductive reactance). To make our numbers a little easier to handle, round off the X_L value to 1000 ohms. One ohm does not make that much difference. It can exist in the tolerance of a resistor.

$$I = \frac{E}{X_L}$$

$$I = \frac{200V}{1000\,\Omega} = .2A$$

Since this circuit contains only an inductor, you can see that the reactance of the inductor behaves just like resistance.

PHASE BETWEEN VOLTAGE AND CURRENT

If you look at one cycle of alternating current on a sine wave, Fig. 13-4, you will see that we have divided the cycle into 360 electrical degrees. This is the same number of degrees that the generator had to turn to produce the cycle of ac.

I.C. SAYS:

"Remember that one full turn or revolution contains 360 degrees."

When studying the sine wave in Fig. 13-4, note that the current flow is changing at the fastest rate at 0 deg., 180 deg. and 360 deg. Also note that at exactly 90 deg., current flow stops increasing and changes direction.

In Chapter 11, we established that an inductor produces a counter electromotive force that depends upon how fast the current is changing. With this in mind, where would you expect counter EMF to be highest? If you said, 0 deg., 180 deg. or 360 deg., you are right. At these particular times in each cycle, the current is changing at its fastest rate and the voltage across the inductor will be at a maximum.

Based on the sine wave shown in Fig. 13-4, where would you expect the voltage to be at a minimum? The answer is: at 90 deg. and 270 deg. At these points, the current stops changing for a moment and the counter EMF of the coil has dropped to zero.

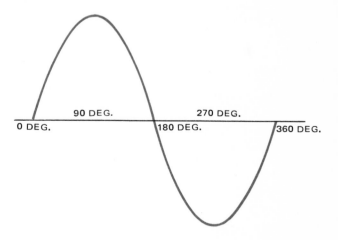

Fig. 13-4. One cycle of alternating current in the form of a sine wave can be divided into 360 electrical degrees.

GRAPHING CURRENT AND VOLTAGE

If you make one graph showing both current flow and EMF produced across the coil, Fig. 13-5, you can see that the current and voltage are both sine waves. While current and voltage waves usually have different amplitudes, at this time we are more concerned with the difference between the phase of the waves.

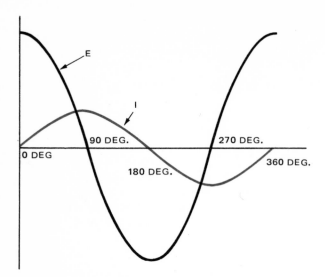

Fig. 13-5. This graph compares out-of-phase sine waves of voltage and current in an inductive circuit.

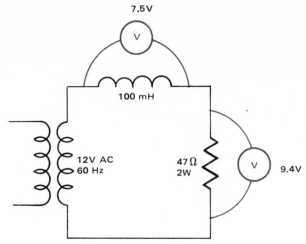

Fig. 13-6. A series "RL circuit" contains both an inductor and a resistor. Note voltage drop readings across these components.

Note that the zero points of the two waves differ by 90 deg. Therefore, we can say that the voltage across the inductor is 90 deg. out of phase with the current through it.

In Fig. 13-5, the voltage wave crosses the reference line (axis) at 90 deg., while the current wave crosses the line at 180 deg. Since current is 90 deg. "later" than voltage, we say that current "lags" voltage by 90 deg. In this circuit, when the voltage across the inductor is at a maximum, current through the inductor drops to zero. *The current lags the voltage in all inductors.*

AC BEHAVIOR OF SERIES RL CIRCUITS

The next step in the study of inductive reactance is to see what happens when we connect an inductor and a resistor in series. This type of circuit is called an "RL circuit" (R for resistance, L for inductance). If you recall your study of series circuits in Chapter 6, you know that when two elements are in series, they have the same amount of current passing through them.

Hook up a series RL circuit as shown in Fig. 13-6. Pick an inductor with a value of around 100 milli-henries and a 47 ohm, 2 watt resistor. Using a voltmeter, measure the voltage across the inductor and the voltage across the resistor. You should get a reading of about 7.5 volts across the inductor and a reading of about 9.4 volts across the resistor.

How can this happen? If these two voltage drops are in series, they should add up to the applied voltage, which is 12 volts. You can see that 7.5 volts + 9.4 volts does not add up to 12 volts.

To answer this question, consider that an inductor has a maximum voltage across it when the current through it drops to zero. A resistor, on the other hand, has zero volts across it when the current drops to zero. In order to add quantities that behave like this, we must use things called "vectors."

VECTORS

Vectors usually are shown as arrows, Fig. 13-7. The length of the arrow represents the magnitude or size of the vector, while the arrowhead shows the direction of the vector. Vectors can be used to show the value of many different things, such as force, movement, voltage and current.

For example, suppose you walked straight north for three miles, then turned right and walked straight east for four miles, Fig. 13-8. If we let vector A represent your northward travel and vector B represent your eastward travel, you can get a good picture of what has occurred.

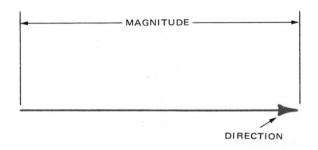

Fig. 13-7. A vector generally is drawn as an arrow. A vector represents direction and magnitude of a given value.

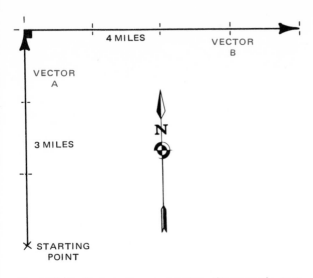

VECTOR
B

4 MILES

VECTOR
A

N

3 MILES

STARTING
POINT

Fig. 13-8. To illustrate the use of vectors, the route of a seven mile walk is charted with vectors A and B.

Note that the little square drawn at the right turn shows that the angle between vector A and vector B is 90 deg. The two most important points you have to think about are:

1. How far have I walked?
2. How far am I from my starting point?

To answer No. 1 is very simple. You have walked a total of seven miles. No. 2 is really quite easy, too. Just join the tail of vector A to the head of vector B, as shown in Fig. 13-9. The figure formed is called a "right triangle," because it has a right angle (90 deg.) in it.

Right triangles have a special relation between their sides. If you take the two shorter sides of the triangle, square them and add these totals, you will get the

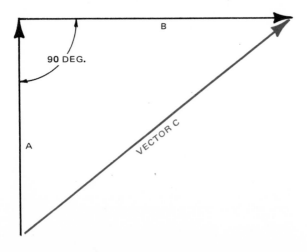

Fig. 13-9. Using vectors A and B from Fig. 13-8, vector C is drawn to complete the right triangle. The formula for solving the length of vector C is $C^2 = A^2 + B^2$.

length of the longest side squared.

In Fig. 13-9, vector C can be found by using the following formula:

$$(\text{Length of C})^2 = (\text{Length of A})^2 + (\text{Length of B})^2$$

$$C^2 = 3^2 + 4^2$$

$$C^2 = 9 + 16$$

$$C^2 = 25$$

$$C = \sqrt{25}$$

$$C = 5 \text{ mi.}$$

Vector C then, gives us the trip distance from start to finish. Note that even though you walked a total of seven miles, you ended up only five miles from your starting point. From this example, we will see how "adding vectors" can help solve the problem of putting an inductor and a resistor in the same series circuit.

RL VOLTAGES

Take another look at the series RL circuit shown in Fig. 13-6. You should recall the following points:

1. The 7.5 volts across the inductor "leads" the current through the inductor by 90 deg.
2. The 9.5 volts across the resistor is in phase with the current through the resistor.
3. The same current flows through both the inductor and the resistor.

If you keep these three points in mind, you can see that the voltage across the inductor is 90 deg. out of phase with the voltage across the resistor. Fig. 13-10 shows these voltages drawn as vectors. If you want to "add" these voltages, you have to use the same formula we worked out in the last example:

$$C^2 = E_L{}^2 + E_R{}^2$$

$$C^2 = 7.5^2 + 9.4^2$$

$$C^2 = 56.25 + 88.36$$

$$C^2 = 144.61$$

$$C = \sqrt{144.61}$$

$$C = 12.0V$$

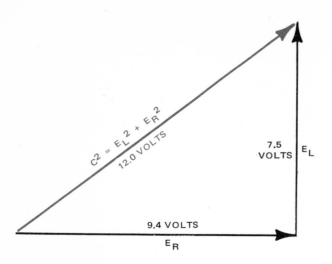

Fig. 13-10. Substituting voltage values for vectors E_R and E_L, use formula shown to solve for total voltage drop in circuit.

From this, you can see that if we add the two voltages like vectors, we get an answer that equals the applied voltage of 12 volts.

IMPEDANCE

By figuring the combination of reactance and resistance as we did in the last example (12.0V), we have "found the impedance" of the circuit. *Impedance, then, is the total opposition that an electrical circuit offers to the flow of varying current at a given frequency.*

You can find the impedance of an RL circuit by adding X_L and R like vectors:

$$\text{Impedance}^2 = X_L{}^2 + R^2$$

Impedance usually is given the symbol Z in formulas and in various forms of circuit analysis. For example, in the series RL circuit shown in Fig. 13-11: the inductor

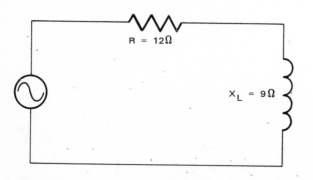

Fig. 13-11. The resistor (R) and the inductor (L) in this series RL circuit offer opposition (impedance) to the flow of alternating current.

has an inductance of 9 ohms; the resistor has a resistance of 12 ohms. Find the total impedance of the circuit:

$$Z^2 = X_L{}^2 + R^2$$

$$Z^2 = 9^2 + 12^2$$

$$Z^2 = 81 + 144 = 225$$

$$Z = \sqrt{225}$$

$$Z = 15\Omega$$

The current will "see" an opposition of 15 ohms to the ac flowing through the circuit.

PHASE ANGLE

Usually a "phase angle" is included with the impedance value. If we draw the two kinds of opposition, Fig. 13-12, we will see that the inductive reactance and the resistance are 90 deg. apart. The angle "A" between the impedance and the resistance is known as the "phase angle." Angle "A" describes the angle (how much out of phase) between the applied voltage and the resulting current.

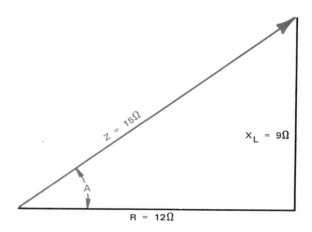

Fig. 13-12. Total impedance (Z) of the circuit shown in Fig. 13-11 is 15 ohms, found by forming vectors X_L and R, and using the formula $Z^2 = X_L{}^2 + R^2$. "A" is the phase angle.

In this text, however, we will not concern ourselves with the phase angle. Rather, we will deal only with the size or magnitude of the impedance. For example, find the magnitude of the current in Fig. 13-13. An ac voltage of 45 volts is applied to the circuit. Total impedance is 15 ohms.

$$I = \frac{E}{Z}$$

$$I = \frac{45}{15}$$

$$I = 3A$$

The current has a magnitude of three amperes.

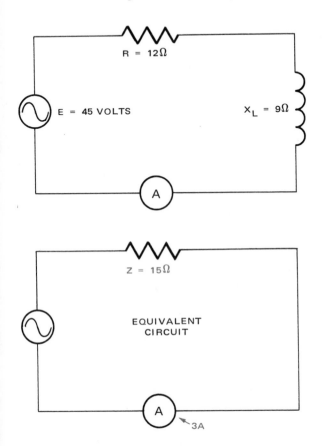

Fig. 13-13. Current flow in this series RL ac circuit is 3 amps, found by figuring impedance (Z) first, then using Ohm's Law $I = \frac{E}{Z}$.

TRANSFORMER LOADING

You recall that we studied transformer operation in some detail in Chapter 11. We found that transformers transfer electrical energy from one circuit to another by electromagnetic induction. When a transformer has no load on the secondary, it will act in a different way than when the secondary is loaded. This is a problem of "phasing." Since we have covered phasing in this chapter, now would be a good time to further explain transformer operation.

UNLOADED TRANSFORMER

When the transformer is not loaded, it is said to have an "open secondary." This means that the primary of the transformer acts just about the same as an inductor. That is: the primary current lags the applied voltage by 90 deg., Fig. 13-14. At the same time, the current will be ahead of the "back EMF" by 90 deg. Back EMF is caused by the magnetic field created when current is flowing in the primary. You can see that the applied voltage and the back EMF have the opposite polarity. Most transformers are built so that the back EMF will be high in the primary when the secondary is open. This is done in an effort to keep the primary current as low as possible when there is no secondary current.

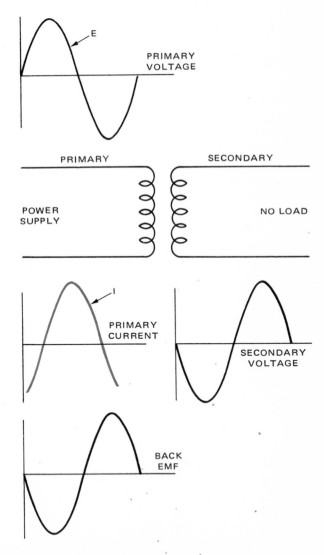

Fig. 13-14. In analyzing the operation of a transformer with no load on the secondary, the primary current lags the applied voltage by 90 deg.

Even though no current flows in the secondary, the expanding and collapsing magnetic field in the primary still cuts through the turns of the secondary winding. This action results in a voltage being induced in the secondary. This induced voltage, however, will be 90 deg. behind the current in the primary. Therefore, the secondary voltage will be a total of 180 deg. out of phase with the primary voltage.

It *is* possible to wind the primary and secondary coils of transformers so that the induced secondary voltage is "in phase" or "out of phase" with the primary. In Fig. 13-15, dots are used to show you whether the terminals are in phase or out of phase.

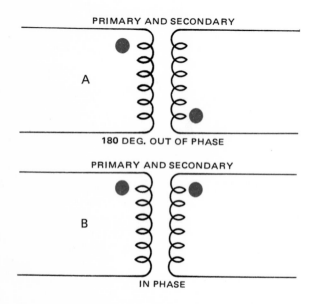

Fig. 13-15. Dots indicate position of terminals on a transformer. A—Primary and secondary windings are out of phase. B—Secondary is wound to bring windings in phase.

LOADED TRANSFORMER

When the transformer's secondary is loaded, we want to get *current* from its windings. As is true in any inductive circuit, the current in the secondary lags the *voltage* in the secondary by 90 deg. See Fig. 13-16. Because of the phase relationship that exists with an unloaded secondary, you can see in Fig. 13-17 that the secondary current is 180 deg. out of phase with the primary current. Therefore, the flux lines of the secondary oppose those of the primary. This will reduce the strength of the primary magnetic field. As a result, there is less back EMF in the primary. With less back EMF, primary current increases, Fig. 13-18. When more current flows in the secondary, more current will flow in the primary.

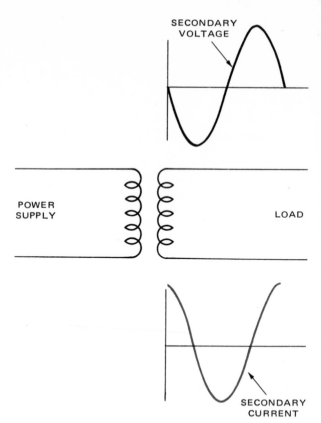

Fig. 13-16. When a transformer's secondary is under load, the current in secondary winding lags secondary voltage by 90 deg.

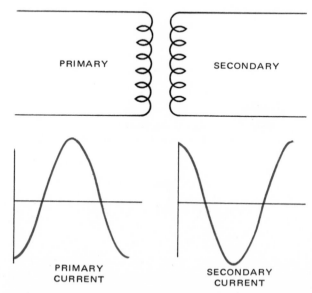

Fig. 13-17. When a transformer's secondary is under load, the secondary current is 180 deg. out of phase with primary current.

From this explanation, you can see that when we load the secondary there is an automatic increase in primary current. If we overload the secondary, or if it

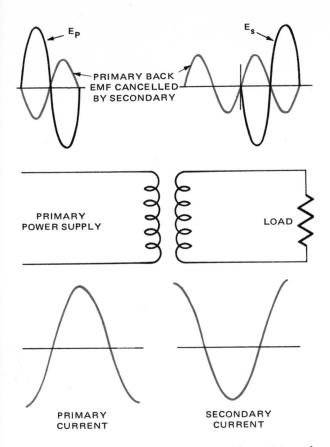

Fig. 13-18. Since secondary current in a transformer is out of phase with primary current, the primary magnetic field and back voltage are weakened and more current will flow.

becomes shorted, the current in the primary will increase greatly. Therefore, the secondary winding must be fused to protect the primary winding. If not, it is possible that the transformer will burn out.

Transformers are rated in terms of primary voltage, secondary voltage and secondary current. The transformer symbolized in Fig. 13-19 has a secondary voltage of 12.6 volts, and a secondary current of 1.2 amps. Therefore, do not go over 1.2 amps when you load the transformer.

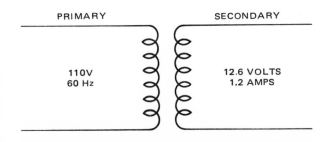

Fig. 13-19. Transformers are rated and labeled for primary current (110V), secondary voltage (12.6V) and secondary current (1.2A).

REPULSION COIL PROJECT

You can demonstrate transformer action by building a repulsion coil. Fig. 13-20 shows the coil and plans for its construction.

Procure the following parts and assemble them according to the construction procedure that follows.

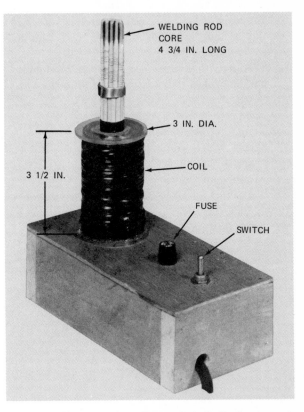

Fig. 13-20. The repulsion coil is built basically from wood, wire and welding rods. See the bill of materials and construction procedure. (Project by Derry Jossefides)

No.	Item
1	Plastic or fiber tubing, 3 1/2 in. long with a 3/4 to 1 in. inside diameter.
2	Plastic or fiber ends for tubing, 3 in. in diameter.
20	Welding rods, cut to 4 3/4 in. lengths.
1	115 volt, 15 amp toggle switch.
1	8 amp fuse.
1	Fuse holder.
1	Male plug and lamp cord.
2	Hookup wires, 4 ft. long.
1	2 1/2 lb. No. 22 magnet wire.
1	6 volt light bulb.
1	18-22 gage magnet wire, 30 ft. long.
1	Aluminum tubing, 1 1/2 in. in diameter. Wood for base.

CONSTRUCTION PROCEDURE

Start by making the coil, core and base, then assemble the project.

1. Cut plastic or fiber ends and glue them to tubing to make a coil form 3 in. in diameter and 3 1/2 in. long.
2. Wind 2 1/2 lb. of No. 22 magnet wire on coil form. Wrap wire in layers. Do not let coils pile up.
3. Cut welding rods (coat hangers, 16-18 gage steel wire or a large bolt) to 4 3/4 in. lengths.
4. Tape rods together or use epoxy at both ends to hold individual rods together and form the core of the repulsion coil.
5. Make base to fit your needs. See Fig. 13-20.
6. Mount coil on base with wood screws.
7. Attach white wire, black wire, switch and fuse holder to the coil. See Fig. 13-21.
8. Attach lamp cord to fuse holder.
9. Wrap 30 ft. of 18-22 gage magnet wire around a 2 in. pipe. Tape wire together and remove from pipe.
10. Solder light bulb to coil.

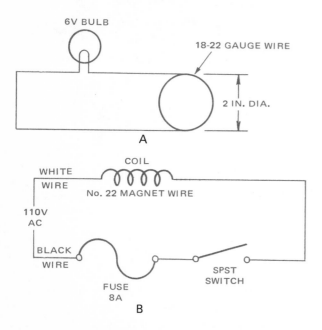

Fig. 13-21. Plans for building the repulsion coil project. A—Layout of coil assembly for this transformer action demonstrator. B—Schematic of parts.

When the project is completed, plug the lamp cord into a 110 volt electrical outlet. Drop the ring of aluminum tubing over the core (welding rods) and flip the switch. The ring will shoot up in the air in a surprising demonstration of transformer action.

To further demonstrate the project, lower the coil of 18-22 gage magnet wire over the core and watch the bulb light up.

CAPACITIVE REACTANCE AND IMPEDANCE

We have studied two of the three main components (parts) in electrical and electronic circuits. These are the resistor and inductor. The third main component we want to cover in depth is the capacitor.

HOW A CAPACITOR WORKS

You recall that we studied capacitor construction and typical applications in Chapter 4. *We found that a capacitor is a device that stores electrons until they are needed by the circuit in which the capacitor is installed.*

One type of capacitor is made of two metal plates separated by a layer of air. If you connect this capacitor in a dc circuit, the battery will force electrons onto one of the plates. The excess of electrons on that plate will cause electrons on the other plate to be repelled and flow towards the positive pole of the battery.

As more and more electrons pile onto one plate and repel electrons from the other plate, an electric field is set up between the two plates. Finally, the voltage across the capacitor will be equal to battery voltage and current flow through the battery will stop.

If we take the capacitor out of the circuit, all the energy gained from the battery will remain stored in the electric field of the capacitor. At this point, the capacitor is "charged," because the battery has moved electric charges from one plate to another.

If larger plates are used, there is room for more electrons on each plate. If the plates are brought closer together, the electrons will have a greater repelling effect upon electrons in the opposite plate, and more charge will be built up. The better the dielectric material insulates, the closer the manufacturer can bring the plates together and get more capacitance. *When making tests, the standard unit of capacitance is called a farad.*

BEHAVIOR OF CAPACITORS IN A DC CIRCUIT

In Chapter 4, we also found that if a capacitor is placed in a dc circuit, no current will flow once the capacitor has charged up. You can verify this fact by setting up the circuit shown in Fig. 13-22 and making certain voltmeter and ammeter tests. This test circuit

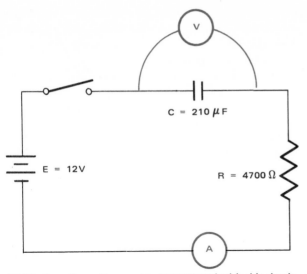

13-22. Capacitor action can be demonstrated with this circuit. Voltmeter/ammeter tests show that once the capacitor is fully charged, no dc will flow.

has a 12 volt dc battery connected in series with a 210 microfarad capacitor, a 4700 ohm resistor and an ammeter. Connect a voltmeter across the capacitor.

To make the tests: Set the voltmeter range selector to the 50 volt scale. Set the ammeter range to read about 10 milliamps. Close the switch and observe the voltage across the capacitor. Voltage builds up slowly, taking about five seconds to reach battery voltage in the circuit. Then, take a look at the ammeter. Once the capacitor has charged up to 12 volts, the ammeter will show no dc flowing in the circuit. A poor quality voltmeter connected across the capacitor could permit a current as high as one milliamp.

I.C. SAYS:

"One of the most important points to remember is that a capacitor blocks the flow of direct current."

TIME CONSTANTS

In the last test setup, you may have wondered what determined how long it took for the capacitor to become fully charged. Actually, the time required for a capacitor to become fully charged depends on two factors:
1. Resistance (R) in the circuit.
2. Capacitance (C) of the capacitor.

If you multiply the resistance (ohms) by the capacitance (farads), you will get a number called the "RC time constant" of that circuit. It takes five of these RC constants for the capacitor to charge up to the voltage of the battery in the circuit:

One RC time constant = R x C
Five RC time constants = Full charge of capacitor

Example:
Connect a 5000 microfarad capacitor and a 2000 ohm resistor in series with a 150 volt battery, as shown in Fig. 13-23. How long will it take the capacitor to charge up to circuit voltage?

$$5000 \ \mu F = \frac{5000}{1,000,000} \ F$$

$$RC = 2000 \ x \ \frac{5000}{1,000,000}$$

$$RC = \frac{10,000,000}{1,000,000}$$

$$RC = 10 \ seconds$$

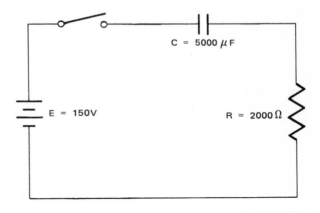

Fig. 13-23. Test circuit presents an "RC time constant" problem. How long will it take for the capacitor to charge up to battery voltage?

If the RC time constant is 10 seconds, then it will take the capacitor 5 x 10 = 50 seconds, Fig. 13-24, to charge up to 150 volts. It takes five RC times constants to do all that charging. The amount of time, of course, will vary according to R and C values.

CAPACITORS AND AC

In our last example, you found that it took a while for the voltage to build up across the capacitor.

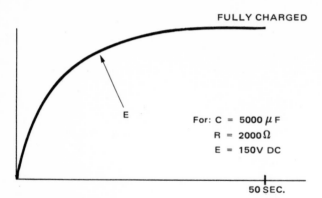

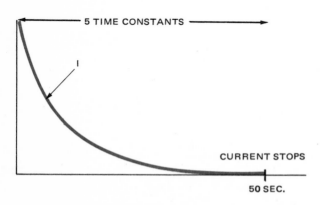

Fig. 13-24. The RC time constant for the circuit in Fig. 13-23 is 10 seconds, or a total of 50 seconds to charge the capacitor.

Although the current started to flow in the circuit immediately after the switch was closed, it took almost a full minute for the capacitor to become fully charged and for the current to die down to zero.

Fig. 13-25 shows a graph of both the voltage across the capacitor and the current in the test circuit. From these sine waves of current (I) and voltage (E), you can see that the current "leads" the voltage. *If you connect a capacitor in an alternating current circuit, the current will lead the voltage by 90 deg.*

NOTE: If you recall from our earlier study of inductors, you will remember that just the opposite condition existed. *With an inductor installed in the circuit, the voltage leads the current.*

CAPACITIVE REACTANCE

Capacitors also show reactance when connected in an alternating current circuit. *The opposition offered by capacitors to a change in voltage is termed "capacitive reactance."*

To demonstrate this type of opposition to ac, connect a 12 volt bulb to a 12 volt, 60 Hz ac source, Fig. 13-26, and check the brightness of the bulb. Then, add a nonpolarized 50 microfarad capacitor in series with the lamp and note that the bulb is less bright. This is caused by capacitive reactance, which bears the symbol X_c.

The formula for figuring capacitive reactance is:

$$X_c = \frac{1}{2\pi fC}$$

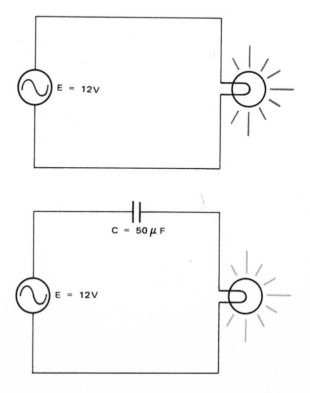

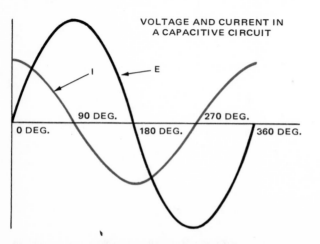

Fig. 13-25. Graph compares 90 deg. out-of-phase sine waves of voltage and current in a capacitive circuit.

Where

f   is the frequency of the applied ac voltage.
C is the capacitance in farads.
π is the constant 3.14.

If the capacitance is given in microfarads, the following formula will make your calculations easier:

$$X_c = \frac{1,000,000}{2\pi fC}$$

VECTOR ADDITION

As with inductors, a capacitor's reactance can be added to the resistance in an ac circuit by using vector addition. Fig. 13-27, for example, shows a series circuit that has a 100 volt, 60 Hz ac power supply, a 220 microfarad capacitor and a 5 ohm resistor.

First, find capacitive reactance (X_c):

$$X_c = \frac{1,000,000}{2\pi fC}$$

$$X_c = \frac{1,000,000}{2 \times 3.14 \times 60 \times 220}$$

$$X_c = 12\,\Omega$$

Next, find impedance (Z):

$$Z^2 = X_c^2 + R^2$$

$$Z^2 = 12^2 + 5^2$$

$$Z^2 = 144 + 25$$

$$Z^2 = 169$$

$$Z = \sqrt{169}$$

$$Z = 13\,\Omega$$

Finally, find the current (I):

$$I = \frac{E}{Z}$$

$$I = \frac{100}{13}$$

$$I = 7.7A$$

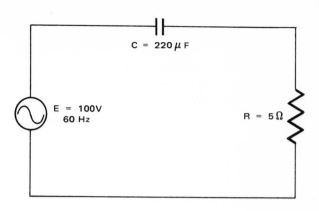

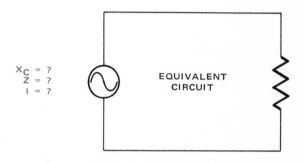

Fig. 13-27. Find the various unknowns in this capacitive circuit. The unknowns include capacitive reactance (X_c), impedance (Z) and current flow (I).

VOLTAGE DROPS

By using vector addition, we found that the impedance of the circuit shown in Fig. 13-27 is 13 ohms. Then, using Ohm's Law and substituting our value of impedance for resistance ($I = \frac{E}{Z}$ instead of $I = \frac{E}{R}$), we found that the current in the circuit is 7.7 amps.

Since this is a series ac circuit, the same 7.7 amps must flow through both the capacitor and the resistor. Therefore, the voltage drops across the capacitor and resistor are:

$$V_c = IX_c$$

$$V_c = 7.7 \times 12$$

$$V_c = 92.4V$$

$$V_R = IR$$

$$V_R = 7.7 \times 5$$

$$V_R = 38.5V$$

Note that if we try to add the voltages across the capacitor and the resistor like ordinary numbers, we do not get 120 volts for an answer. This is because the voltages are vectors and must be added like vectors.

Try adding these two voltages like vectors and see if we get the correct answer:

$$V^2 = V_C^2 + V_R^2$$

$$V^2 = (92.4)^2 + (38.5)^2$$

$$V^2 = 8537.8 + 1482.3$$

$$V^2 = 10{,}020.1$$

$$V = \sqrt{10{,}020.1}$$

$$V = 100.1V$$

Note that this answer is the same as the applied voltage of 100 volts.

TEST YOUR KNOWLEDGE

1. The opposition to ac by an inductor is called _____.
2. The symbol for an inductor is _____.
3. How many electrical degrees are there in one complete revolution?
4. What is meant by impedance?
5. Draw a vector diagram and tell how vectors can be used in ac circuits.
6. What does a capacitor do to current in a dc circuit?
7. Find the RC time constant for a 30,000 ohm resistor and a 200 microfarad capacitor connected in series. Show your work.
8. The value for inductive reactance usually is given in what terms?
9. The voltage on a sine wave would be at its maximum at _____ degrees and _____ degrees.
10. At what number of degrees would you expect to find the counter EMF to be the highest on a sine wave?
11. In order to add voltages in inductive and capacitive reactance circuits, we must use _____.
12. Impedance in a circuit is usually labeled _____ in a vector diagram.
13. Dots are used on schematics with transformers to show _____.
14. It takes about _____ time constants to charge a capacitor up to the voltage of a dc circuit power supply.
15. The RC time constant of all capacitors is approximately equal. True or False?
16. Write the formula for the impedance of a circuit with inductive reactance.

SOMETHING TO DO

1. The electronic calculator has made the analysis of circuits with impedance a lot easier. How is a calculator able to show you the square of a number or square root of a number, multiply, divide and do all of the other functions so easily?
2. See if you can locate a nomograph for inductive reactance and capacitive reactance circuits. Show how they work.
3. What is a saturable core reactor? How does it work?
4. How do you determine the value of a resistor which can be safely used to discharge a capacitor? What stops the resistor from overheating if you exceed its wattage rating when discharging capacitors?
5. An electric shaver makes use of a transformer. How? What is done to its windings in terms of phasing? What stops it from overheating when it is not loaded?
6. Perform some experiments which prove what a capacitor does to the current in a dc circuit.
7. Take three different sizes of capacitors and charge them. Measure the amount of time necessary for the capacitors to lose their charge. Use three different sizes of resistors to discharge them and measure the time needed. What do the results tell you?

Chapter 14
LCR CIRCUITS

In this chapter, we will look at LCR circuits. These are circuits that contain inductance (L), capacitance (C) and resistance (R). We will study their behavior and look at some of the useful things we can do with LCR circuits.

SIMPLE LCR CIRCUIT

Fig. 14-1 shows an LCR circuit connected to a 50 volt ac power supply with a frequency of 100 Hz. LCR values in the circuit include: inductor = .159 henry; capacitor = 21.2 microfarads; resistor = 7 ohms. Since this is an ac circuit, we will have to find the reactance of both the capacitor and the inductor in order to calculate the impedance of the circuit.

To solve the circuit for inductive reactance:

$$X_L = 2 \pi fL$$

$$X_L = 2 \times 3.14 \times 100 \times .159$$

$$X_L = 100 \, \Omega$$

When solving for capacitive reactance, the formula must be changed because the value of the capacitor is given in microfarads:

$$X_C = \frac{1,000,000}{2 \pi fC}$$

$$X_C = \frac{1,000,000}{2 \times 3.14 \times 100 \times 21.2}$$

$$X_C = \frac{1,000,000}{13,314}$$

$$X_C = 75 \, \Omega$$

FINDING IMPEDANCE

To find the total impedance of the circuit shown in Fig. 14-1, we have to add the reactances and the resistance like vectors. Note in Fig. 14-2 that the vectors for X_L and X_C are drawn in opposite directions. You can see the reason for this if you recall our earlier study of capacitors and inductors.

In Chapter 13, we found that in an inductor, the

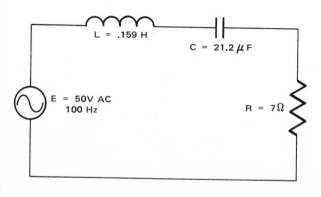

Fig. 14-1. A typical LCR circuit contains an inductor (L), a capacitor (C) and a resistor (R). Formulas for finding inductive reactance and capacitive reactance are given in the text.

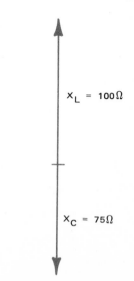

Fig. 14-2. Vectors drawn for inductive reactance (X_L) and capacitive reactance (X_C) extend away from each other because they represent opposing actions.

voltage leads the current by 90 deg. (a right angle). In a capacitor, the current leads the voltage by 90 deg. See Fig. 14-3. These two events really are opposing actions, which is the reason why the vectors for X_L and X_C are drawn in opposite directions.

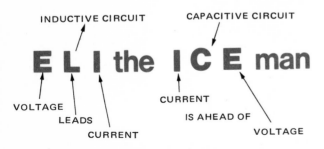

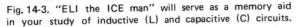

Fig. 14-3. "ELI the ICE man" will serve as a memory aid in your study of inductive (L) and capacitive (C) circuits.

SOLVING X_L AND X_C

Look at the two vectors representing X_L and X_C in Fig. 14-2. The best way to add them is to imagine that each vector represents a displacement. That is, imagine yourself at the bottom of a high ladder. If you climb ten steps straight up, then three steps straight down, you would end up seven steps above your starting point. See Fig. 14-4. As you can see, you just subtract one number from the other to get the result.

ADDING VECTORS

To relate the "ladder" method of adding vectors to our example, Fig. 14-5, first add the two vectors for X_L and X_C. This will give us a vector 25 units long that points in the same direction as X_L (because X_L is

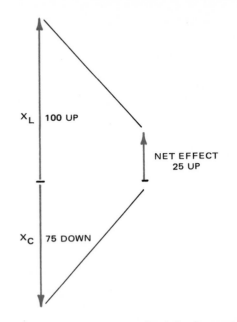

Fig. 14-5. When adding the vectors for inductive reactance (X_L) and capacitive reactance (X_C), make use of the "ladder" method shown in Fig. 14-4.

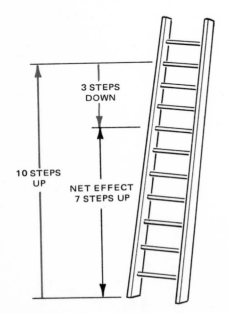

Fig. 14-4. When adding positive and negative vectors, think of climbing up (positive) and down (negative) the steps or rungs of a long ladder.

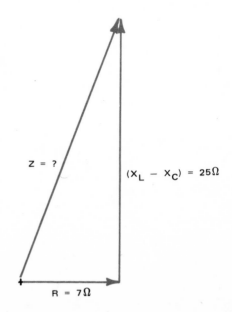

Fig. 14-6. To find total impedance of a circuit, draw the vector diagram shown and use the formula given in the text. Total impedance is 26 ohms.

longer than X_C). We now have two vectors left, Fig. 14-6. The vector with a length of 25 units represents the net reactance of the circuit. The vector with a length of seven units represents the resistive part of the circuit.

Next, place the tail of one vector at the head of the other (do not change the direction of the vector when you move it). This is the same type of vector problem that you solved for total impedance in Chapter 13.

$$Z^2 = 7^2 + 25^2$$

$$Z^2 = 49 + 625$$

$$Z^2 = 674$$

$$Z = \sqrt{674}$$

$$Z = 25.96 \ \Omega \text{ (round off to 26)}$$

FINDING CURRENT

You have found the total impedance of the circuit, Fig. 14-7. Knowing the impedance (26 ohms) and the applied voltage (50 volts), you can use the following formula to find the current flowing through the circuit:

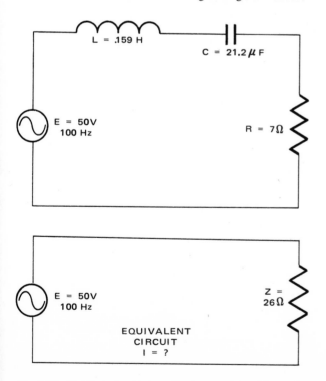

$$I = \frac{E}{Z}$$

$$I = \frac{50}{26}$$

$$I = 1.92A$$

FINDING VOLTAGE

Once you know the current in the circuit, you can use the following formula to find the voltage across each part of the circuit:

$$V_L = I \times X_L$$

$$V_L = 1.92 \times 100$$

$$V_L = 192V$$

$$V_C = I \times X_C$$

$$V_C = 1.92 \times 75$$

$$V_C = 144V$$

$$V_R = I \times R$$

$$V_R = 1.92 \times 7$$

$$V_R = 13.4V$$

Note that circuits containing reactance often produce voltages that are higher than the applied voltage. Remember that the voltage across the inductor and the voltage across the capacitor are opposite to each other (180 deg. out of phase).

Fig. 14-8 shows the voltages across the components. If you try to add all the voltages like ordinary numbers, you will get an answer that is much higher than 50. Remember, these voltages must be added like vectors.

LCR SAMPLE CIRCUITS

Try solving another LCR circuit to make sure you understand how to do this type of problem. Fig. 14-9 shows a series LCR circuit with a 150 volt, 200 Hz ac power supply. It also has a .219 henry inductor, a 10.6 microfarad capacitor and a 20 ohm resistor. Find the voltages across each of the components in the circuit on a step-by-step basis.

Fig. 14-7. To find the current flowing in this LCR circuit, divide the applied voltage by the total impedance. The current is 1.92 amps.

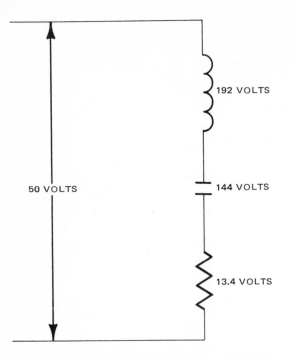

50 VOLTS

192 VOLTS

144 VOLTS

13.4 VOLTS

Fig. 14-8. Voltages across LCR components total more than the applied voltage. Individual voltages must be added like vectors to match the applied voltage.

Find reactance of inductor:

$$X_L = 2 \pi fL$$

$$X_L = 2 \times 3.14 \times 200 \times .219$$

$$X_L = 275\Omega$$

Figure reactance of capacitor, based on formula for use when C is given in microfarads:

$$X_C = \frac{1,000,000}{2 \pi fC}$$

$$X_C = \frac{1,000,000}{2 \times 3.14 \times 200 \times 10.6}$$

$$X_C = \frac{1,000,000}{13,314}$$

$$X_C = 75.1\Omega \quad \text{(Round off at 75)}$$

Find net reactance offered by inductor and capacitor:

$$\text{Net reactance} = X_L - X_C$$

$$\text{Net reactance} = 275 - 75$$

$$\text{Net reactance} = 200\Omega$$

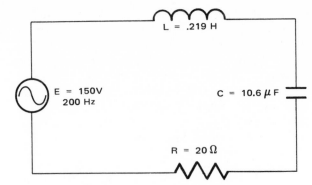

L = .219 H

E = 150V
200 Hz

C = 10.6 μ F

R = 20 Ω

Fig. 14-9. This sample LCR circuit must be solved for reactance, impedance, current through the circuit and voltages across the components.

Draw vector diagram, Fig. 14-10.
Add vectors to get impedance:

$$Z^2 = 200^2 + 20^2$$

$$Z^2 = 40,000 + 400$$

$$Z^2 = 40,400$$

$$Z = \sqrt{40,400}$$

$$Z = 201\Omega$$

Using the impedance value, figure current in the circuit:

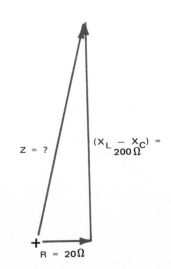

Z = ?

$(X_L - X_C) = 200\Omega$

R = 20Ω

Fig. 14-10. To obtain the total impedance of the circuit shown in Fig. 14-9, draw a vector diagram and add the vectors for net reactance and resistance.

$$I = \frac{E}{Z}$$

$$I = \frac{150}{201}$$

$$I = .75A$$

Find voltage across each component:

$$V_L = I \times X_L$$

$$V_L = .75 \times 275$$

$$V_L = 206V$$

$$V_C = I \times X_C$$

$$V_C = .75 \times 75$$

$$V_C = 56.25V$$

$$V_R = I \times R$$

$$V_R = .75 \times 20$$

$$V_R = 15V$$

Using methods for adding vectors and Ohm's Law, we have taken an LCR circuit and solved for the current through the circuit and the voltage across each component in the circuit.

RESONANCE

In analyzing circuits that contain both an inductor and a capacitor, the net reactance of the two can be found by subtracting the capacitor's reactance from the inductor's reactance. A good question to ask is "What happens when the reactance of the capacitor equals the reactance of the inductor?"

You can see that the net reactance of the circuit is zero, Fig. 14-11. Also, the impedance of the circuit is equal to the resistance of the resistor in the circuit. *In an LCR circuit, when the reactance of the capacitor equals the reactance of the inductor, the circuit is called a " resonant" circuit.*

For example, our last circuit had a .219 henry inductor, a 10.6 microfarad capacitor and a 20 ohm resistor in series fed by a 150 volt, 200 Hz signal. To pose this problem of a resonant circuit, we will leave

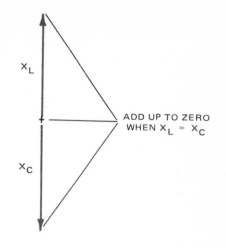

Fig. 14-11. Resonance occurs when inductive reactance (X_L) equals capacitive reactance (X_C). Using the "ladder" method for solving net reactance, the result is zero.

the voltage of the applied signal at 150 volts, but change the frequency to 104.5 Hz. See Fig. 14-12. Then, we will analyze the circuit step-by-step:

$$X_L = 2\pi fL$$

$$X_L = 2 \times 3.14 \times 104.5 \times .219$$

$$X_L = 143.7\Omega$$

$$X_C = \frac{1,000,000}{2\pi fC}$$

$$X_C = \frac{1,000,000}{2 \times 3.14 \times 104.5 \times 10.6}$$

$$X_C = \frac{1,000,000}{6956.5}$$

$$X_C = 143.7\Omega$$

Net reactance $= X_L - X_C$

Net reactance $= 143.7 - 143.7$

Net reactance $= 0\Omega$

Since $X_L = X_C$, we say that this circuit "resonates" at a frequency of 104.5 Hz.

Carry on and find the impedance of the circuit at resonance and the voltages developed across the inductor and the capacitor.

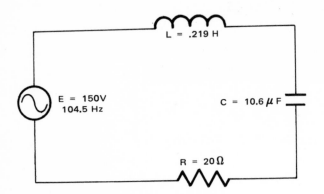

Fig. 14-12. This sample resonant circuit has a net reactance of zero. Also solve for impedance, current through the circuit and voltage across the components.

$$Z^2 = (\text{net reactance})^2 + R^2$$

$$Z^2 = 0^2 + R^2$$

$$Z = R = 20\,\Omega$$

$$I = \frac{E}{Z}$$

$$I = \frac{150}{20} = 7.5A$$

$$V_L = I \times X_L$$

$$V_L = 7.5 \times 143.7$$

$$V_L = 1078V$$

$$V_C = I \times X_C$$

$$V_C = 7.5 \times 143.7$$

$$V_C = 1078V$$

$$V_R = I \times R$$

$$V_R = 7.5 \times 20$$

$$V_R = 150V$$

RESONANT FREQUENCY

At the resonant frequency of 104.5 Hz, the voltages across the inductor and the capacitor are much larger than at other frequencies. For example, compare voltages at a frequency of 104.5 Hz with voltages at 200 Hz.

Frequency	Voltage	
(Hz)	V_L	V_C
104.5	1078	1078
200	206	56.25

Actually, at resonant frequency, the voltages across the inductor and the capacitor are at their highest value. See Fig. 14-13. Note how quickly the voltage drops off as we get farther from resonant frequency.

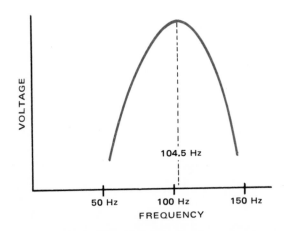

Fig. 14-13. Chart shows voltage across individual components in the circuit at different frequencies. Voltage is at a maximum at the resonant frequency of 104.5 Hz.

TUNING

The most common and useful way of using resonance is in radio circuits. The circuit shown in Fig. 14-12 is resonant at a frequency of 104.5 Hz. You might say that this circuit is "tuned" to 104.5 Hz.

Therefore, when you turn the dial on your radio to listen to your favorite station, you are tuning it to the frequency where it is resonant, Fig. 14-14.

CHANGING CAPACITANCE IN A TUNED CIRCUIT

If we change the value of the capacitance in a tuned circuit, we will find that the circuit is no longer resonant at the same frequency. To experiment, change the value of the capacitor from 10.6 microfarads to 20 microfarads, Fig. 14-15. *Remember, when the capaci-*

Fig. 14-14. The tuning control on a radio (see arrow) adjusts a variable capacitor to tune the circuit to a frequency that is resonant.
(McIntosh Laboratory, Inc.)

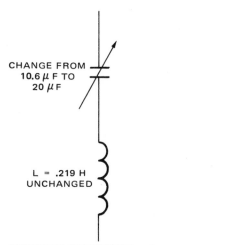

CHANGE FROM 10.6 μ F TO 20 μ F

L = .219 H UNCHANGED

RESONANT FREQUENCY = ?

Fig. 14-15. By changing the value of the capacitor in Fig. 14-12, capacitive reactance changes. Therefore, the circuit will not resonate.

tance changes, so does the capacitive reactance. Since the inductor remains the same, X_C no longer equals X_L, and the circuit will no longer resonate at this frequency.

Work the following formulas to find the frequency at which this new circuit will resonate. Remember, $X_L = X_C$.

$$X_L = X_C$$

$$2 \pi fL = \frac{1}{2 \pi fC} \quad \text{(f is resonant frequency)}$$

$$2 \times 3.14 \times f \times .219 = \frac{1,000,000}{2 \times 3.14 \times f \times 20}$$

$$1.38 \times f = \frac{1,000,000}{125.6 \times f}$$

$$1.38 \times f = \frac{7962}{f}$$

Next, multiply both sides by f:

$$1.38 \times f \times f = 7962$$

$$1.38 \times f^2 = 7962$$

$$f^2 = \frac{7962}{1.38}$$

$$f^2 = 5770$$

$$f = \sqrt{5770}$$

$$f = 76 \text{ Hz}$$

This circuit analysis proves that if we change the capacitor from 10.6 microfarads to 20 microfarads, we can make the circuit resonate at a frequency of 76 Hz. In effect, we have "tuned" the circuit to a frequency of 76 Hz. Therefore, if we want to tune a circuit to a certain frequency, all we have to do is change the value of the capacitor. See Fig. 14-16. NOTE: You can do the same thing by changing the value of the inductor.

TUNING A RADIO

To see how "tuning" works, take a look at the tuning circuit of a radio, Fig. 14-17. The antenna picks up small voltages from radio waves transmitted through the air by many different stations. Each of these stations sends out waves at a different frequency. NOTE: The antenna is shown as an inductor because most modern radio antennas are wound in a coil.

The purpose of the tuning circuit is to make the circuit resonate at the frequency you desire, so that only one station is received in the radio. For example, if your favorite radio station broadcasts signals at a frequency of 680 kHz, you would adjust the capacitor

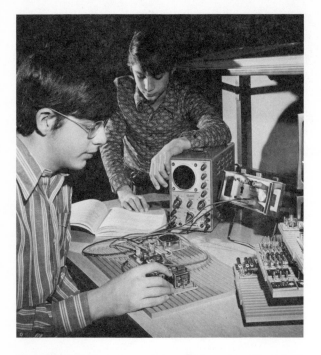

Fig. 14-16. Students use a power supply and oscilloscope to "tune" a circuit. (Hickok Teaching Systems, Inc.)

VARIABLE CAPACITORS

"Tuning" usually is performed by varying the capacitor in the circuit, rather than the inductor. This is done because it is relatively easy to make a variable capacitor.

Fig. 14-18 shows a typical variable capacitor. It is made of a number of metal plates attached to a shaft called a "rotor." Another set of plates called the "stator" is attached to the frame of the capacitor. Air is generally used as the dielectric (insulating) material.

When the rotor shaft is turned, Fig. 14-18, the plates on the rotor move between the plates on the stator. As the rotor is turned farther, more area of the rotor plates is exposed to the stator plates. You recall from Chapter 4 that the greater the area of the plates, the higher the capacitance. To increase the capacitance, then, you have to turn the rotor so that a larger area of the plates overlap.

When a capacitor is used in a radio, the rotor shaft is extended through the front cover. A tuning control is attached to the shaft to move the pointer on a dial marked for many different frequencies.

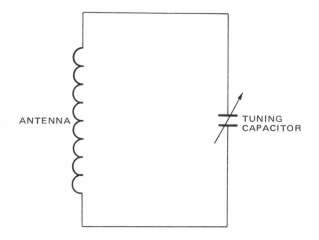

Fig. 14-17. Simplified tuning circuit. Radio waves are picked up by the antenna. The capacitor is tuned to a station's assigned broadcasting frequency.

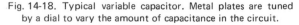

Fig. 14-18. Typical variable capacitor. Metal plates are tuned by a dial to vary the amount of capacitance in the circuit.

in the tuning circuit until the circuit resonates at 680 kHz. When this happens, you have "tuned" the radio to receive your station's signal.

At the same time, small voltages whose frequency is 680 kHz develop very large voltages across both the antenna and the capacitor. These voltages are amplified by other components of the radio to give you your favorite program.

METAL LOCATOR PROJECT

You can build a simple metal locator by attaching a search coil to a wooden handle and connecting the coil to a circuit board and parts. Used in conjunction with a "tuned" transistor radio, the metal locator will detect coins, keys, watches and other metal objects lost in sand, dirt or grass.

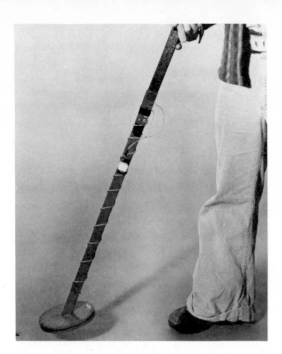

Fig. 14-19. Metal locator utilizes a shop-made search coil and circuit board mounted on a wooden handle. This device works in conjunction with a transistor radio to signal the location of hidden metal. (Project by Ed Belliveau)

Study Figs. 14-19 and 14-20. Then, obtain the following items and assemble them according to the construction procedure given.

No.	Item
1	9V battery.
1	680 pF capacitor.
1	.001 μ F capacitor
1	NPN silicon transistor (general purpose).
1	5 K potentiometer.
1	1 K resistor.
1	4.7 K resistor.
1	SPST switch.
1	Coil made from 30 ft. No. 26 wire.
1	Wooden handle (broken hockey stick, etc.).
1	1/4 in. plywood disk, 6 in. in diameter.
2	1/4 in. plywood disks, 7 in. in diameter.
	Circuit board.
	Hookup wire, AWG 20.

CONSTRUCTION PROCEDURE

To build the metal locator:

1. Using 1/4 in. plywood, cut out one 6 in. disk and two 7 in. disks.
2. To make the coil form for the search coil, glue the three disks together (small coil in the middle).
3. Glue the handle to the disks.
4. Wind 30 ft. No. 26 wire on the coil form and wrap the lead wire up the handle, Fig. 14-19.

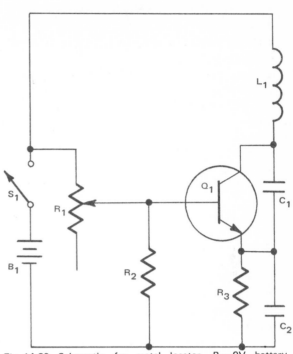

Fig. 14-20. Schematic for metal locator. B_1—9V battery. C_1—680 pF capacitor. C_2—.001 mF capacitor. L_1—inductor (search coil). Q_1—transistor. R_1—5 K potentiometer R_2—1 K resistor. R_3—4.7 K resistor. S_1—SPST switch.

5. Build circuit as shown in Fig. 14-20. Mount parts on a perforated board and solder all connections. Q_1 can be any general purpose transistor.
6. Connect search coil lead wires to circuit board.
7. Tape circuit board and 9V battery to handle.

PUTTING THE PROJECT TO WORK

To use the metal locator:

1. Place metal locator on surface to be tested.
2. Turn on transistor radio and tune it to a high frequency station. Tune it slightly off the station, yet still audible.
3. Turn on the metal locator and move the volume control until you can hear a squeal from the radio.
4. If no squeal is heard, repeat step 2 at the next lower station.
5. Then, repeat step 3.
6. Keep doing this until you find the frequency with the loudest squeal.
7. Pass the coil slowly over the area to be searched. As the coil passes over buried metal, the inductance of the coil will change. This, in turn, will change the frequency of the squeal and tell you where to start digging for your metal you located.

TEST YOUR KNOWLEDGE

1. What do the letters LCR stand for in an electrical formula?
2. If you added the voltage drops across the components in a reactive circuit, your result would be different than the voltage of your power supply. Why?
3. When the capacitive reactance equals the inductive reactance, we have a condition where the circuit is said to be in _____.
4. An antenna usually is shown in a circuit schematic as a _____ or an _____.
5. What happens when we tune a circuit in a radio?
6. In a resonant circuit, X_L will be equal to _____.
7. When a circuit is at a resonate frequency, will the voltage across individual components be at a maximum or minimum?
8. What is the formula for finding impedance (Z) of a circuit with one inductor and one resistor?
9. In a series RL circuit containing one inductor and one resistor, if $X_L{}^2$ is equal to 9 and R_2 is equal to 216, what is the value of Z? Show your work.
10. Complete the following list by using the base unit for each element of the circuit.
    a. _____ Resistance.
    b. _____ Impedance.
    c. _____ Power.
    d. _____ Capacitance.
    e. _____ Current.
    f. _____ Frequency.
    g. _____ Inductance.
    h. _____ EMF.
    i. _____ RMS.
11. In an LCR circuit, when the reactance of the capacitor equals the reactance of the inductor, the circuit is called a resonant circuit. True or False?

THINGS TO DO

1. Find as many different kinds of radio coils as you can. Does the size or cost of a radio change the number of coils used? Why?
2. Take apart an old television set. Separate the various components into piles. How many different components are used? What are their range of sizes?
3. There was some talk that TV sets cause X rays. Where do these rays come from? How are people watching TV protected from these X rays? How did the rays get out of the TV set so that they would affect people?
4. What are the sizes of the tuning capacitors used in radios?
5. Make a diagram of the tuning circuits used in a radio. What are the values for each?
6. Why do most radios use variable capacitors instead of variable inductors for tuning?
7. Does each city in the U.S. use the same set of frequencies for radio transmission? Are any of them the same or are they all different? Would the distance between cities make any difference which frequencies are used?
8. There are some nights when you can pick up radio stations hundreds of miles away from your home. How is that possible? Why does it happen at night? Why does it happen only on certain nights?
9. What agency of the government controls radio stations, frequencies used, etc.? Why must there be some type of control?
10. What international agency controls radio station frequencies when other countries are involved?
11. If you were to operate a "ham" radio station, how would you go about learning what to do? Who assigns call letters?

Chapter 15
FILTERS

In Chapter 14, we looked at LCR circuits and found that they can be used as tuning circuits in radios. Tuning, however, is just one of the many different ways we can make use of LCR circuits.

In this chapter, we will make a close study of LCR circuits and see how they are used to "filter" out unwanted frequencies.

The circuit shown in Fig. 15-1 is a series LCR circuit. As an experiment, it would be interesting to chart what this circuit would do if a voltage of one millivolt of seven different frequencies were applied to it. The results of this experiment are worked out for you in Fig. 15-2.

As you can see from the chart, the resonant frequency of the circuit is 1000 kHz. Note that at resonant frequency, the impedance (Z) of the circuit is at a minimum, and it is equal to the resistance in the circuit. Also, at resonance, voltage across the inductor and the capacitor is at a maximum.

Fig. 15-3 shows how the current through the circuit changes with frequency. The curve in Fig. 15-4 indicates voltage across the inductor. It illustrates, maximum voltage at the resonant frequency of the circuit

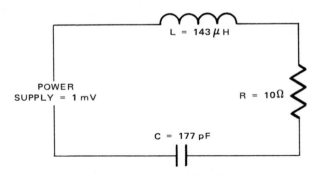

Fig. 15-1. This series LCR circuit is used as a filter. It has a resonant frequency of 1000 kilohertz (kHz) as indicated in the chart in Fig. 15-2.

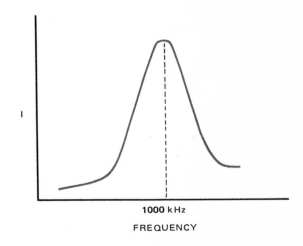

Fig. 15-3. Current flow curve shows how current through a series LCR circuit changes with frequency.

VALUE OF ELEMENTS IN CIRCUIT

		E	X_L (Ω)	X_C (Ω)	R (Ω)	Z (Ω)	V_L (μV)	V_C (μV)
FREQUENCY	400 kHz	1 mV	360	2250	10	1890	190	1190
	600 kHz	1 mV	540	1500	10	960	562	1562
	800 kHz	1 mV	720	1125	10	405	1777	2777
	1000 kHz	1 mV	900	900	10	10	90,000	90,000
	1200 kHz	1 mV	1080	750	10	330	3272	2273
	1400 kHz	1 mV	1260	640	10	620	2032	1032
	1600 kHz	1 mV	1440	560	10	880	1636	636

Fig. 15-2. Chart based on the LCR circuit in Fig. 15-1 lists values of elements under the applied voltage of 1 millivolt at various frequencies.

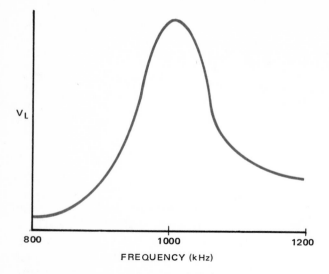

Fig. 15-4. Voltage curve shows how voltage across an inductor in a series LCR circuit reaches a maximum at the resonant frequency of the circuit.

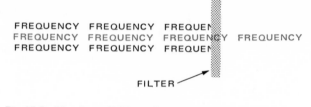

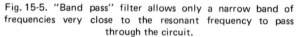

Fig. 15-5. "Band pass" filter allows only a narrow band of frequencies very close to the resonant frequency to pass through the circuit.

BANDSTOP FILTER

A "bandstop" filter, on the other hand, stops any frequencies very close to the resonant frequency from passing through the circuit. *Remember, then, that the band pass filter allows a narrow band of frequencies to pass; the bandstop filter prevents a narrow band of frequencies from passing.*

We would use a bandstop filter if we wanted to make sure that one certain frequency would not get into a given circuit. For example, the frequency of the alternating current in your home is 60 Hz. This frequency lies in the audio range, which falls between 20 Hz and 20,000 Hz. If, by chance, any small voltages at this frequency did "sneak" into a radio circuit, they would produce an annoying humming sound at the speaker.

The amplifier shown in Fig. 15-6 illustrates the use of a bandstop filter. The amplifier has been protected against any 60 Hz voltage getting into its input because the LCR circuit has been "tuned" to 60 Hz. It works like this: The input impedance of the amplifier is about 1000 ohms. That is, any signal coming into the

and very small voltages at frequencies far from the resonant frequency. This information can be very useful. All you have to do is to choose the proper values of inductance (L) and capacitance (C) in order to make the circuit resonate at the frequency you desire.

BAND PASS FILTER

If we use a circuit to separate different frequencies, this action is called "filtering." If we use a circuit to filter out frequencies above and below the resonant frequency, this circuit is called a "band pass" filter. The effect of this type of circuit is shown in Fig. 15-5.

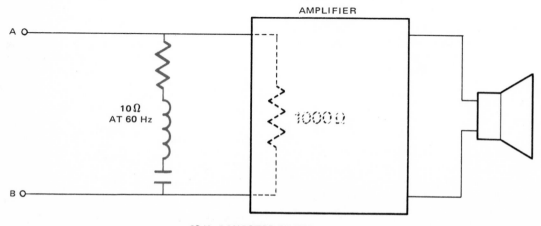

60 Hz BANDSTOP FILTER

Fig. 15-6. The part of this circuit shown in color is a "bandstop" filter designed to intercept a signal at 60 Hz frequency before it gets to the amplifier.

amplifier "sees" a resistance of 1000 ohms. If any 60 Hz signals get to terminals A and B, these signals will travel through the 10 ohm path and very little current will flow into the amplifier. Signals at any other frequency "see" a high impedance through the filter circuit and flow directly into the amplifier instead of the 10 ohm path.

In the amplifier circuit shown in Fig. 15-6, we have made a bandstop filter, using the same components as in Fig. 15-1 (L, C and R). This filter circuit will stop any frequencies close to 60 Hz from entering the system.

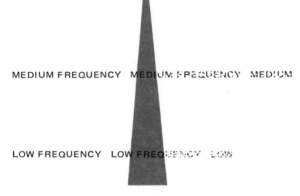

Fig. 15-7. The idea behind a "high pass" filter is to make up a circuit that will pass a high frequency signal only.

HIGH PASS FILTER

A "high pass" filter is a circuit that will allow only high frequencies to pass through it. See Fig. 15-7. Any low frequencies that try to pass through this type of circuit will be attenuated (reduced in amplitude). A very simple high pass filter can be made using only a capacitor, Fig. 15-8.

In order to understand how a high pass filter works, you have to remember how capacitors and inductors behave at different frequencies. Note in Fig. 15-8 that as frequency gets higher, the reactance of the capacitor gets lower.

If we wanted to pass high frequencies and stop low frequencies from getting to our load resistor, we would use a capacitor whose reactance was very high. Suppose we have a load resistor with a value of 100 ohms. If we want to stop frequencies lower than 500 Hz from getting through, all we have to do is pick a capacitor whose reactance will be higher than the value of the load resistor when the frequency is under 500 Hz.

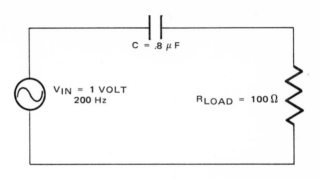

Fig. 15-8. This "high pass" filter will filter out almost all of the low frequency (200 Hz) signal. Only about one-tenth of the voltage will get through to the load resistor.

Suppose the capacitor has a capacitance of .8 μ f. If a signal with a frequency of 200 Hz enters the filter, the value of X_c will be:

$$X_c = \frac{1,000,000}{2 \pi fC}$$

$$X_c = \frac{1,000,000}{2 \times 3.14 \times 200 \times .8}$$

$$X_c = 995\Omega$$

To keep the numbers simple, we will say that X_c is equal to 1000 ohms. Also, suppose that the signal applied to the circuit has an amplitude of one volt and a frequency of 200 Hz. How much of that voltage gets to the load resistor (R_L)?

Using Ohm's Law to solve this problem, first find the current in the circuit:

$$Z^2 = X_c^2 + R_L^2$$

$$Z^2 = (1000)^2 + 100^2$$

$$Z^2 = 1,000,000 + 10,000$$

$$Z^2 = 1,010,000$$

$$Z = \sqrt{1,010,000}$$

$$Z = 1005\Omega$$

$$I = \frac{E}{Z}$$

$$I = \frac{1}{1005}$$

$$I = .00099A \ (.99 \ mA)$$

Then, check and compare the voltages across the capacitor (V_C) and load resistor (V_R):

$$V_C = I \times X_C$$

$$V_C = .00099 \times 1000$$

$$V_C = .99V$$

$$V_R = I \times R$$

$$V_R = .00099 \times 100$$

$$V_R = .099V$$

Most of the voltage at this frequency has been dropped across the capacitor. Only about one-tenth of the voltage will reach the load resistor.

By circuit analysis, then, we have found that low frequencies are "attenuated" or cut down by a high pass filter. Next, we will see what is happening in this circuit at a high frequency. In this setup, we will use the same filter. However, we *will change* the frequency of our signal to 5000 Hz, Fig. 15-9.

Now, analyze this circuit and see how much of the signal gets across the load resistor (R_L):

$$X_C = \frac{1,000,000}{2 \times 3.14 \times 5000 \times .8}$$

$$X_C = \frac{1,000,000}{25,120}$$

$$X_C = 40\Omega$$

Now, find the impedance of the circuit:

$$Z^2 = 40^2 + 100^2$$

$$Z^2 = 1600 + 10,000$$

$$Z^2 = 11,600$$

$$Z = \sqrt{11,600}$$

$$Z = 108\Omega$$

Then, figure the current:

$$I = \frac{E}{Z}$$

$$I = \frac{1}{108}$$

$$I = .0093A$$

Finally, calculate the voltages across the capacitor and load resistance:

$$V_C = I \times X_C$$

$$V_C = .0093 \times 40$$

$$V_C = .37V$$

$$V_R = I \times R$$

$$V_R = .0093 \times 100$$

$$V_R = .93V$$

Note that if the frequency of the signal is 5000 Hz, we will get .93 volts across the load resistor. This is almost

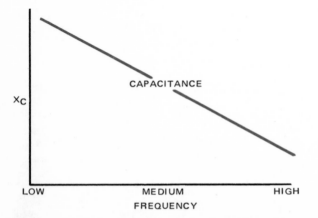

Fig. 15-9. As the frequency of the signal gets higher, capacitive reactance goes down. Therefore, a capacitor of the proper value can be selected to filter out the right range of frequencies.

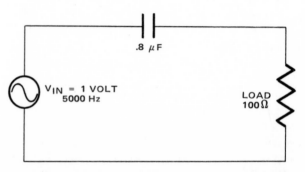

Fig. 15-10. Using the same "high pass" filter shown in Fig. 15-7, but with the signal at high frequency (5000 Hz), much more of the signal will get through to the load resistor.

all of the signal which was applied, Fig. 15-10. In other words, this circuit has passed a frequency of 5000 Hz.

IMPROVED HIGH PASS FILTER

If you want to improve on the last circuit, add an inductor in parallel with the load resistor, Fig. 15-11. Without going into actual numbers, you can see the results.

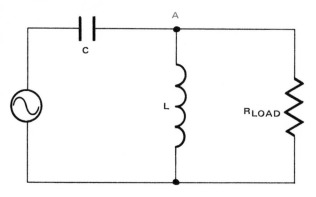

Fig. 15-11. Add an inductor to a "high pass" filter to improve performance. At low frequency, most of the voltage will be dropped across the capacitor and inductor. At high frequency, most of the voltage will get through to the load resistor.

At high frequencies, capacitive reactance (X_C) is very low. This means that most of the signal will get through the capacitor (C) to point A. From A, the signal "sees" two possible paths to follow, either by way of the inductor (L) or through the load resistor (R_L). Since inductive reactance (X_L) goes up as frequency rises, Fig. 15-12, inductive reactance will be

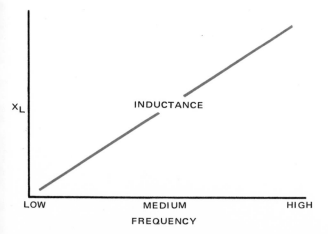

Fig. 15-12. Inductance is low at low frequencies and increases at high frequencies. This characteristic makes an inductor suitable for use in a "low pass" filtering circuit.

very high. Therefore, most of the signal will take the easier path and go through the load resistor and develop voltage across it.

If we connect low frequency voltage to the filter shown in Fig. 15-11, X_C will be very high and most of the signal will be dropped across X_C. The small amount of signal that does get to point A now has a choice of going through R_L or through X_L, which is very low at low frequencies. Therefore, most of the signal will go through X_L, and very little will develop across R_L.

The filter just explained is the normal type of high pass filter. The values of the capacitor and inductor will depend on the frequency you want to pass and on the size of the load resistance. In later courses, you will learn how to design your own filters.

LOW PASS FILTER

A low pass filter is a circuit that will allow only low frequencies to pass through it. In order to make up a low pass filter, simply interchange the location of the inductor and capacitor in a high pass circuit. Compare the high pass filter in Fig. 15-11 with the low pass filter in Fig. 15-13.

Note in Fig. 15-13 that if a low frequency is applied to the low pass filter, it will:
1. "See" a low value of inductive reactance (X_L).
2. "See" a high value of capacitive reactance (X_C) in parallel with the load resistor (R_L).

This means that low frequency signals will pass through the filter to the load resistor.

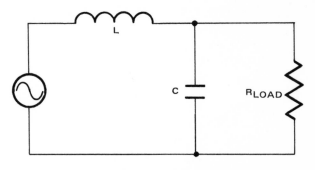

Fig. 15-13. In this "low pass" filtering circuit, the inductor allows low frequencies to pass, while the capacitor blocks passage. Most of the voltage passes through the load resistor.

High frequency signals, on the other hand, "see" a high value of inductive reactance. Therefore, the small amount of signal that does get by X_L will pass through the low value of X_C instead of through the load resistor (R_L). The high frequency signals are blocked.

POWER SUPPLY FILTER

Some devices, such as hi-fi amplifiers and television sets, need a source of high voltage direct current. A power transformer usually is used, Fig. 15-14, to step up the 120 volts found in your home. A device called a "rectifier," Fig. 15-15, also is used to change the ac voltage to dc voltage.

POWER TRANSFORMER

Fig. 15-14. A power transformer is used in this stereo power amplifier. (McIntosh Laboratory, Inc.)

Fig. 15-15. Rectifiers (circled) are used in hi-fi amplifiers and television sets to convert ac voltage to dc voltage.

You will learn more about rectifiers in Chapter 16. All you need to know now is that after the ac has passed through the rectifier, it is called "pulsating direct current." See Fig. 15-16. While the rectified current is dc, the "ripple" at the top of the waves acts like ac. If this ac ripple gets into the amplifier section, it could cause an annoying hum.

RIPPLE

In order to get rid of the ac ripple, a capacitor, Fig. 15-17, usually is added to the power supply output. When the ripple rises to its peak value point A in Fig. 15-18, the capacitor charges to the peak value of the output voltage. When the ripple starts to drop in value (point B), the capacitor discharges through the load resistor and helps keep the voltage across R_L at or near the peak value of the direct current. This cycle is repeated every time the pulsating current rises and falls.

The output of this circuit is dc with only a small amount of ripple, Fig. 15-19. What the circuit has done is to filter out the ac ripple from the pulsating dc. This is the most common type of filter used in power supplies. Fig. 15-20 shows other, slightly different forms of filters which do the same type of power supply filtering.

BANDWIDTH AND Q

When we looked at band pass and bandstop filters, we found that these filters would pass or stop a narrow band of frequencies around the resonant frequency (f_R) of the circuit. Now we will find the biggest factor that affects bandwidth, called the "Q" of the circuit.

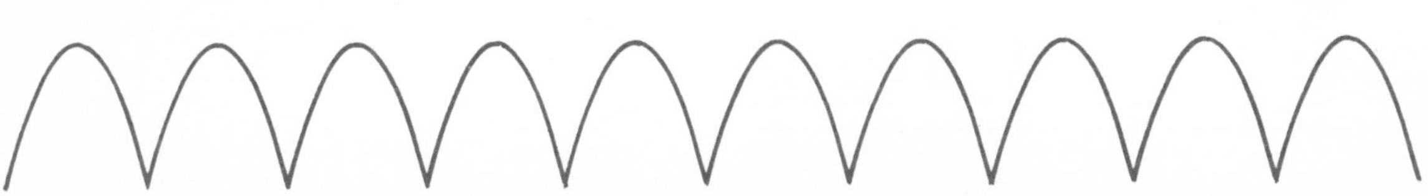

PULSES OF DIRECT CURRENT AT TOPS OF SINE WAVES

Fig. 15-16. When the rectifier changes alternating current to pulsating direct current, an "ac ripple" is formed on the top of the sine waves.

Fig. 15-17. A capacitor is the most common form of filter used in power supplies. When ac is changed to dc, the capacitor helps filter out the "ac ripple." (Fisher Radio)

WHAT THE CAPACITOR WILL DO TO THE POWER SUPPLY OUTPUT

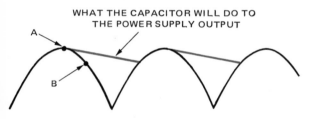

Fig. 15-18. The charging (A) and discharging (B) of the capacitor serves to filter the "ac ripple" from the pulsating dc.

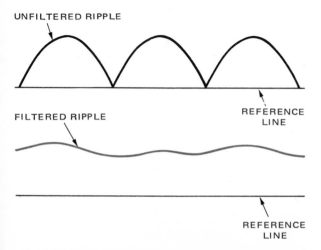

Fig. 15-19. This filtered ripple results from the action of the capacitor shown in Fig. 15-18. Note how the highs and lows of the unfiltered and filtered ripples coincide.

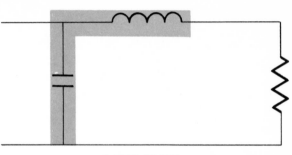

L TYPE FILTER

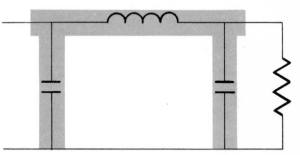

π TYPE FILTER

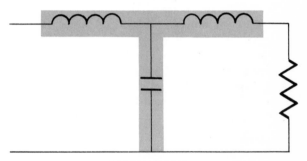

T TYPE FILTER

Fig. 15-20. Special types of power supply filters are shown. Basically, they are named for the arrangement of components in the circuit.

The Q of the circuit is defined as the ratio of the reactance of an inductor or capacitor to circuit resistance. Therefore, the value of Q can be found by dividing the reactance of the inductor at resonant frequency by the resistance in the circuit:

$$Q = \frac{X_L}{R}$$

As the formulas will prove, the higher the value of Q, the narrower the band of frequencies around the resonant frequency, Fig. 15-21.

The value of Q is very useful. If we know both the Q value and the resonant frequency of a circuit, we can find the width of the band of frequencies that is passed

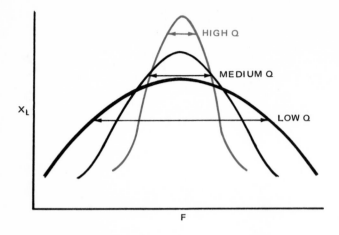

Fig. 15-21. These curves show the varied effects of "Q" at resonant frequency. Note that the higher the "Q" value, the narrower the bandwidth will be.

or stopped. The bandwidth, Q and the resonant frequency (f_R) are related by the following formula:

$$BW = \frac{f_R}{Q}$$

Take a look at a sample circuit. Fig. 15-22 shows a band pass filter. See if you can find what band of frequencies is passed by this filter. Suppose that the resonant frequency is 10 kHz and the inductive reactance (X_L) is 500 ohms at this frequency:

$$Q = \frac{X_L}{R}$$

$$Q = \frac{500}{10}$$

$$Q = 50$$

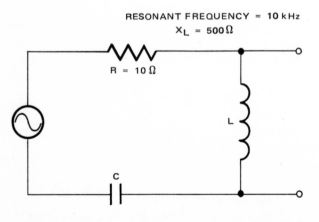

RESONANT FREQUENCY = 10 kHz
$X_L = 500\,\Omega$

$R = 10\,\Omega$

L

C

Fig. 15-22. Sample circuit illustrates the use of a "band pass" filter designed to pass a narrow band of frequencies around the resonant frequency of the filter.

Once we have found the value of Q, we can figure out the bandwidth of the filter:

$$BW = \frac{f_R}{Q}$$

$$BW = \frac{10,000}{50}$$

$$BW = 200\ Hz$$

This means that the filter will pass a 200 Hz band of frequencies whose center frequency is the resonant frequency of the filter, Fig. 15-23. To state this another way, the filter will pass frequencies from 9900 Hz to 10,100 Hz. The idea of bandwidth and Q is important in radio circuits, as you shall see as you continue your study of electronics.

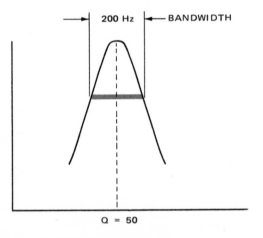

Fig. 15-23. The "band pass" filter shown in Fig. 15-22 will pass this narrow band of frequencies. Note that the center frequency is the resonant frequency of the circuit.

SMALL STROBE LIGHT PROJECT

Build a simple strobe light, Fig. 15-24, and freeze the motion of any mechanism that turns 60 times per second. Under the flashing light, the rotating part will appear to be standing still. NE 2P and five percent tolerance resistors may be difficult to obtain in some locations.

Build the small strobe light from the following parts:

No.	Item
1	HEP R0053 diode.
1	GE NE 2P neon lamp.
1	16 K resistor (R_1).
1	110 K resistor (R_2).
1	110V ac lamp cord.
	Connecting wires.

Assemble the parts and make the connections shown in the schematic in Fig. 15-24.

Use the strobe light in a dimly lit room. Hold it close to the motor, fan, clock or whatever rotating unit you want to check. Be careful. If you get too close to moving parts, an accident may occur.

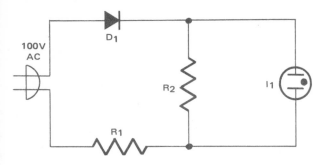

Fig. 15-24. Schematic gives the arrangement of components in a shop-made strobe light circuit. D_1 — Diode. I_1 — Neon lamp. R_1 and R_2 — Resistors.

TEST YOUR KNOWLEDGE

1. At the resonant frequency, the impedance of an LCR circuit is at _____ value.
2. What do you call a filter circuit which is used to stop one certain frequency?
3. Define what is meant by ripple.
4. One way to cut down on ripple is to use _____.
5. Match the following:
   a. _____ Passes low frequencies.
   b. _____ Passes high frequencies.
   c. _____ Stops a group of frequencies.
   d. _____ Passes a group of frequencies.
   1. Bandstop filter.
   2. Low pass filter.
   3. Power supply filter.
   4. High pass filter.
   5. Band pass filter.
6. In an inductive circuit, the voltage _____ the current.
7. Draw the symbols for the following:
   Ammeter.
   Antenna.
   Electrolytic capacitor.
   Buzzer.
   Ground.
   Incandescent lamp.
8. What is the range of frequencies that can be heard by the human ear?

SOMETHING TO THINK ABOUT

1. How does a radio station transmit frequencies? You cannot hear the message in the air as it goes to a radio receiver. Why?
2. What is the range of hearing for other animals?
3. Why is it possible to hear radio signals coming from the moon easier than from some spot on earth which is only a few thousand miles away?
4. There are bands that surround the earth that block radio signals. What are they called? Where are they located?
5. What are microwave signals? How are they transmitted?
6. How is it possible for the telephone company to use the same wire for more than one conversation? Sometimes you can hear other voices when you are talking to someone on the telephone. How can that be possible?
7. In some cities, you can pick up the audio from a local TV station on the radio. How is this possible?
8. All countries do not generate 60 Hz power for use in the home. Which ones do and which ones do not? What would happen if you took an electrical appliance designed for 60 Hz to a country operating at 50 Hz? What would happen if someone from a 50 Hz country brought an electrical appliance to a country which used 60 Hz.
9. How do ultrasonic machines work? How are they able to do things like welding without creating all the heat found in normal welding operations?
10. If you had a coil of wire, how do you determine the amount of inductance? If you wanted a certain inductance, how would you determine how much wire is needed? What determines the size of cores used?
11. What kind of fire extinguisher should be used on electrical fires? Why?
12. Why are so many electrolytic capacitors used in filtering circuits?

Chapter 16

DIODES

As we move along in this course from electricity to basic electronics, we will study the so-called "solid state devices." Solid state is a term given to electronic components that have replaced vacuum tubes. These include diodes, transistors, SCRs (silicon controlled rectifiers) and many others.

Basically, solid state devices do one of three things:
1. Stop the flow of electrons.
2. Start the flow of electrons.
3. Control the amount of electron flow.

In this chapter, we will cover "diodes." *A diode is a solid state device usually designed to permit electron flow in one direction and block flow from the other direction.*

Some people have trouble understanding how solid state devices work. Some of the reasons for the difficulty are: there are quite a few different devices; new ones are being discovered every day. To help solve this problem, some companies have published manuals that show the symbols for each device. They also publish graphs that explain:
1. What each component can do.
2. What amounts of current and voltage can be used without overloading the device.

OVERLOADS

Overloads ruin many solid state devices. To help you visualize what happens, think of how easily a string can be used to pull a small toy. Yet, if this same string is used to pull a large rock, the string will snap.

In another situation, consider that the chain on a swing set will support a child with no trouble. However, if this same chain is used in an attempt to pull a car out of a ditch, the chain will break.

Each thing – the string, chain and solid state device – can be loaded only with what it is built to carry. Remember, even a fuse can carry only so much load before it will "open." The big difference is that

fuses are made to be replaced. Solid state devices usually are more expensive and much more difficult to replace.

DIODES AND HOLE FLOW

The first solid state device we will cover is the diode. Its symbol is shown in Fig. 16-1. To get a better idea how diodes work, think back to the flow of electrons explained in Chapter 1. If necessary, review the electron theory before continuing with this chapter.

DIODE SYMBOL

Fig. 16-1. The symbol for a diode is used in drawings to show where the diode is located in the circuit.

Solid state devices, including diodes, work on the principle of "hole flow." Hole flow is in the opposite direction of electron flow. In Fig. 16-2, hole flow is demonstrated with marbles instead of electrons to help you understand the principle.

Note in Fig. 16-2 that as a marble moves to the right, it leaves a hole to the left. Then, the next marble moves up to fill the hole. This, in turn, leaves a hole in place of the second marble. As this action continues, you can see that the marbles (electrons) are moving to the right. At the same time, however, the hole is moving to the left. This is the hole flow we will be referring to whenever we discuss solid state devices.

In earlier chapters, we mentioned the use of rectifiers or diodes. Many people think that these are two different names for the same thing. Actually, there is a difference between them, but it is really just a matter of

Diodes

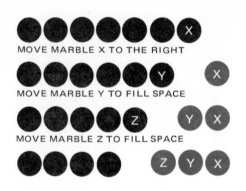

MOVE MARBLE X TO THE RIGHT

MOVE MARBLE Y TO FILL SPACE

MOVE MARBLE Z TO FILL SPACE

Fig. 16-2. Marbles may be used to show the concept of moving electrons. Marbles (electrons) move to the right; "hole flow" is to the left.

how much current they will carry without burning up. A diode usually is thought of as carrying low current, while a rectifier generally means high current.

I.C. SAYS:

"When you hear the term 'diode,' you should think of current in the milliamp range. A 'rectifier' usually means current in the amps range."

Both diodes and rectifiers allow flow in only one direction. To understand how this is possible, you need to study the function of a "PN junction." See Fig. 16-3.

PN JUNCTION

A diode is made of P type material and N type material. Most people think of the P material as positive. The N material, as you may have guessed, stands for negative. A PN junction, then, is the line of separation between P and N material in a semiconductor (diode, in this case). See Fig. 16-3.

In order to get electron flow across a diode, the electrons must travel through the P and N material. This function can be related to water trapped behind a dam, Fig. 16-4. When the water reaches a certain high level, it can be "run out" through gates put there for that purpose.

You may recall that this was the way water power was used to generate electricity. With this setup, the water can only travel out through the gates. It cannot go backward. When the water is at a level lower than the gates, it cannot flow in either direction. A diode acts almost the same way. *When a certain level of electron flow is reached, it can pass through the diode.*

However, to understand how a diode works, you must know more about the PN junction. If we wanted to put P material and N material together, we could fuse them in special ovens. However, to make it easy,

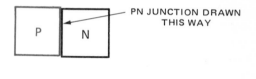

PN JUNCTION DRAWN THIS WAY

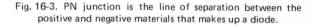

ACTUAL SIZE SMALLER THAN THIS

Fig. 16-3. PN junction is the line of separation between the positive and negative materials that makes up a diode.

Fig. 16-4. Water flowing out through the gates of a dam can be compared with current flow (one way only) through a diode. (Consumers Power Co.)

just think that both materials are heated and stuck together, Fig. 16-5. This is somewhat similar to the way you would solder two components together.

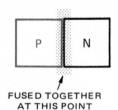

FUSED TOGETHER
AT THIS POINT

Fig. 16-5. Positive and negative materials must be fused together to form a diode.

BIASING

Once our P and N materials are put together, we will want to get current flowing in the circuit, Fig. 16-6. To do this, we will need some type of power supply. A battery with the polarities shown in Fig. 16-7 would be a good source of power. By connecting the N material to the negative side of the battery and the P material to the positive side of the battery, we will get current flow.

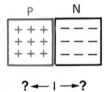

? ←— I —→ ?

Fig. 16-6. Once formed, the diode is ready to pass current (I) in one direction only. Which way will it flow?

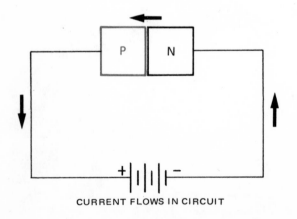

CURRENT FLOWS IN CIRCUIT

Fig. 16-7. When a battery is connected to the diode, positive to P material and negative to N material, current will flow.

Then, since like charges repel, Fig. 16-8, the *electrons* in the N material are repelled toward the junction. On the other side of the diode, the *holes* (plus signs) in the P type material are repelled toward the junction. When these actions occur, the electrons and holes meet at the junction and current flows through the diode. This is what is known as "forward biasing" of a diode, which is just another way of saying that the diode *will* pass current. Also, as you might guess, we can do just the opposite to stop the flow.

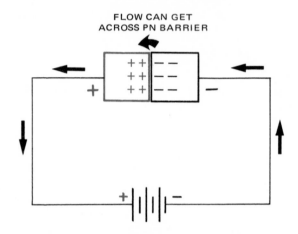

FLOW CAN GET
ACROSS PN BARRIER

Fig. 16-8. With this setup, electrons at right are repelled toward the PN junction. At the same time, "hole flow" at the left is toward PN junction.

REVERSE BIAS

To reverse bias the diode, Fig. 16-9, put the positive terminal of the battery to the N material. Also, hook up the negative terminal of the battery to the P material. This will draw electrons and holes away from the junction. When this happens, we cannot get any current flow through the diode. You will soon see that it is possible to get flow in either direction if we put a high enough voltage on the diode. In some cases, you will find there are diodes made just for that purpose. Other times, this will ruin them.

PEAK INVERSE VOLTAGE

Current going through a diode when it is forward biased is written I_f. Voltage is written V_F. A diode, however, has two important ratings. One is the current in the forward direction (I_f). The second is the amount of voltage that it can take in the reverse direction without being damaged. This reverse, or inverse, voltage is called the "peak inverse voltage," which is written

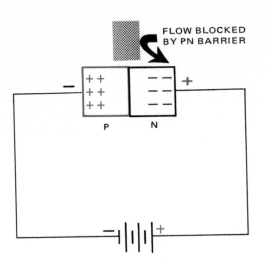

Fig. 16-9. When a battery is connected to the diode, negative to P material and positive to N material, current flow is blocked (reverse bias).

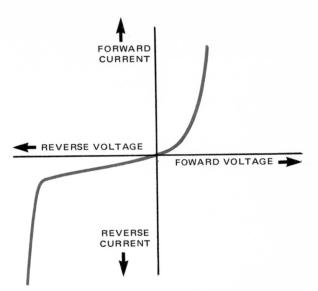

Fig. 16-11. Forward and reverse current and voltage values for a diode can be plotted on a graph.

PIV. These ratings differ for all the different kinds of diodes. To show what the ratings are, many people plot them on a graph, Fig. 16-10.

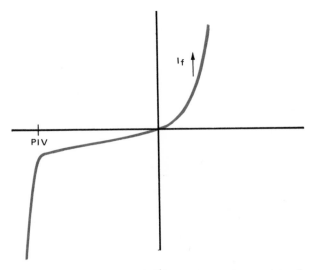

Fig. 16-10. This graph is plotted from two important ratings of a diode, peak inverse voltage (PIV) and forward bias (I_f).

GRAPHING

In plotting a voltage and current graph for diodes, Fig. 16-11, the line going from left to right (on the horizon) is called the "horizontal axis." The line going up and down is called the "vertical axis." Voltage is shown on the horizontal axis; current is shown on the vertical axis. By studying the plotted curves, you can tell how much voltage and current each diode can stand without being ruined.

DIODES IN SERIES

Some circuits use two diodes (rectifiers) in series, Fig. 16-12. If one of these diodes has a rating I_f equal to 3 amps, and the second diode has a rating I_f equal to 1 amp, what is the safe amount of current that you could pass through the diode? Note that one of the diodes can handle only 1 amp. If we tried to pass 3 amps through it, this diode would be ruined. Therefore,

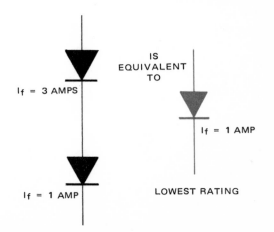

Fig. 16-12. Series-connected diodes will pass an amount of current that is equivalent to the diode of the lowest rating.

figure that two diodes in series can handle current equivalent to the diode with the lower rating.

Next, take these same diodes and look at their PIV, Fig. 16-13. Also consider their PIV in series. Bear in mind that when two diodes or rectifiers are connected in series, their PIV is the sum of the two. In this case, our PIV will increase to 900 (600 + 300).

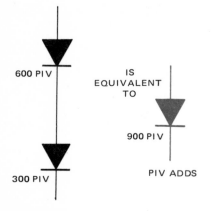

Fig. 16-13. "PIV adds" when diodes are connected in series.

DIODES IN PARALLEL

If we hooked those same 1 amp and 3 amp diodes in parallel instead of in series, we would get different results. The I_f value for the parallel hookup would add up to 4 amps, as shown in Fig. 16-14. This means that if there is a need for more current than a given diode can take, we can hook two diodes in parallel to handle the current load.

Looking at the PIV for this parallel hookup, Fig. 16-15, you can see it now has the rating of the diode

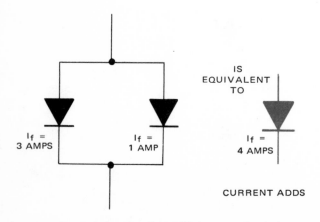

Fig. 16-14. "Current adds" when diodes are in parallel.

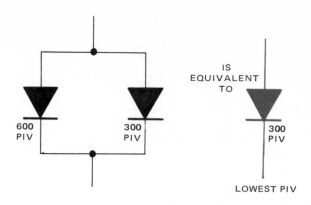

Fig. 16-15. Diodes connected in parallel will have peak inverse voltage equivalent to the diode with the lower PIV rating.

with the lower value. Anytime we hook two diodes in parallel, the hookup can handle more current, but we must watch for low PIV. *If the circuit voltage is higher than the PIV rating, the diode will pass current in the reverse direction.* Relating this to the dam we discussed earlier, it is as if the water is pumped back up to the top of the dam.

SYMBOL POLARITY

The symbol for a diode, Fig. 16-16, is labeled plus and minus. The minus side is known as the "cathode." The plus side is called the "anode." You might ask, "What difference does it make?" It *is* important because not all diodes or rectifiers are marked with a plus or a minus.

Fig. 16-16. Diode symbols should always be drawn with polarity labels in mind. The wide end of the symbol faces the direction of foward bias.

However, by using an ohmmeter, you can find out which end of the diode is cathode and which is anode. With this information, you can put the diode into the circuit the right way. Otherwise, you might hook the diode the wrong way and block current flow.

DIODE POLARITY

If a diode is not marked for polarity, hook the positive lead from the ohmmeter to one end of the diode and the negative lead to the other end. See Fig. 16-17. If you get a low resistance reading, you are forward biasing the diode. Polarity is correct. Mark a minus sign on the end of the diode hooked to the negative lead of the ohmmeter. Mark a plus (anode) sign on the other end of the diode.

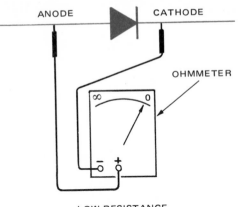

Fig. 16-17. To check polarity, make an ohmmeter test across the diode. Low resistance indicates that diode is forward biased.

If the test reading shows high resistance, Fig. 16-18, you are reverse biasing the diode. Polarity is incorrect. Simply reverse the lead connections and retest for forward biasing. Then, mark the diode ends for correct polarity.

It is important when making ohmmeter tests that you know which of the terminals is plus and which is minus. Although the outside of the meter is marked with a plus and minus, this polarity might not be correct when compared with the battery. The safest way to find out is to open the back of the meter and see where the leads from the battery are connected (plus to plus, minus to minus).

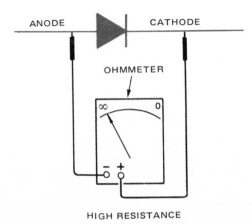

Fig. 16-18. If the ohmmeter test for diode polarity results in high resistance, the diode is reverse biased.

MEASURING RESISTANCE

Diodes do not have a fixed resistance. Resistance will change, depending on the amount of voltage put on the diode. To help you understand this reaction to applied voltage, simply measure three or four different diodes for their resistance value at various voltages. If you use a **VTVM** (vacuum tube voltmeter), instead of a **VOM** (volt ohm milliammeter), you will have more resistance scales to work with in plotting the resistance of each diode.

Fig. 16-19 shows how to connect the ohmmeter test leads across the diodes or rectifiers to make resistance tests. As you move the range selector switch from R x 1 to R x 10, you are decreasing the voltage applied by the ohmmeter to the diode. Note that each diode or rectifier has different resistance values, depending on where the range selector switch is set.

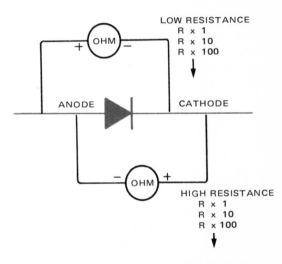

Fig. 16-19. To make resistance checks on diodes, connect ohmmeter test leads as shown and set the range selector switch to various positions given on a VTVM.

The results of resistance tests on 1N604, 1N519 and 408A diodes on all the ohmmeter ranges are plotted on the chart shown in Fig. 16-20. This chart proves that solid state devices are not fixed resistors. Therefore, current will increase or decrease, depending on the amount of voltage being applied. How much the current will change is not fixed by Ohm's Law because, in some cases, the law does not apply to solid state devices.

SMALL SIGNAL VS POWER DIODES

Under test, small signal diodes show a big difference between forward resistance and reverse resistance. Diodes used in high current circuits do not. Therefore,

	FORWARD	REVERSE	FORWARD	REVERSE	FORWARD	REVERSE
R x 1	13	∞	7	∞	10	∞
R x 10	45	∞	37	∞	72	∞
R x 100	200	∞	230	∞	560	∞
R x 1000	1100	700K	1.5K	∞	4.1K	∞
R x 10K	5K	420K	9K	ALMOST ∞	30K	∞
R x 100K		160K	50K	7 MEG	200K	ALMOST ∞
R x 1 MEG				6 MEG	1.1 MEG	70 MEG
	1N604		1N519		408A	

Fig. 16-20. Chart shows results of resistance tests made by forward biasing, then reverse biasing three different diodes.

when tests on the 408A unit resulted in only small differences in forward and reverse resistance values, it was determined to be a power type diode (rectifier). The 1N604 and 1N519 units, on the other hand, had big differences in resistance values, so they were considered to be small signal diodes. These typical differences in resistance are normal for diodes of both types. See Fig. 16-21.

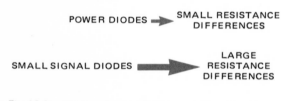

Fig. 16-21. Diode resistance differences help to identify power diodes and small signal diodes.

IDENTIFICATION NUMBERS

The level of electron flow depends on the size of the diode. However, there are many sizes of diodes; so many, in fact, that they are given numbers rather than names. Most diodes start with 1N, followed by other numbers. Examples include: 1N40, 1N363, 1N946 and 1N3750.

SUBSTITUTION

There are thousands of diodes used in circuits. Since many companies make diodes, there are many ways used to label them. In addition, you must be able to find a subsitute for each diode number. This is necessary because some diodes shown in circuits are not available in all electronic stores. Therefore, most companies publish a cross-reference (substitution)

catalog, Fig. 16-22. This catalog makes it easier for you to change from one diode manufacturer to another and still have your circuit work. Be aware, however, that there may be a slight difference in I_f and PIV.

I.C. SAYS:

"If you want to substitute a diode and cannot find one with values exactly as the one in your circuit, make sure that you use one with higher I_f and PIV ratings."

If you go to diodes with smaller ratings, you run the risk of ruining them with too much current or too high voltage once they are in the circuit.

COLOR CODE

Basically, diodes are made from one of two materials. One type of diode uses germanium and the other uses silicon. Germanium carries smaller amounts of current than silicon. The silicon diodes (sometimes called "rectifiers") are larger and carry heavier loads. Because of this, they even look different.

The smaller germanium diodes will often have a color code painted on the side, Fig. 16-23. This color code is the same one used for resistors. To read the code, start with 1N, followed by whatever the colors tell you "by the numbers." If the colors on our diode are brown, orange and red, Fig. 16-24, this diode is a 1N132 germanium diode. By following our resistor color code, brown is 1, orange is 3 and red is 2. You can see how this color code helps you to label diodes.

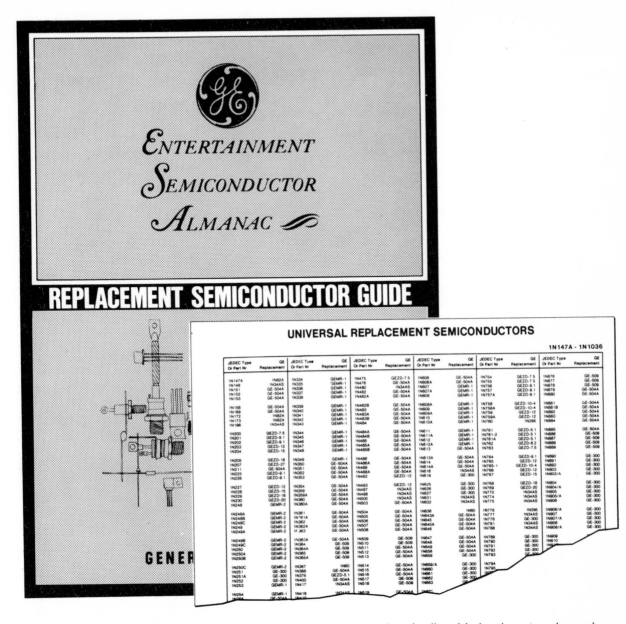

Fig. 16-22. Parts manufacturers and suppliers publish cross reference catalogs that list original equipment numbers and replacement parts numbers.

GERMANIUM DIODE

Fig. 16-23. Small germanium diodes usually have bands of color that help to identify the diodes by number.

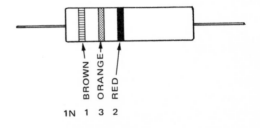

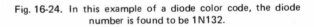

1N 1 3 2

Fig. 16-24. In this example of a diode color code, the diode number is found to be 1N132.

USES

Now we will see some of the things that diodes can do when hooked into a circuit. In Fig. 16-25, we have a

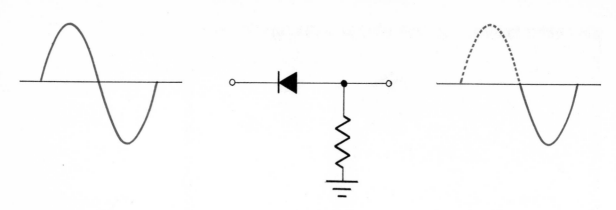

Fig. 16-25. If a diode hooked into a circuit is followed by a resistor to ground, it will "clip" off the top half of the sine wave.

diode followed by a resistor going to ground. This will cut off the top half of our sine wave and let only the bottom half through. If we reverse the direction of the diode, Fig. 16-26, we will get the top half of our sine wave and cut off the bottom half. Because each of these diodes is in series with the resistor going to ground, it will clip off one-half of the sine wave. For this reason, diodes used in this way are called "clippers."

As we go through more circuits in solid state, you will be seeing more letters used under both I and V. For example, in Fig. 16-27, the forward current through the diode in series with this resistor will result in a clipper potential (V_{CP}), which is about 0.6 volts for a silicon diode.

FULL WAVE RECTIFIER

If you change the clipper circuit shown in Fig. 16-27, by adding another diode to ground in the opposite direction, we will get a square wave. See Fig.

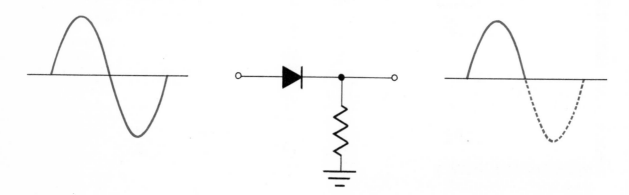

Fig. 16-26. If the diode in Fig. 16-25 is connected in reverse, it will "clip" off the bottom half of the sine wave.

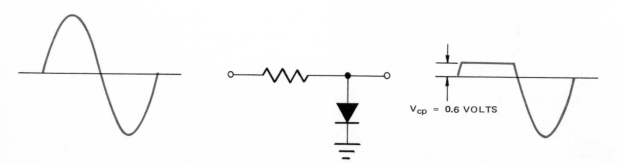

$V_{CP} = 0.6$ VOLTS

Fig. 16-27. The term V_{CP} is related to drawing sine waves and writing formulas. It stands for "voltage/clipper potential."

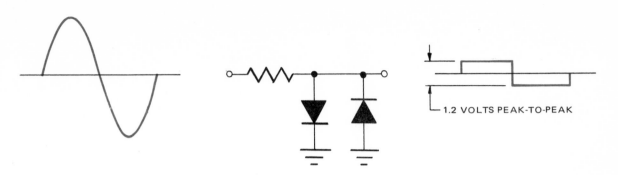

Fig. 16-28. If we make a change in Fig. 16-27 by adding another diode to ground in the opposite direction, we will get a square wave.

16-28. This double shunt clipper circuit gives us a clipped wave of about 1.2 volts peak-to-peak when using a silicon diode.

However, one of the biggest uses for diodes is found in a circuit used to change ac to dc. This circuit usually is known as a "full-wave rectifier." Some people also call it a "bridge rectifier circuit." As you can see in Fig. 16-29, there are four diodes (rectifiers) connected from the secondary of a transformer. Note that this transformer has a 12.6 volts, 1.2 amp output (same type mentioned in earlier chapters). To change ac to dc, the rectifiers act in the way shown in Figs. 16-30, 16-31 and 16-32. Actually, there are hundreds of other ways in which you can use diodes in circuits.

SOLDERING

You must be very careful when installing diodes, especially germanium diodes. A soldering iron can be used to connect a diode in a circuit, but it could give

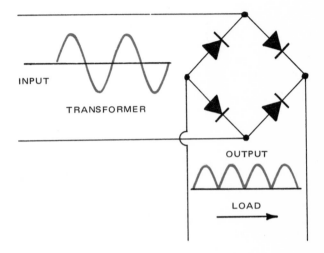

Fig. 16-30. Schematic shows full wave rectifier in a circuit. Note the direction of forward bias through the four diodes.

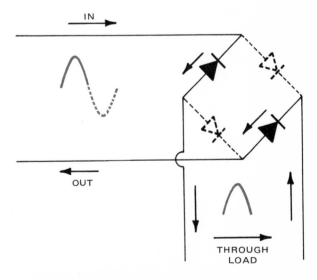

Fig. 16-31. The first half cycle of the conversion from ac to dc is illustrated. Arrows indicate current flow. Note that two diodes are blocked.

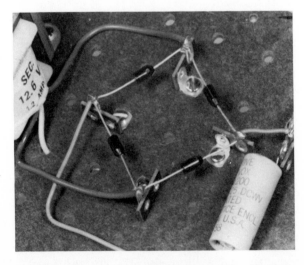

Fig. 16-29. This full wave rectifier (bridge rectifier circuit) uses four diodes to convert alternating current to direct current.

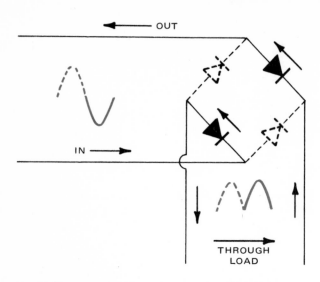

Fig. 16-32. The second half cycle of the ac/dc conversion is presented. Note blocked diodes and new path of current flow.

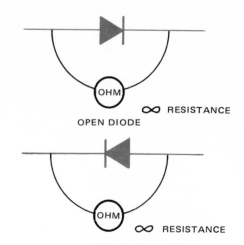

Fig. 16-33. If resistance checks with an ohmmeter indicate high (∞) resistance in both directions, the diode is open.

ZENER DIODE

Fig. 16-34. This symbol is used in drawings to represent a zener diode. Again, the wide end of the symbol faces direction of forward bias.

off enough heat to destroy the diode. To avoid this, use a heat sink (metal device that dissipates heat) between the diode and the soldered connection. This little trick will save many dollars and much time.

Even the so-called expert who tries to solder a diode connection without using a heat sink has ruined many diodes in the process. Very few packaged electronic components or plans for building circuits will warn you about this possibility of overheating the diodes. *Always use a heat sink when soldering diode connections.*

Overheating diodes or rectifiers can cause one of two things to happen:

1. It may open the diode.

How would you know this? Remember how we measured the resistance of a diode in two directions? In one direction, we found the resistance to be very high. In the opposite direction, the resistance was very low. If your diode or rectifier is open, the resistance will be *very high (∞) in both directions,* Fig. 16-33.

2. It may short the diode.

Again, your resistance check will prove this by measuring *low resistance in both directions.* Generally, however, when a diode connection is soldered without a heat sink, the diode will open. High currents in a circuit will do the same thing.

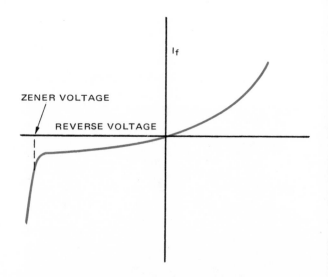

Fig. 16-35. The curve on this graph shows that reverse voltage falls sharply at zener voltage.

ZENER DIODE

Another fairly popular type of diode is one called a "zener diode." Its symbol is shown in Fig. 16-34. This type of diode is used to regulate the voltage. Fig. 16-35

illustrates the reason it can be used in that way. Note that the voltage in the reverse direction reaches a point (zener voltage) where the current in the reverse direction will go to a very high level. Up until that time, the reverse current was almost constant.

As we reverse bias this zener diode, the current will remain at a low level. Finally, you will get to the point where any increase in voltage will cause the current to go way up. This "zener point" is often called the "saturation point" of the diode. Actually, the voltage level where saturation of the diode results will range from a few volts to over 400 volts.

BREAKDOWN VOLTAGE

A zener diode has two important specs (specifications):
1. Zener breakdown voltage.
2. Power rating.

You recall from Chapter 4 that power is equal to the voltage times the current ($P = E \times I$). We can use this information to find the maximum current we can use in a circuit with a zener diode. First, look up the power rating for the diode. Next, check the operating voltage of our circuit, and we can use the following formula:

$$\text{maximum current} = \frac{\text{power rating}}{\text{operating voltage}}$$

I.C. SAYS:

"Never go above the wattage that a zener diode can handle, otherwise it will be ruined."

To use the formula for finding the maximum current, assume that we have a 1 watt zener diode with a breakdown voltage of 40 volts:

$$I_{MAX} = \frac{\text{power rating}}{\text{operating voltage}}$$

$$I_{MAX} = \frac{1 \text{ watt}}{40 \text{ volts}}$$

$$I_{MAX} = 25 \text{ mA}$$

As a follow-up, you can hook a milliammeter in series with the diode to check the maximum current for a diode having this power rating.

ZENER DIODE USED AS A VOLTAGE REGULATOR

We can use a zener diode as a voltage regulator, Fig. 16-36. Note that the zener diode is used as a shunt.

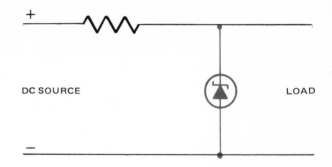

Fig. 16-36. Schematic reveals how a zener diode can be used as a voltage regulator.

That is, it is hooked up in parallel with our load. To fix the operating load, we must place the resistor ahead of the diode. This will also control the range of load current and any changes in voltage from the input without going over the power rating of the zener diode.

When voltage from the input increases, current through both the reverse biased zener diode and load will also increase. However, as the zener current increases, its resistance decreases and still more current is passed through it. See Fig. 16-37. The voltage drop across our resistor will also go up. This lets us keep the voltage across our load at an almost constant level.

If the voltage across the load should drop, the resistance of the zener would increase and keep the

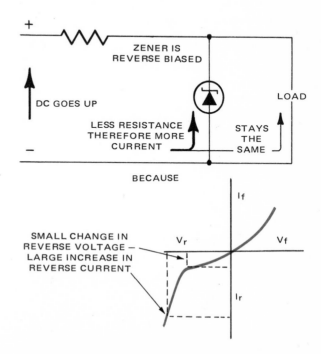

Fig. 16-37. This schematic serves to explain how a zener diode keeps constant voltage across the load.

voltage across the load almost constant. In this way, the zener diode can keep an almost constant output to the load even though the input is always changing.

CHECKING ZENER DIODES

You cannot test a zener diode the same way you checked other diodes. Since the resistance of a zener changes in a different way, we must use an oscilloscope to check it.

RESEARCH

There are other types of diodes besides the ones mentioned in this chapter. They have names like tunnel diode, light emitting diode, Fig. 16-38, and many others. Research work is always going on to find new

Fig. 16-39. An engineer is doing research work in electronics. (Autonetics Div. of North American Rockwell)

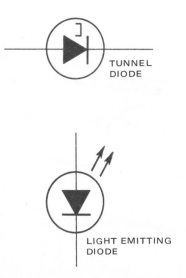

Fig. 16-38. These symbols are used in circuit drawings to represent tunnel diodes and light emitting diodes.

and better solid state devices. For example, in Fig. 16-39, work is being done on a 10 micron photo-detector. It is this type of research that has made electronics such a rapid moving field.

TEST YOUR KNOWLEDGE

1. A diode is identified by labeling one end the _____ and the other end the _____.
2. In order for a diode to conduct, it must be properly _____.
3. How much current will two diodes in series be able to conduct?
4. PIV stands for _____.

5. What would the color code of a 1N420 diode look like?
6. Draw the symbols for a diode and a zener diode.
7. Before soldering, a diode should be protected with a _____.
8. Diodes are usually thought of as carrying current in the _____ range, while rectifiers carry current in the _____ range.
9. The anode is _____.
10. What results would you expect to get if you were to make a resistance check on a diode?
11. If a diode must be replaced, what is the best thing to do if the exact values could not be found for a replacement part?
12. Give one use for a zener diode.

THINGS TO DO

1. Get as many different types of diodes as possible. Label each lead. Are all of them identified the same way?
2. Why is a diode called a solid state device?
3. Take a diode apart. Can you see how it has been made? If you were to use a microscope, what other things could you see?
4. Why are there so many different types of diodes? How is a manufacturer able to control the process for making so many different types?
5. How do you know which type of diode to use in a circuit? Make a list ranging from those carrying low current to those carrying high current.

Diodes

6. How do they make diodes with the glass cases? How do they get the glass around it? How do they put the colored bands on the glass without breaking it?

7. Try an experiment with a couple of diodes. Measure the amount of heat it takes to ruin each type. Plot the results to show how much difference there is in each type.

8. How much overload current will a diode take before it is ruined? Does it make any difference whether it is silicon or germanium?

9. Using rectifiers, build a full wave bridge. How many ways can you filter the output?

10. How many different circuits can you find using zener diodes? Do they fall into any one type?

11. Take a zener diode apart. Does it look any different from a rectifier or standard type diode?

12. How much voltage is needed to break down a zener diode? How much overload will it take before it is ruined?

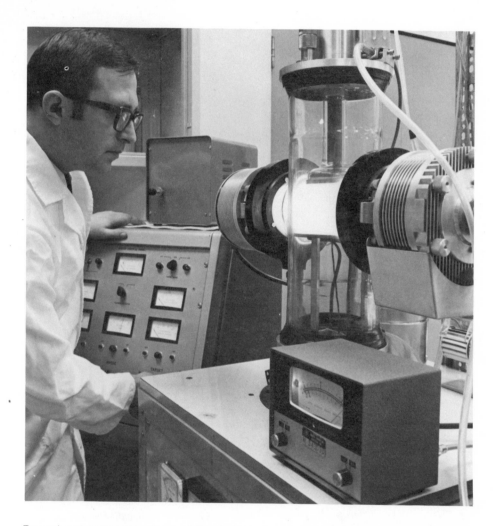

Research and development of new solid state devices is a continuing process that results in improved electronics. (Autonetics Div. of North American Rockwell)

227

Chapter 17
TRANSISTORS

In this chapter, we will continue our study of solid state devices. We covered diodes in Chapter 16. A diode, you recall, will permit electron flow in one direction and block the flow from the other direction. Now, we will study transistors. *A transistor is a solid state device used for switching and/or amplifying the flow of electrons in a circuit.*

Research done by three scientists not too long ago led to the discovery of the transistor, Fig. 17-1. Although the first transistor looks rough by today's standards, its principle of operation is still the basic idea behind many of the solid state devices which we now use, Fig. 17-2.

TRANSISTOR TYPES

There are two basic types of transistors, the "switching" type and the "amplifying" type. No matter which of the thousands of transistors you use, they either switch or amplify. Some transistors can do both, switch *and* amplify.

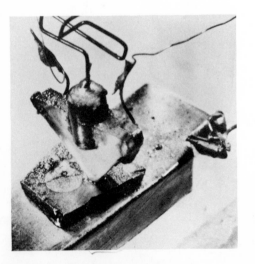

Fig. 17-1. The first transistor was designed to "amplify" by passing electrical signals through a solid, semiconductor material.

Fig. 17-2. Today's transistors look more compact and attractive, but the operating principle is still based on the idea behind the first transistor.

SWITCHING

To help you understand what we mean by switching and amplifying, we will use the light switches in your home as examples of these functions.

In the first example (switching), Fig. 17-3, a light switch must be flipped in order to let the current flow to the light bulb or to stop current flow across the switch terminals. Normally, there are only two possible results from flipping the switch: the light bulb will light up or it will shut off. Any other result would be caused by an electrical problem.

ANSWER BOX PROJECT

Your second example (switching) is in building an answer box, Fig. 17-4, similar to the ones you see on television quiz shows. A question is asked of two people. The first one to push the button is the one who gets the chance to answer the question.

By its construction, the answer box can be used to prove who pushed the button first. The circuit for the answer box has a transistor connected to each of two buttons, but only one transistor can conduct at any given time. Therefore, the person with the faster finger will always go first. Even if the second person hits the

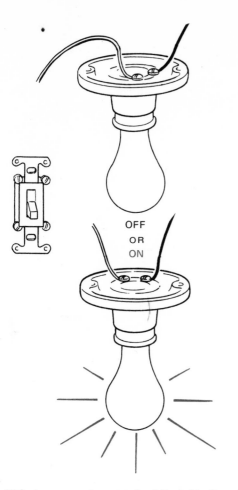

Fig. 17-3. In an everyday example of "switching" an electrical signal, a light switch must be flipped to make the bulb light up or shut off.

Fig. 17-4. This quiz show answer box is an interesting project that illustrates how transistors can switch electrical signals. (Project by Larry Wood)

button a fraction of a second later, the light will not come on.

Obtain the following materials and assemble them according to Fig. 17-5 and the construction procedure given:

No.	Item
1	6-9 volt battery.
2	6-9 volt light bulbs.
2	NPN transistors.
2	2.2K resistors.
2	Push button switches.
1	4 in. x 6 in., 18 gage steel box.
1	4 in. x 6 in. circuit board.
	Connecting wires.

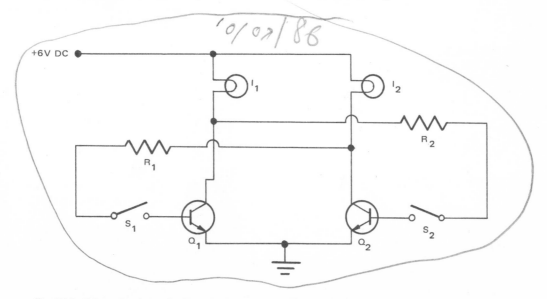

Fig. 17-5. Schematic shows hookup for answer box. Parts include: B_1—6-9V battery. $I_{1,2}$—6-9V bulbs. $Q_{1,2}$—NPN transistors. $R_{1,2}$—2.2 K resistors. $S_{1,2}$—Switches.

CONSTRUCTION PROCEDURE

To build the answer box:

1. Connect positive leads of lights, using wire or alligator clips.
2. Connect a resistor to negative lead of each light.
3. Connect each resistor to one terminal of a switch.
4. Connect other leads of switches to bases of two NPN transistors.
5. Connect emitters together; also attach a wire to the emitters, forming a ground.
6. Connect collector of transistors to point where resistor meets light of other wire (see Fig. 17-5).
7. Attach power supply's positive terminal to either light; connect negative terminal to ground.
8. Make a metal box large enough to hold circuit board and complete circuit (approximately 4 in. x 6 in. x 1 1/2 in.).
9. Drill four 5/8 in. holes in box and install lights and push buttons (see Fig. 17-4).
10. Solder all terminals on circuit board and secure board in box.

Use the answer box in conjunction with flash cards, questions from the teacher or any other quiz program. If this project is mass produced, it will be possible to have a class-wide contest. The teacher asks the questions and the students compete in pairs to guess the answers.

AMPLIFYING

In your third example of switching and amplifying, a dimmer switch, Fig. 17-6, is used in conjunction with the light bulb to demonstrate "amplification of an electrical signal." With the dimmer switch, you can turn the control knob so that the amount of light given off by the light bulb will vary from dim to very bright.

You also can turn the dimmer switch control knob to shut off the light bulb. Then, as you turn back the knob, you will reach a point where the light bulb will come on. From that point on, the more you turn the control knob, the brighter the light bulb will get.

SWITCHING APPLICATIONS

Switching transistors are used in circuits which are either "on" or "off" (similar to lighting circuit just described). These transistors are found in computers that perform "counting" operations. In counting circuits, the signal is either present or not present. There are no "in-betweens."

The computer signal can be in the form of a hole

Fig. 17-6. In an example of "amplifying" an electrical signal, turning the dimmer switch control knob makes the bulb give off light ranging from dim to very bright.

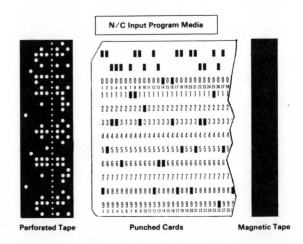

Fig. 17-7. In computer work, the signal can be in the form of holes punched in a tape, or holes in a card, or electronic pulses on a magnetic tape.

punched in tape or a hole in a card or an electronic pulse on a piece of magnetic tape, Fig. 17-7. This last application uses almost the same type of magnetic tape that you use in a tape recorder. Although you cannot see the hole or pulse, it *is* there.

We can take the punched tape, for example, and run it through a tape reader, Fig. 17-8, to count the holes.

Fig. 17-8. "Switching transistors" are used in tape readers. The signal is created by way of holes punched in a tape. (Superior Electric Co.)

These holes could represent people in a census, money in an account or signals sent to a turret drill, Fig. 17-9. Can you imagine drilling a piece of steel because a hole in a tape sent the signal by means of a transistor? The signal ordered the machine to drill a hole or not to drill a hole. The transistor performed the switching operation.

AMPLIFYING APPLICATIONS

Amplifying transistors are used in places where the amount of signal is important. You would use this type of transistor for a stereo, where you want to amplify (strengthen) a radio signal, Fig. 17-10. Because these signals are small, they must be made larger so you can hear them over a speaker.

The amplifying transistor must be able to go from "off" through an "on" range of increasing volume, similar to increasing bulb brightness with a dimmer switch. The radio signals must go through a few transistors to increase the size of the signals. These transistors are the final power type which feed the signal into the speaker for the music or voices (audio) that we hear.

Fig. 17-9. This tape-controlled turret drill will perform specific drilling operations ordered by the computer tape. (Superior Electric Co.)

Fig. 17-10. Amplifying transistors (circled) are used in radios to build up weak signals. (McIntosh Laboratory Inc.)

PNP AND NPN

Usually, when you hear the word "transistor," certain letters are used with it to identify the type of transistor being discussed. The first type to be considered is the PNP transistor. In working with circuit drawings, you will find that there are two ways of drawing the PNP transistor symbol, Fig. 17-11. Both symbols are in popular use, and they stand for the same device.

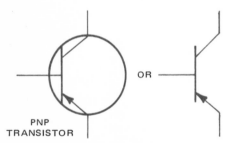

Fig. 17-11. These two PNP transistor symbols are commonly used in circuit drawings.

The second transistor for study is the NPN type. Its symbol also can be drawn two ways, Fig. 17-12. You recall that these same letters, PN, were used with reference to diodes in Chapter 16. Although the transistors are drawn as large pieces of P and N materials, Fig. 17-13, these junctions are so small that it takes a microscope to see them.

TRANSISTOR CASES

When you see transistors on circuit boards, Fig. 17-14, you will see many different kinds. Some look like small thimbles. Others have shiny metal cans. Still

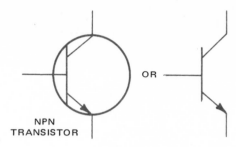

Fig. 17-12. Either of these two ways of drawing NPN transistor symbols is acceptable.

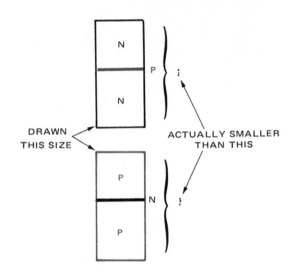

Fig. 17-13. Drawings of transistor P and N materials are much larger than the actual materials.

Fig. 17-14. Circuit boards with transistors and other solid state devices are pictured in a stockpile at this assembler's station. (Acme Electric Corp.)

others have black containers. This presents a real problem for someone just starting out in electronics. Before you can solder transistors into a circuit (using a heat sink, of course), you must identify the leads.

LEAD IDENTIFICATION (SYMBOL)

Transistor leads are known as the emitter, base and collector. If you have a PNP transistor, its leads (wires) will be labeled as shown in Fig. 17-15. The letters PNP also let you know which way the arrow is pointing. In this case, PN could mean <u>P</u>OINTING <u>IN</u>.

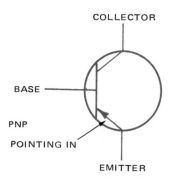

Fig. 17-15. Marking the leads on a PNP transistor symbol.

If you look at an NPN transistor, Fig. 17-16, think of <u>N</u>OT <u>P</u>OINTING <u>IN</u>. Note also that the lead with the arrow is always the emitter. Knowing this should be a big help when you start working with transistors.

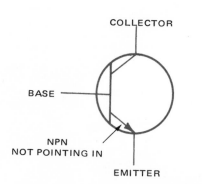

Fig. 17-16. Marking the leads on an NPN transistor symbol.

DIRECTION OF FLOW

Since a transistor can either switch or amplify, we need to know what to do to the leads so they can conduct. The first example, Fig. 17-17, uses an idea similar to the light switch. In order to turn on the light,

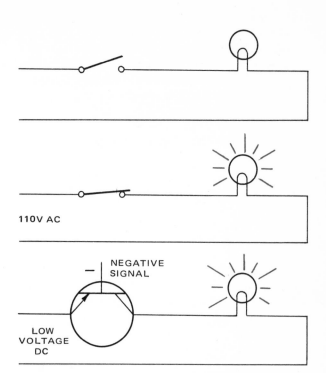

110V AC

Fig. 17-17. Examples are given for various switching operations. 1—Switch open, light bulb off. 2—Switch closed, light bulb on. 3—Transistor on, light bulb on.

you have to flick the switch with your hand. If you use a PNP transistor in place of the light switch, you can do the same thing.

In order to turn on the PNP transistor, we must put a negative signal on the base. When this happens, the electrons will flow through the transistor. When working with a PNP transistor, this flow of electrons is from the collector to the emitter, Fig. 17-18. As with the light switch, we must use some outside source to allow the flow of electrons. Note that the flow of electrons from the collector to the emitter is against the direction

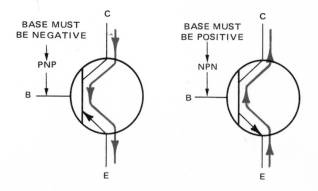

Fig. 17-18. When working with transistors, the flow of electrons is opposite the way the emitter arrow points.

indicated by the arrow on the emitter.

With an NPN transistor, we must put a positive signal on the base to make it conduct. This will let the electrons flow from emitter to collector. Again, the flow of electrons is against the direction in which the arrow on the emitter points.

LEAD IDENTIFICATION (ACTUAL)

The next thing you will want to know is how to identify the leads on an actual transistor. The top two transistors shown in Fig. 17-19 are the small signal type. The first step toward identifying them is to turn the transistor upside down so you can see the leads at the bottom. Next, try to identify the leads as emitter (E), base (B) and collector (C).

If the transistor has a small tab, it usually is made with the emitter lead closest to the tab. See Fig. 17-19. Then, going around the transistor in clockwise direction, the next lead is the base and the last lead is the collector.

If the transistor you are trying to identify is laid out in a half round shape with a flat part, Fig. 17-19, place the flat part up. In this position, the leads are arranged as emitter, base and collector in left to right order. We say "usually" because this is not always true. Some manufacturers label their leads in different ways. The safest and best way to check transistor leads is by using an ohmmeter (to be explained later in this chapter).

The third type of transistor, shown in Figs. 17-19 and 17-20, is a power transistor, with a metal plate on the back to take away the heat. In order to label these leads, lay down the transistor with the metal plate up

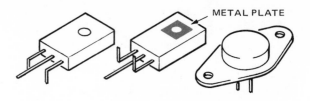

Fig. 17-20. Power transistors usually are the flat, plastic type, having leads on the bottom or back. The metal plate dissipates excess heat.

and away from you. Then, the left hand lead is the emitter, the middle one is the collector and the right hand lead is the base.

The other power transistor shown in Figs. 17-19 and 17-20 also must be turned upside down to identify its leads. Then, position the transistor so that the two leads are closer to the left hand side. Now, you can label them. The top lead is the emitter, the bottom one is the base. What happened to the collector? This type of transistor has its collector connected to the case. Therefore, the whole bottom of the transistor is the collector.

FLASHING LIGHT PROJECT

There are a lot of uses for a small flashing light, Fig. 17-21. Build the circuit shown in Fig. 17-22 and place it in a small box or soap dish. The flashing light can be used inside a paper mache animal and placed on a table. Halloween will give you another chance to use this flasher. Turn it on and drop it inside a cut-out pumpkin or jack-o-lantern.

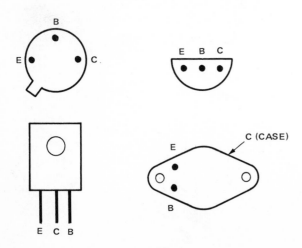

Fig. 17-19. Typical transistor shapes include: Top—Two small signal types. Bottom—Two power types. Labels are: E—Emitter. B—Base. C—Collector.

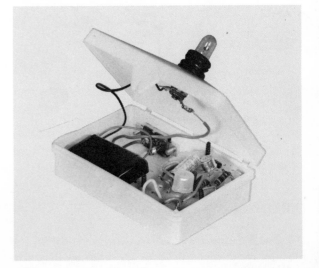

Fig. 17-21. Circuitry for a small blinking light is contained in a soap dish. (Project by Dick Butcke)

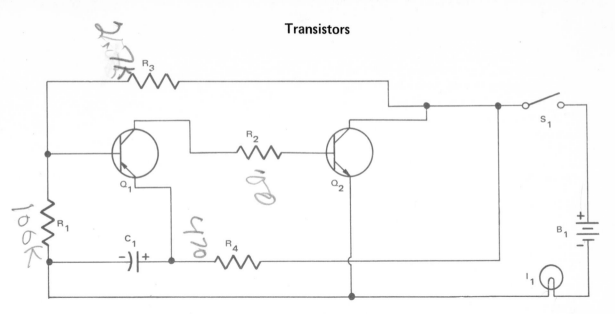

Fig. 17-22. Schematic shows hookup for a small blinking light. Parts include: B_1—Battery. C_1—Capacitor. I_1—Lamp. $Q_{1,2}$—Transistors. $R_{1,2,3,4}$—Resistors. S_1—Switch.

Obtain the following materials and assemble them according to Fig. 17-22 and the construction procedure given:

No.	Item
1	9V dc battery.
1	100-μF 15WVDC electrolytic capacitor.
1	4 1/2V dc lamp.
1	GE 2 transistor (Q_1).
1	GE 8 transistor (Q_2).
1	100 K resistor (R_1).
1	100 Ω resistor (R_2).
1	2.7 K resistor (R_3).
1	470 resistor (R_4).
1	SPST switch.
1	Perforated circuit board.

CONSTRUCTION PROCEDURE

To build the flashing light:
1. Lay out wiring of schematic on a breadboard.
2. Make sure that flasher works.
3. Transfer wiring to perforated board.
4. Solder all joints. Tape all joints that might short out flasher.
5. Place flasher in desired container.

HEAT SINK

Heat sinks are used in many solid state applications to dissipate (take away) heat from heat-sensitive areas where damage might otherwise occur. Power transistors are one of those applications.

Power transistors can carry much more current than the small signal type. However, the larger current causes extra heat, so we must do something to help the transistor get rid of that heat. We can do this by attaching the transistor to a heat sink, Fig. 17-23. This is not the same kind of heat sink we used when soldering diode leads, but it does the same job. It takes heat away from the transistor so it will not burn up.

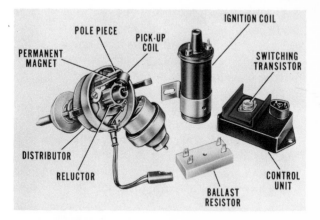

Fig. 17-23. Chrysler's electronic ignition system uses a power switching transistor mounted on a heat sink (in color).

Fig. 17-24 shows how to mount a power transistor and mica washer on a heat sink. To install the washer:
1. Place a small dab of silicone grease on underside of transistor.
2. Install mica washer on transistor.
3. Put a small dab of silicone grease on washer.
4. Attach assembly to heat sink.

If you use screws to fasten the assembly to the heat sink, be sure to insulate the screws with extruded fiber or plastic washers, or use nylon screws and nuts. Otherwise, the circuit will be affected when the assembly is installed.

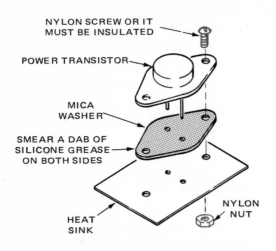

Fig. 17-24. When installing a power transistor on a heat sink, a mica insulator must be used between the transistor and heat sink. Attaching screws also must be insulated.

PNP IDENTIFICATION

Suppose you have a transistor to install, but cannot tell whether it is a PNP or NPN type. First, make sure that you know which is the positive lead and which is the negative lead from your ohmmeter. If necessary, take off the back of the instrument and check the polarity of the battery against the lead connections (positive-to-positive, negative-to-negative).

To test the transistor for type:

1. Hook the positive lead from the ohmmeter to the base of the transistor, Fig. 17-25.
2. Connect the negative lead from the ohmmeter first to one transistor lead, then to the other.
3. If both readings show high resistance, hook the negative ohmmeter lead to the base of the transistor, Fig. 17-25.
4. Connect the positive lead from the ohmmeter first to one transistor lead, then to the other.
5. If both readings show low resistance, you have a PNP transistor.

NPN IDENTIFICATION

Suppose your ohmmeter tests show high resistance with the negative ohmmeter lead connected to the base of the transistor and the other lead is switched from transistor lead to transistor lead. See Fig. 17-26.

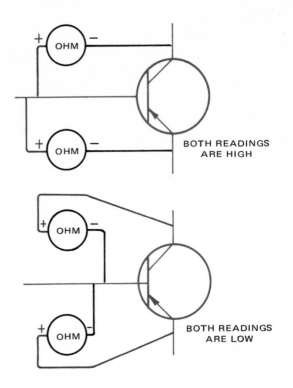

Fig. 17-25. Results of these ohmmeter tests prove that transistor is the PNP type.

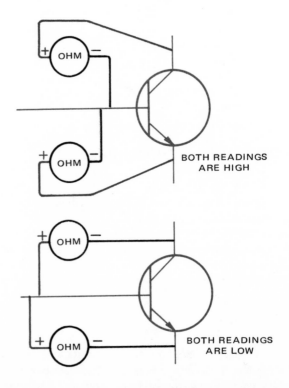

Fig. 17-26. Results of these ohmmeter tests prove that transistor is the NPN type.

Continue testing as follows:
1. Reverse the ohmmeter leads, connecting the positive lead to the base of the transistor.
2. Connect the negative lead from the ohmmeter first to one transistor lead, then to the other.
3. If both readings show low resistance, you have an NPN transistor.

GERMANIUM OR SILICON?

After you have determined whether your transistor is PNP or NPN, the next thing you need to know is whether it is silicon or germanium. Bear in mind that a germanium transistor will carry less current than a silicon transistor.

I.C. SAYS:

"Do not mix up the terms 'silicon' and 'silicone.' Silicon is an element used in making transistors. Silicone is the grease used when installing heat sinks."

A quick check can be made with an ohmmeter to determine whether a transistor is made from silicon or germanium. In the test of a PNP transistor shown in Fig. 17-27, first connect the ohmmeter negative lead to the collector and the positive lead to the emitter. With this hookup, you will get a high resistance reading from emitter to collector.

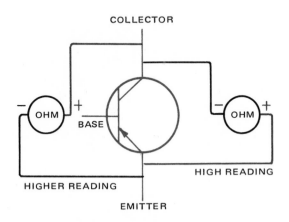

Fig. 17-27. Ohmmeter test results prove whether the PNP transistor under test is made from germanium or silicon.

Then, reverse the ohmmeter lead connections and the resistance reading will go even higher. If you can read the ohms on the meter scale, you have a germanium transistor. If the reading is in the megohms-to-infinity (∞) range, you have a silicon transistor.

To make these same checks on an NPN transistor, connect the ohmmeter leads as shown in Fig. 17-28. With the ohmmeter positive lead on the collector and the negative lead on the emitter, the resistance will be high. Switching the leads, then, will make the resistance even higher. Again, if you can read the ohms on the meter scale, you have a germanium transistor. If you cannot read the resistance value because it is near infinity, you have a silicon transistor.

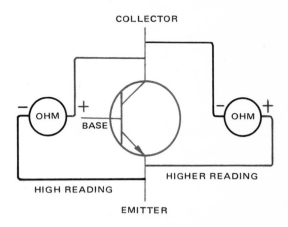
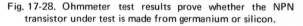

Fig. 17-28. Ohmmeter test results prove whether the NPN transistor under test is made from germanium or silicon.

Do not mix up transistor tests with diode tests. Though the procedures are similar, the results do not mean the same thing. From the transistor test examples given, you can see why it is important to identify the transistor as PNP type or NPN type.

MANUFACTURING TRANSISTORS

At this point of study, you probably would like to know how transistors are made. There are many steps in their construction. Back in the 1950s, transistors were assembled by hand operation, Fig. 17-29. It was a slow process, which is why transistors were very expensive to manufacture.

Today, transistors have their start as a huge silicon rod grown in a special furnace, Fig. 17-30. Then, wafers are cut from the silicon ingot, Fig. 17-31, and placed in a special machine where they are polished to a mirror finish. Following this, the wafers are put into special reactors, Fig. 17-32, where controlled vapors apply very thin coating (doping) on the wafers. These vapor coatings "make up" the transistor, diffusing (distributing evenly) the P and N material to make either PNP or NPN silicon transistors.

Next, the wafers are cut up and completely covered

Fig. 17-29. In early 1950s, workers used hand-operated fixtures to assemble transistors. (Western Electric Co.)

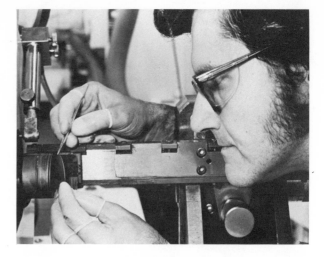

Fig. 17-31. Wafers are cut from the silicon ingot by means of a diamond saw. (Western Electric Co.)

Fig. 17-30. Using modern transistor manufacturing techniques, a single silicon rod is "grown" in a special furnace that maintains precise temperatures for about 12 hours. (Western Electric Co.)

Fig. 17-32. After polishing, the wafers are placed in a special reactor where controlled amounts of gas vapors are applied. These coatings "make up" PNP or NPN transistors. (Autonetics Div. of North American Rockwell)

with plastic. This is called "encapsulating" the transistor. During this operation, the transistor chips are handled by vacuum pick up needles and connected to .001 in. diameter gold wire. See Fig. 17-33.

Then, the encapsulated transistors are put into a miniature "Ferris wheel" to be power aged, Fig. 17-34. Here, current flows through the transistor for about

Fig. 17-33. Operators using microscopes feed gold wire through a hollow point needle. This assembly procedure bonds the leads and coats the transistor with plastic. (Western Electric Co.)

two hours at a level above their power rating. After this break-in operation, a technician tests each transistor with a scope to plot its curves. See Fig. 17-35. These tests tell them what number to put on the transistor. Then, the transistors are sorted into bins according to type, ready for careful packaging and shipment.

Fig. 17-34. An aging rack rotates transistors for two hours of above-rated power aging to meet set levels of reliability. (Western Electric Co.)

Fig. 17-35. Technician connects scope test leads to transistor and plots curve of transistor performance. Results of tests permit transistor to be "numbered." (Acme Electric Corp.)

This brief description of transistor production helps explain why transistors must be handled with care. The PN junctions are so thin, they can be ruined very easily if you get careless.

CURRENT PATHS

Earlier in this chapter, you learned how to label the leads and identify the two types of transistors. Then, you had a quick look at a transistor manufacturing process. Now, you will want to know how to use transistors.

PNP CURRENT

If we hook a PNP transistor into a simple circuit, Fig. 17-36, we can see how the current flows. Note that there are two current paths. The first path is through

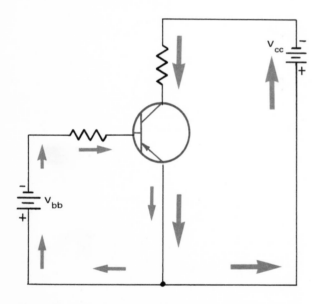

Fig. 17-36. PNP transistor circuit has two batteries and two current paths, both passing through emitter. V_{cc} path carries larger current flow than V_{bb} path.

the base and into the emitter, then back to the battery. This current usually is very small, but the amount depends on the circuit in which it is used. The resistor in this path is similar to a small load resistor. It is there to control the amount of current.

BATTERY LABELS

The battery voltage in the first path of current flow is labeled V_{bb} because it is connected to the base. We are using a system with a double set of letters to avoid

confusion with labels for voltage drops and voltages at various points in the circuit. Note that the other battery connected from emitter to collector is labeled V_{cc} for the same reason.

TOTAL CURRENT

The second path for current in Fig. 17-36 is from the collector through the emitter. This path is the main flow of current in the circuit. Note that the emitter handles the current flow from both the base and collector.

The transistor in this particular circuit is said to be forward biased because current can flow through the junctions. Forward bias means "direction of electron flow," just as it did in our study of diodes.

The amount of current in the emitter can be found by using the following formula and doing some simple addition:

$$I_e = I_b + I_c$$

This formula states that the emitter current is equal to the base current plus the collector current. For example, if the base current is 40 microamps and the collector current is 3 milliamps, find the emitter current. Using the formula, you have:

$$I_e = I_b + I_c$$

$$I_e = 40 \ \mu A + 3 \ mA$$

$$I_e = 40 \ \mu A + 3000 \ \mu A$$

$$I_e = 3040 \ \mu A \text{ or } 3.04 \ mA$$

GAIN

Current flow information is really important when we get into amplifying transistors. We started this chapter by stating that the two types of transistors used are: switching and amplifying. When we want to consider by how much a transistor can amplify a current, we call it the "gain" of the transistor.

Gain is labeled "beta" and has a letter which looks similar to a fancy capital B (β) as its symbol. The gain of a transistor is nothing more than the ratio of the collector current, I_c, and the base current, I_b:

$$\beta = \frac{I_c}{I_b}$$

Using the same current values, what is the gain for the PNP transistor in our last example? Using the formula:

$$\beta = \frac{I_c}{I_b}$$

$$\beta = \frac{3 \text{ mA}}{40 \ \mu\text{A}}$$

$$\beta = \frac{3000 \ \mu\text{A}}{40 \ \mu\text{A}}$$

$$\beta = 75$$

If you look through a transistor catalog, you will find gain or β listings from 30 to 50,000. However, the most common gain falls in the range of 30 to 200. For example, when you hear someone say that a transistor has a gain of 75, it refers to the ratio of collector current (75) to base current (1).

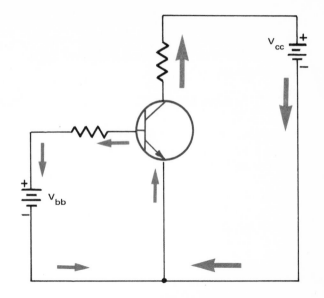

Fig. 17-37. NPN transistor circuit also has two batteries and two current paths. NPN has opposite direction of current flow than PNP.

NPN CURRENT

If we look at an NPN transistor in use, Fig. 17-37, we will see its current paths. Note that again there are two circuits:

1. From the first battery (V_{bb}), through the emitter, the base and back to the battery.
2. From the second battery (V_{cc}) through the emitter, the collector and back to the battery.

The second circuit is the main path for current. Note in both examples, Fig. 17-36 and 17-37, that the smaller current flow is through the base of the transistor.

The base turns the transistor "on" or "off." To do this, it must receive either a negative signal (PNP) or a positive signal (NPN). These signals cause the transistor to turn on, which relates to flipping the switch to turn on the lights in your home.

Depending on how much we can amplify the signal (β), we can figure out how much current we can get through the collector to emitter safely. This relates to using the dimmer switch and turning it on full, so that your house light is as bright as you can get it.

SIGNAL FLOW

Note in Figs. 17-36 and 17-37 that the emitter is used in both the base circuit and in the collector circuit. Therefore, it is called a "common emitter circuit."

There are two other ways you can wire a transistor. One is known as "common base," the second is

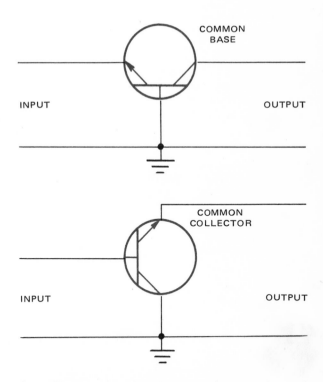

Fig. 17-38. Signal flow through the use of common base and common collector hookups are shown for a PNP transistor.

"common collector," Fig. 17-38. These terms are mentioned a lot. Now you can see why they are labeled this way. Signal flow is through one particular transistor connection which is common to both the input circuit and the output circuit of the transistor.

DATA SHEETS

The last thing we will cover on transistors is other data you might find in parts catalogs. Transistor voltages, for example, are marked as CBO, CEO or EBO. After manufacture, the transistors are tested and marked this way.

CBO, then, means voltage from collector to base with the emitter open, Fig. 17-39. The letter missing in the label is the lead that is open.

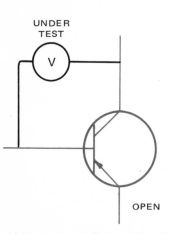

Fig. 17-39. Voltmeter test indicates voltage from collector to base with the emitter open. This function is listed as V_{CBO} under transistor voltages on data sheets.

In the case of CEO, which letter is missing? Since B is missing, this means that there is collector to emitter voltage with the base open. See Fig. 17-40.

The last marking, EBO, is emitter to base voltage with the collector open. See Fig. 17-41.

Gain (β) is also listed in catalogs, along with operating temperature, cutoff frequency and collector cutoff current (I_{CBO}) at some collector-to-base voltage (V_{CB}).

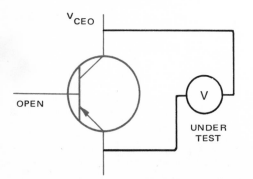

Fig. 17-40. This voltmeter test shows voltage from collector to emitter with the base open (V_{CEO}).

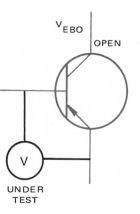

Fig. 17-41. Final voltmeter test reveals voltage from emitter to base with the collector open (V_{EBO}).

You will begin to recognize these letters in circuit drawings as you get a little deeper into electronics. Some catalogs also tell you the type of case the transistor comes in. The shapes we used to illustrate how to find the labels for our leads are just four of the many case types. Note in Fig. 17-42 that cases have numbers such as TO-5 and TO-39, TO-41 and TO-66. There are many others.

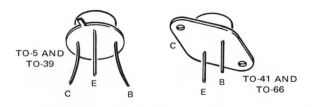

Fig. 17-42. Transistor cases come in a variety of shapes and leads arrangements. Two common case designs are shown.

TEST YOUR KNOWLEDGE

1. The two basic functions of transistors are ____ and ____
2. Draw the symbols for PNP and NPN transistors. Label each one for type.
3. Label the three leads on each transistor drawn in answer to question 2.
4. The lead of the transistor that has the arrow is the ____
5. "Doping" is a term used to describe ____
6. When a transistor is completely coated with plastic, it is said to be ____
7. "Beta" is a term used to describe the ____
8. What is meant by the term EBO in a parts catalog for semiconductors?

9. All silicon transistors have the same "Beta," and all germanium transistors have different "Betas." True or False? *FALSE*

10. Which type of transistor, silicon or germanium, will carry more current? *GERMANIUM*

THINGS TO DO

1. Take apart NPN and PNP transistors. Is there any difference in construction between the two types?

2. Find as many different shapes of transistors as you can. Identify the lead location and name for each type.

3. Solder a couple of transistors without using heat sinks. How much does it take before they are ruined? Do the leads discolor in any way? Can you "see" when they are ruined?

4. Visit a plant or try to get a film that shows how a transistor is encapsulated. How is the process done? How do they identify which type transistor is inside the plastic? Are they processed in batches or one at a time?

5. Why do manufacturers give guarantees on solid state components? Do they last longer than those with tubes?

6. You often hear that if a solid state device is going to fail, it will do so in the first 30 days. If not, it will last for years. Why?

7. Find some circuits that have common emitter, base and collector wiring. Are there any advantages for each type?

8. Make a list of the properties listed in a replacement catalog. What do each of the items mean or stand for? Why are there so many different categories?

9. Communications satellites are affected by the altitude. Some have solid state devices which were ruined by "things" in the atmosphere. How is that possible?

10. You often see a weather report given where clouds are shown on a radar screen. What causes the blips on a radar screen? How can you tell clouds from airplanes?

11. Holography is becoming more important in science. What is it? How does it work? What can you do with it?

12. How is it possible to send communications messages to submarines? Does it make any difference if they are under ice?

An aerospace worker wires the power supply into an electronics assembly.
(Goodyear Aerospace Corp.)

Chapter 18

INTEGRATED CIRCUITS AND OTHER SOLID STATE DEVICES

In the last chapter, we studied transistors. Now, we will look at some of the other types of solid state devices: how they are constructed; how they function; how they are used.

FIELD EFFECT TRANSISTORS

A field effect transistor (FET) is another solid state device that will do many things that a transistor does. FETs usually are made from a small block of silicon called the "channel."

In Fig. 18-1, for example, a chunk of N type silicon is used as the channel. Then, a small piece of P type material is attached to the side of the channel to form a

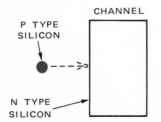

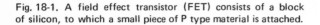

Fig. 18-1. A field effect transistor (FET) consists of a block of silicon, to which a small piece of P type material is attached.

P-N junction. The P type material is called the "gate." Finally, two leads are added at either end of the channel. These are known as the "source" and the "drain."

The P-N junction of the FET is very similar to the type of junctions you dealt with in the last two chapters. Fig. 18-2 illustrates a completed field effect transistor.

Looking back to Chapter 17, you could compare the *source* of the FET with the *emitter* of a transistor, the *drain* with the *collector* and the *gate* with the *base*, Fig. 18-3. Unlike the transistor, however, the FET is not made of PNP or NPN junctions, Fig. 18-4. Instead, it is

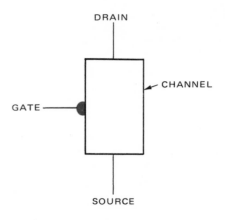

Fig. 18-2. A completed FET is shown with the channel made from a block of silicon. Gate material and leads are attached.

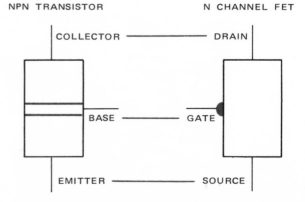

Fig. 18-3. In comparing transistor and FET leads, note that the emitter, base and collector of the transistor equals the source, gate and drain of the FET.

made from a solid piece of N material with P material as its gate. It also could have been made from a solid piece of P material with N material as its gate. See Fig. 18-5.

Why the FET *does not* work on the junction idea will be explained later.

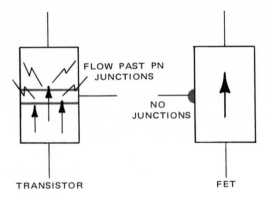

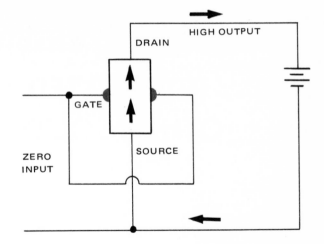

Fig. 18-4. Comparing flow of electrons past the PN junction of a transistor with flow through the channel of the FET.

Fig. 18-6. With zero input on the FET, current output is high because nothing is blocking the flow through the channel.

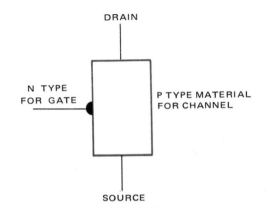

Fig. 18-5. Making the FET from P type material combined with N type material to serve as the gate.

18-6. In order to let the FET conduct, we must not block the flow of electrons. This is a fact because the FET is made from a solid piece of material. The FET does not require biasing to make it conduct (as a conventional transistor does). Instead, the gates are used to block the flow of electrons, Fig. 18-7. As the gates are made more negative, fewer electrons can flow from source to drain.

TYPES OF FETS

There are several types of FETs.
1. Junction field effect transistors.
   a. JFET — N channel.
   b. JFET — P channel.
2. Metal oxide semiconductor field effect transistors.
   a. MOSFET — N channel.
   b. MOSFET — P channel.

As indicated, both JFETs and MOSFETs can be made of either N type material or P type material.

JFET

We will examine the JFET first, Fig. 18-6. Note that we have another gate on the opposite side of the channel and the two gates are tied together. In this way, we can use the gates to block the flow of electrons through the channel.

The input circuit is from the gates to the source, and the output circuit is from the source to the drain, Fig.

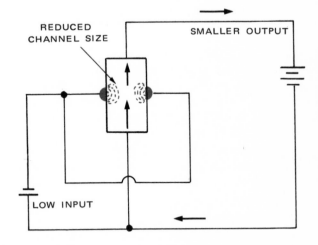

Fig. 18-7. Putting a low voltage on the input makes the size of the channel smaller and reduces output current.

DEPLETION LAYER

With voltage applied to the gates of the JFET, a depletion layer is set up in the channel. This layer resists or opposes the flow of electrons through the channel. Fewer electrons can get past the depletion layer than when no signal was put on the gates. Then, when increased voltage is applied to the gates, current in the output circuit drops to a very low value. Finally, when the voltage on the gates reaches a high negative value, the depletion layer extends all the way across the channel, Fig. 18-8. The electrons cannot flow and the current falls to zero.

Note that we used voltage on the gates to stop the flow of current. With FETs, current does not flow across a junction as it does with conventional tran-

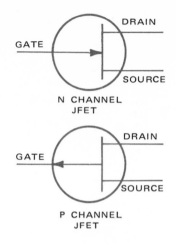

Fig. 18-9. Symbols are shown for N channel and P channel JFETs. Only the arrow changes direction.

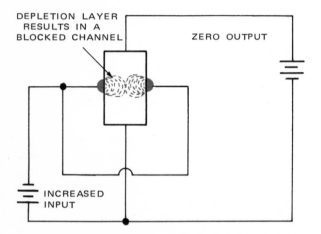

Fig. 18-8. Increasing voltage on the input sets up a depletion layer that blocks the flow of current through the channel.

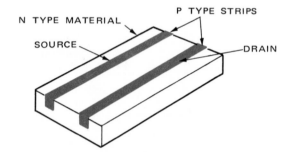

Fig. 18-10. Diffusing two P type strips on a substrate of N type material for a MOSFET.

sistors. Therefore, there is much less noise with FETs. This characteristic makes them very useful in amplifier circuits. Since there is little noise on the input, it means there is less noise amplified at the output.

Although we have shown the JFET as an N channel (made of N type material), we could have used P type material. In this case, the JFET would work with polarities being just the reverse of N channel usage.

The symbols for N channel and P channel JFETs are shown in Fig. 18-9. The only difference between these two symbols is the direction in which the arrow points.

MOSFET

The metal oxide semiconductor FET (MOSFET) is similar to the JFET. The major difference is in how they are made. With a MOSFET, Fig. 18-10, a piece of N type material (substrate) is used. On it, two P type strips are diffused. One of the strips will become our source, and the other becomes our drain. Between them, the P channel is diffused. This layer, shown in Fig. 18-11, is not as thick as the first two.

NOTE: The drawings in Figs. 18-10, 18-11 and 18-12 are meant to give you some idea how a MOSFET

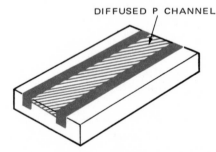

Fig. 18-11. Diffusing the P channel between the two P type strips of a MOSFET.

is made. Do not think that this is what a MOSFET would look like if you took one apart. The diffused layers on the substrate are very thin and are measured in microns (only a few thousandths of an inch thick).

Looking at an end view of the MOSFET in Fig. 18-12, a thin film of dielectric (insulating) material is grown over the channel. On top of this, another film is deposited to complete the gate of the MOSFET. The

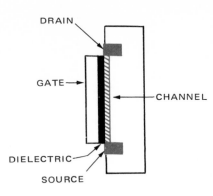

Fig. 18-12. Finished MOSFET showing gate, drain, source and channel. The dielectric creates a capacitor between the gate and the channel.

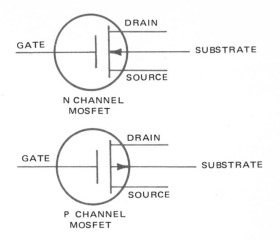

Fig. 18-13. Symbols are shown for N channel and P channel MOSFETs. The substrate is the block of material on which the various parts were diffused.

last layer applied is metal, which is why the unit is called a "Metal Oxide Semiconductor FET."

The dielectric separates the metal from the channel. Since it acts as an insulator, this particular FET is also called an IGFET (Insulated Gate Field Effect Transistor).

In a MOSFET, then, the gate is separated from the channel by a dielectric. As you recall from Chapter 4, when two conductors are separated by an insulator, you have a capacitor. The small capacitor in a MOSFET can be damaged very easily by spikes (sudden, unintended pulses of electricity) and even static charges. Even the capacitance of your body or stray voltages when soldering can ruin a MOSFET. For this reason, manufacturers ship them with their leads shorted.

I.C. SAYS:

"Use extreme caution when working with a MOSFET. Solder all other components first and save the MOSFET until last. Use heat sinks and avoid touching the leads. Your body capacitance can ruin them. Ground all sources of static electricity before contacting a MOSFET."

There is a whole series of MOSFETs. No attempt will be made to describe all of them. However, it *is* important to know that P type material (instead of N type) could have been used in our example of how a MOSFET is made. Again, they would operate with opposite polarities. The N channel MOSFET which we used in our example has a symbol with the arrow pointing in, Fig. 18-13. The P channel MOSFET has its arrow pointing out.

To make our N channel MOSFET conduct, hook it up the same way you hooked up the JFET. See Fig. 18-6. As the gate is made more negative, the current between the source and drain will again decrease

because of the depletion region, Fig. 18-7. The main difference between the JFET and MOSFET is that the MOSFET has an extremely high input impedance; much more than the JFET.

Once the FET has been made, it is placed in a small plastic or metal case with three leads. The completed FET looks just like a transistor. In fact, the only way you can tell some of them apart is by looking up the code numbers printed on them. These numbers usually can be found in a semiconductor data book.

ADVANTAGES OF THE FET

The main advantage that FETs have over conventional transistors is the fact that any signal trying to enter the FET "sees" a very high impedance. Fig. 18-14 shows a FET used as an amplifier connected to a radio tuning circuit. When you tune in a station, the tuning circuit will develop the signal voltages of that station. These voltages, in turn, will be amplified by the FET, which uses a small voltage at the gate to control a large current through the channel.

If you think back to Chapter 15, you will recall that we studied the "Q" of a resonant circuit. We found that a high Q circuit has a narrow bandwidth. If you attach a low resistance in parallel with the resonant circuit, it will lower the Q of the circuit and make the bandwidth much wider.

However, if the bandwidth of the tuning circuit of a radio is made too wide, you might pick up the signals from two stations at once. FETs help avoid this because they have a very high input impedance. So high, in fact, that when you attach a FET to a tank (parallel resonant) circuit, the Q of the circuit shows little effect.

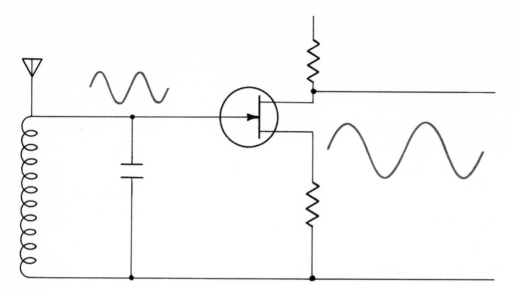

Fig. 18-14. Using the FET in a circuit to amplify the radio signal.

SILICON CONTROLLED RECTIFIERS

Another solid state device found in the electronics field is the silicon controlled rectifier (SCR). This device usually is used in circuits where the current is very high. You will find SCRs in circuits carrying as much as 400 amps.

By using an SCR, it is possible to switch a circuit thousands of times in one second. Consider the fact that during one cycle of a 60 Hz sine wave, there are 16,667 microseconds. See Fig. 18-15. Since the SCR can be switched in about one microsecond, it can be

used to switch high current circuits in extremely short periods of time. This makes the SCR very valuable in switching circuits.

TURNING ON THE SCR

The symbol for an SCR, Fig. 18-16, looks like a diode symbol except that it has one additional line. This line is known as its "gate." The anode and cathode are the same as for a diode.

Since the SCR looks like a diode and is turned on like a transistor, many people say it is a combination of both of them. However, there is one big difference. To turn on the SCR, the gate must be given a pulse that is positive with respect to the cathode. Once it is turned on, the SCR will conduct from cathode to anode. Then, comes the difference. *If you open the circuit to the gate, the SCR will keep on conducting.* Removing the gate current does nothing.

Therefore, the SCR is like a latching switch. Once it

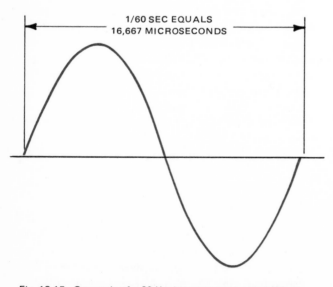

Fig. 18-15. One cycle of a 60 Hz sine wave takes only 1/60 of a second. This is equal to 16,667 microseconds.

Fig. 18-16. Symbol is shown for the SCR. It looks the same as the diode symbol except that another lead known as the "gate" is added.

is triggered, it remains on as long as there is current flowing through it. Only a short pulse to the gate was needed to make it conduct.

TURNING OFF THE SCR IN DC CIRCUITS

Most SCRs need less than 20 milliamps to keep them turned on. One way to turn off the SCR is to lower the load current to a value of less than 20 milliamps. There are many ways to do this. One way would be to use a switch, Fig. 18-17, in the load circuit. If the load is

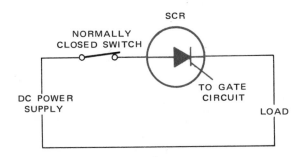

Fig. 18-17. To turn off the SCR in a dc circuit, wire a normally closed switch in series with the SCR. Opening the switch will shut off the SCR.

drawing a heavy current (200 amps or more), it will cause a heavy arcing when you open the switch. However, if the circuit is carrying small current, this would not be a problem.

A second way to turn off the SCR (open the circuit) is to wire a push button switch in parallel to the SCR, Fig. 18-18. When the switch is closed, the SCR is shorted out and will stop conducting. When the push button is released, the gate must be triggered again to turn on the SCR.

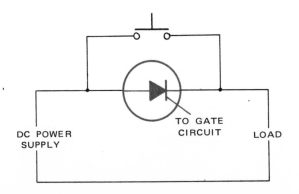

Fig. 18-18. Another way to turn off the SCR in a dc circuit is by wiring a normally open push button switch in parallel with the SCR. Pushing down the button while the SCR is conducting will short it out.

TURNING OFF THE SCR IN AC CIRCUITS

When an SCR is used in an ac circuit, it is an easy matter to switch it off. Actually, the SCR will automatically shut off during each negative alternation (half cycle) when the current from the positive alternation falls to zero. Therefore, it must be triggered "on" every positive alternation. This is not a problem, however, all that is needed is another pulse at the gate.

Since the SCR can be controlled anywhere on the positive alternation, it has many uses in circuits that control heavy currents. You can trigger the SCR early in the cycle, Fig. 18-19, by sending a pulse to the gate at that time. If you want to delay its firing, the SCR can be pulsed "on" later in the alternation.

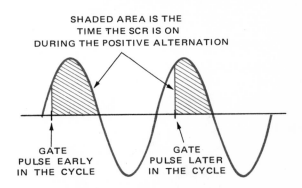

Fig. 18-19. There are over 8000 microseconds in a 60 Hz signal during one alternation. Since the gate can be pulsed anywhere in the positive alternation, the SCR will conduct for as much of the positive half cycle as you wish.

There are over 8000 microseconds in one half cycle of a 60 Hz signal. Since only a short time is needed to trigger the SCR, it can be turned on anytime during the positive alternation. For this reason, SCRs are used in motor speed control circuits, switching circuits, lighting controls and welding circuits.

POWER TOOL SPEED CONTROL PROJECT

You can build the speed control unit shown in Fig. 18-20 as a valuable exercise in working with ac circuits having resistors, rectifiers and SCRs. Then, with the twist of a knob, you can vary the speed of any series universal (brush type) motor that draws less than 400 watts. This is the type of motor found in most drills, table saws, hand grinders and sewing machines.

CAUTION: Do not use the speed control unit for other types of motors.

This speed control will let you slow down your tools

when working with different kinds of material. Suppose you tried to drill holes in a piece of steel and your drill bit squealed and burned. This indicates that your drill is running too fast, so install the speed control and slow it down.

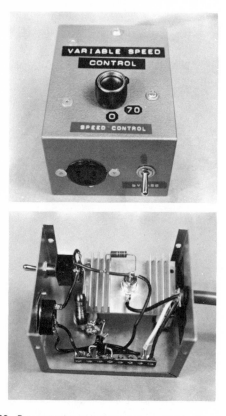

Fig. 18-20. Power tool speed control. Top—Finished project. Bottom—Assembled components. (Project by Jim Shelton)

To build the speed control unit, obtain the items given in the parts list and assemble them according to the construction procedure given.

No.	Item
2	Motorola HEP 162 silicon rectifiers.
1	2.5K resistor, 5 watts (R_1).
1	500 ohm potentiometer, 2 watts (R_2).
1	1.5K resistor, 1/2 watt (R_3).
1	SPST switch, 5 amps.
1	Motorola HEP 302 silicon controlled rectifier.
1	Aluminum heat sink for SCR, 3 1/2 x 1 1/2 in.
1	Three prong panel mount ac receptacle.
1	Three wire grounded line cord.
1	Aluminum minibox, 3 in. x 4 in. x 5 in. Connecting wiring and cord strain relief.

CONSTRUCTION PROCEDURE

To build the power tool speed control:

1. Discard the insulating washers furnished with the SCR, then install SCR on heat sink. See Fig. 18-21.
2. Apply a thin coat of heat sink compound to surface of minibox that will contact top of heat sink.
3. Mount heat sink/SCR assembly in minibox. Make a direct electrical connection between minibox and heat sink, but do not overtighten nut.
4. Mount and connect other components as shown in Fig. 18-20. Make certain that rectifiers and other parts do not short-circuit against bottom half of minibox. Receptacle wired to insure against incorrect connection which could be a safety hazard.
5. Connect line cord.

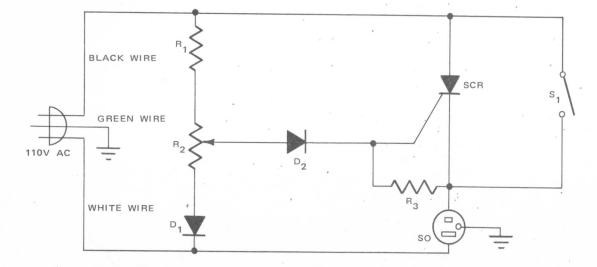

Fig. 18-21. Schematic for power tool speed control: $D_{1,2}$—Silicon rectifiers. $R_{1,3}$—Resistors. R_2—Potentiometer. S_1—SPST switch. SCR—Silicon controlled rectifier. SO—Receptacle.

Fig. 18-22. Color organ with an SCR. Left. Finished project. Right. Circuitry inside case. (Project by Ed Belliveau)

Be sure to use a three wire grounded line cord and three prong panel mount socket to prevent electrical shock in case the insulation within the power tool should break down. Also see that your bypass switch and wiring are able to carry at least 5 amps ac.

After you have completed the speed control unit, plug the line cord of the control unit into a 110 volt ac receptacle. Plug your power tool line cord into the panel mount socket of the control unit. You should be able to control the speed of the power tool up to 70 percent of maximum speed. If you need full speed, simply flip the bypass switch.

COLOR ORGAN PROJECT

Do you want to experiment with SCRs? The color organ shown in Fig. 18-22 is an easy-to-build project that uses one SCR. With it, you are able to vary the brightness of a low wattage bulb in keeping with changes in the audio level from your radio.

You can mount the simple circuit, Fig. 18-23, in any handy plastic box and use a knob on the outside to control the potentiometer. Obtain the following items:

No.	Item
1	Plastic box, approximately 3 in. x 6 in. x 1 1/2 in.
1	Circuit board, about 2 in. x 5 in.
1	Audio output transformer.
1	5K potentiometer.
1	Control knob.
1	1K resistor.
1	.001 μF capacitor.
1	SCR.
1	110 volt receptacle.
1	Lamp cord with plug.
	Connecting wiring.

CONSTRUCTION PROCEDURE

To build the color organ:
1. Mount transformer, resistor, capacitor and SCR on circuit board, Fig. 18-22.
2. Place board in plastic box.
3. Install receptacle in plastic box.
4. Install potentiometer and control knob.
5. Attach lamp cord and connecting wiring, Fig. 18-23.
6. Solder connections.

The transformer output, as usually used, will be the input for this circuit.

Connect the audio input side of the transformer into the speaker of your radio. Connect the lamp cord for a low wattage bulb into the receptacle in the side of the

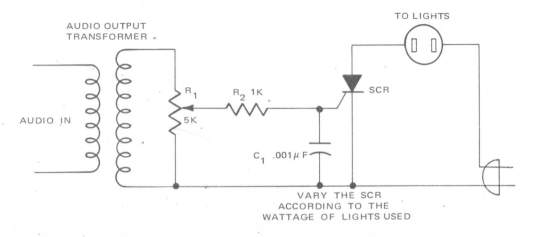

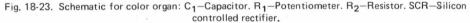

Fig. 18-23. Schematic for color organ: C_1—Capacitor. R_1—Potentiometer. R_2—Resistor. SCR—Silicon controlled rectifier.

plastic box. Plug your color organ lamp cord into a 110 volt receptacle. Turn on the radio and watch the results.

If you want to experiment further, mount the lamp behind different colored plastic sheets. You can even make a four-sided, four-color lamp for this use.

There are a number of other color organ circuits you can build. Some will have six to eight transistors. To decide which one you should build, consider the cost and how fancy you want to make your color organ.

OTHER SOLID STATE DEVICES

Another type of transistor you may work with is called a "phototransistor," Fig. 18-24. The base of this type of transistor is right underneath a tiny lens. As the amount of light through the lens changes, the current

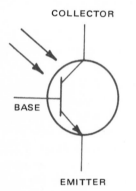

Fig. 18-24. The phototransistor is activated by light.

through the transistor changes. The only difference between the phototransistor and a conventional transistor is the fact that the light on the base controls current flow in a phototransistor. As you recall, current on the base of a conventional transistor controls current flow in the emitter-collector direction.

A "light emitting diode" (LED) is a semiconductor device that gives off a tiny bit of light when current flows through it. Because a LED is a diode, current will flow only when it is forward biased. The symbol for a LED is shown in Fig. 18-25.

The main use of LEDs is in the display of numbers in test instruments or calculators, Fig. 18-26. The LEDs are arranged in seven-segments, Fig. 18-27. When the proper voltages are applied to the LEDs, one or more of the segments will light up to form a number. For example, if five of the segments are forward biased, they will form the number 3. Small bulbs could be used, but LEDs are much cheaper and can be turned on and off thousands of times without failing.

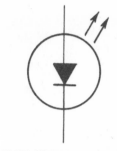

LIGHT EMITTING DIODE

Fig. 18-25. Symbol for a light emitting diode (LED).

Fig. 18-26. On-vehicle computer uses LEDs for digital readouts. (National Semiconductor)

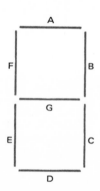

Fig. 18-27. LEDs in seven segments can be used to display all numbers from 0 to 9.

INTEGRATED CIRCUITS

In reading electronics magazines, no doubt you have seen a large number of projects that use integrated circuits (ICs). An IC is a circuit containing resistors, transistors and capacitors made very small, then placed in a compact package, Fig. 18-28.

THICK FILM CIRCUITS

Certain types of ICs are made by reducing the size of a circuit, attaching a number of terminals and encasing

Fig. 18-28. Integrated circuits are manufactured in several different shapes.

the entire circuit in a plastic package. Fig. 18-29 shows the various steps involved in making this "thick film" type of integrated circuit:

1. Material which is to be the "frame" or "substrate" for finished IC.
2. Flat wires called "conductors" are deposited on substrate. These will be wires for circuit.
3. An insulating glaze or film is put over conductors.
4. Layers of carbon are deposited on film. These will become resistors in the circuit. Values are set by erasing bits of carbon until we have the right amount of resistance.

5. Leads are added and conductors are "tuned." This means that conductors are covered by a layer of solder to allow them to carry more current.
6. Small transistors and diodes are added to circuit.
7. Whole circuit is coated with plastic and marked so that it can be identified.

Fig. 18-30 shows some "thick films" before they have been coated with plastic.

ADVANTAGE OF INTEGRATED CIRCUITS

The main advantage of integrated circuits is that so much electronic circuitry can be put in a very small package, Fig 18-31. There are many ICs today that contain a whole AM radio. To make it work, all you need to do is attach an antenna, a battery and a speaker. The IC alone is smaller than a book of matches and the price is just a few dollars.

STEPS IN MAKING AN IC

Another type of integrated circuit is made from a single block, or wafer, of silicon. First, a circuit drawing is made much larger than the finished circuit. This is done so that the drawing can be reduced and still maintain accurate spacing between parts. No part of any line can be allowed to touch another line. If it did, some part of the circuit would be shorted out.

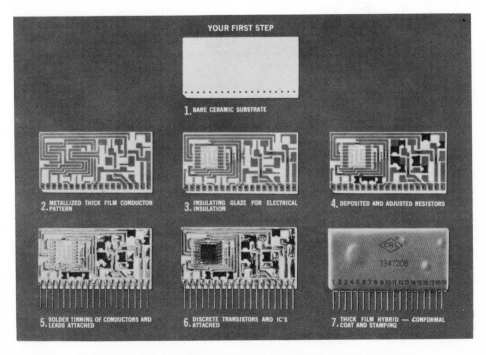

Fig. 18-29. Steps are shown in making thick film integrated circuits.
(Centralab Electronics Div. of Globe-Union, Inc.)

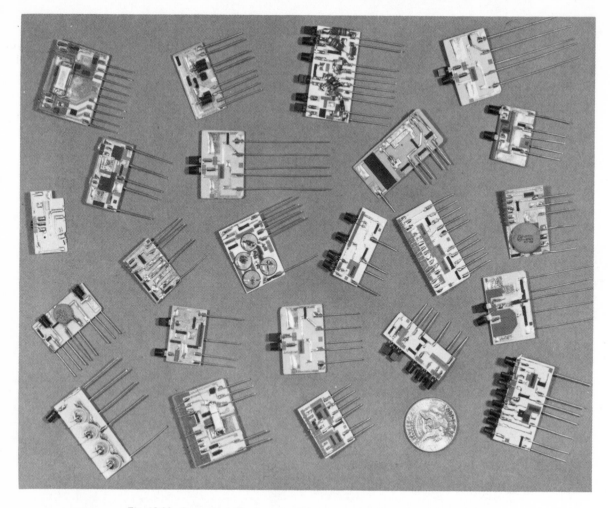

Fig. 18-30. Some thick film integrated circuits are shown before encapsulation.
(Sprague Electric Co.)

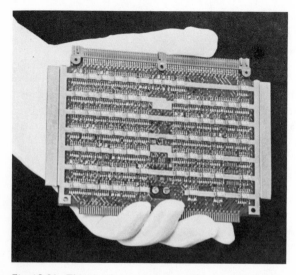

Fig. 18-31. This circuit board has over 70 integrated circuits.
(Autonetics Div. of North American Rockwell)

Next, a special camera, Fig. 18-32, is used to reduce the drawing down to 400 times smaller than original size. It also can shrink lines down to .0001 in. (1/10,000 of an inch). This allows several hundred completed circuits to be put on a piece of silicon only 1 1/4 inches in diameter, see Fig. 18-33.

MONOLITHIC IC

Special furnaces (similar to those used for making transistors) add the impurities (N and P), Fig. 18-34, to make the completed unit. Integrated circuits manufactured by this process are called "monolithic" ICs, because they are made from a single piece of silicon (mono means one). Since ICs are so small, the only way to see what the circuit looks like is to use a microscope, Fig. 18-35.

Before shipment, ICs are checked on a special

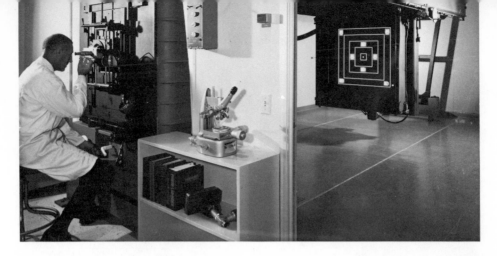

Fig. 18-32. A special reducing camera is used in making integrated circuit drawings much smaller in size. (The Boeing Co.)

Fig. 18-33. A silicon wafer is filled with integrated circuits. This 1 1/4 in. diameter wafer holds several hundred circuits. (Sprague Electric Co.)

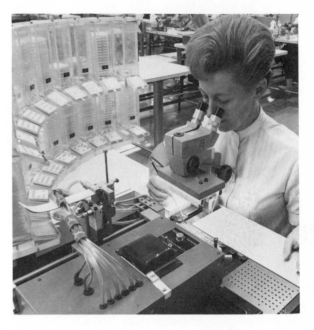

Fig. 18-35. Assembly worker examines an IC with a microscope. (Electronics Group Div. of North American Rockwell Corp.)

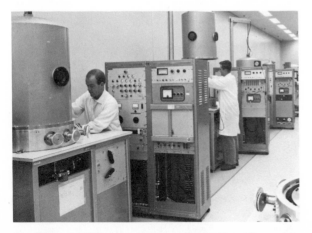

Fig. 18-34. Special ovens are used to deposit N and P materials on integrated circuits. (The Boeing Co.)

machine. The one shown in Fig. 18-36 can test hundreds of ICs in a single day. This once was a very difficult job. Now, it is a simple matter of hooking up the IC and automatically checking it. Since the IC is such a small part, it is very important that its circuit works properly.

IC SIZE

You may wonder why integrated circuits are so low priced when they are so hard to manufacture. It is because the manufacturer makes ICs by the thousands. Special machines produce ICs so fast and so accurately that the price can be kept quite low.

The development of ICs looks promising. Their small size and low price mean that you will be seeing more

Fig. 18-36. This special testing machine automatically tests ICs and prints the results on punched cards so circuit performance can be analyzed. (The Boeing Co.)

and more of these circuits in the next few years. Because they are so small, ICs can be used where, in the past, even a single component would not fit. The IC shown in Fig. 18-37 has over 100 transistors on it. Think of some of the places where it can be used. Hearing aids, for example, are an ideal IC application.

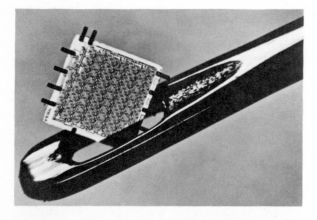

Fig. 18-37. This integrated circuit containing 100 transistors, resistors and diodes is interconnected by the etched pattern of the printed circuit. It is small enough to pass through the eye of a needle. (Western Electric Co.)

Research has made all these solid state devices possible. The discovery of the transistor, for example, really sparked the growth we have in electronics today.

IC SYMBOLS

Several of the symbols for integrated circuits are shown in Fig. 18-38. Because the actual circuit inside the container is so complex, circuits are rarely drawn. Instead, a circuit will give you a symbol and some numbers. Other times, they will tell you what the IC can do, such as the 3 input AND gate and the 3 input OR gate. With this information, you can wire the IC into the circuit to do what you want. You can imagine the time it would take to draw the circuit for the IC having 100 transistors that you saw earlier.

THE "AND" SYMBOL

The "AND" symbol is used when drawing a circuit where signals must be present at *all* inputs in order to get an output. For example, there must be a signal at A and B and C before you get an output at D. You can see why AND deserves its name.

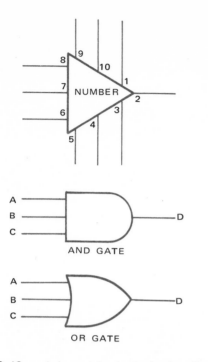

Fig. 18-38. IC symbols used in circuit drawings have many different shapes, rather than forming true electrical schematics.

A simple circuit can be used to show how the AND symbol might be used. Suppose you wanted to make a circuit in which three things must take place before the next step can occur, Fig. 18-39. In this diagram, we will say that before the machine can start moving safely, the machine operator must do three things.
1. Close the switch for input A.
2. Turn selector switch to terminal 3 for input B.
3. Use the push button switch for input C.

There must be input signals to all three inputs before we can get an output. When all three steps have been taken, the machine will start moving. If any one of the three steps is left out, the machine will not move.

If we made a table of inputs to see if an output resulted, we would have the following:

INPUT			OUTPUT
A SPST SWITCH	B SELECTOR SWITCH	C PUSH BUTTON SWITCH	D MACHINE MOVES
No	No	No	No
No	No	Yes	No
No	Yes	No	No
No	Yes	Yes	No
Yes	No	No	No
Yes	No	Yes	No
Yes	Yes	No	No
Yes	Yes	Yes	Yes

The table gives all possible ways in which we could have the circuit. It is only when we have signals to all the inputs (last line) that we finally get an output.

TRUTH TABLE

The preceding table can be done in a number of different ways. Instead of the NO and YES listings, we could use 0 (zero) to tell us that the signal was missing (not present) and number 1 to indicate that the signal was there. If we use 0 and 1, our table would look like this:

INPUT			OUTPUT
A SPST SWITCH	B SELECTOR SWITCH	C PUSH BUTTON SWITCH	D MACHINE MOVES
0	0	0	0
0	0	1	0
0	1	0	0
0	1	1	0
1	0	0	0
1	0	1	0
1	1	0	0
1	1	1	1

This is known as a "TRUTH TABLE." It tells us what we must do to the inputs in order to get an output.

THE "OR" SYMBOL

The TRUTH TABLE for the "OR" in Fig. 18-38 would look like this:

INPUT			OUTPUT
A	B	C	D
0	0	0	0
0	0	1	1
0	1	0	1
0	1	1	1
1	0	0	1
1	0	1	1
1	1	0	1
1	1	1	1

What does this table tell you? It indicates that if a signal is present at any input, there is an output. That is: if there is a signal at A or a signal at B or a signal at C, there is an output at D.

You can see how well the name "OR" fits this setup. There are a whole series of other solid state symbols that can be handled in the same way. These are known as "LOGIC" symbols, and each will have a TRUTH TABLE.

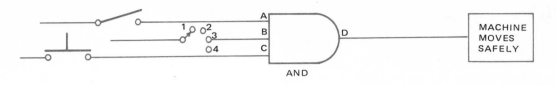

Fig. 18-39. Three things must be done to the input (A, B and C) of this "AND" to get the output (D), which will result in the machine moving safely.

PRINTED CIRCUIT BOARDS

When it is not necessary to make the circuit as small as an IC, you can make your own printed circuits. Simply obtain a board, lay out the circuit, mount your transistors and capacitors and solder them into position.

To make the printed circuit, start with a special board coated with copper. Lay out the circuit you wish to make, using special ink or tape, Fig. 18-40. This is necessary because of the etching process used in making the circuit board. In this process, chemicals will "eat away" the copper that is not covered with ink or tape. If the ink does not protect the copper, there is no way to make the circuits.

Put the ink on the copper in the shape that you want your finished conductors to take. Most of these conductors are a series of paths which will act as wires for your circuit. The next important step is to lay out each component exactly where you want to locate it. Once the copper is removed, it is difficult to make repairs.

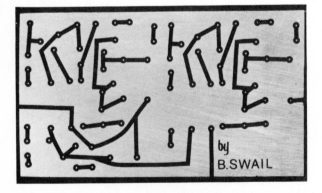

Fig. 18-40. Special tape is used to cover the copper on this clad printed circuit board. You can even mark your name on the board with special ink as the student did.

ETCHING THE PC BOARD

When you finish your printed circuit board layout, the next step is to remove the copper from the places where it is not wanted. This is done by placing the PC board in chemicals that will etch (eat away) unwanted copper, Fig. 18-41. Therefore, the copper that you need to form your circuit must be covered with a layer of material that will resist being etched away. Once the copper is etched, wash it off, remove the tape or ink, clean it, drill the holes, and you have your finished PC board, Fig. 18-42.

Fig. 18-41. Students etching a printed circuit board to remove the unwanted copper. (Hickok Teaching Systems, Inc.)

UNDERCUTTING

The PC board is left in the chemicals only until the uncovered areas of copper are etched away. Do not leave the board in the solution too long or the conductors will be "undercut." See Fig. 18-43. Under-

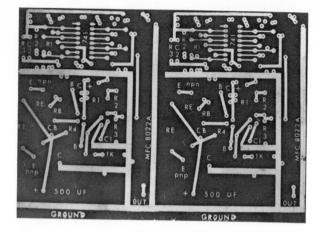

Fig. 18-42. After the ink is removed from the etched printed circuit board, only the copper that the ink protected is left on the board.

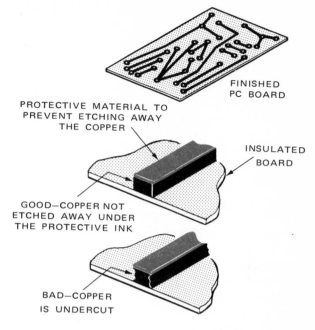

FINISHED
PC BOARD

PROTECTIVE MATERIAL TO
PREVENT ETCHING AWAY
THE COPPER

INSULATED
BOARD

GOOD—COPPER NOT
ETCHED AWAY UNDER
THE PROTECTIVE INK

BAD—COPPER
IS UNDERCUT

Fig. 18-43. Close-up views show two conditions of a finished PC board. The undercut copper was left in the etching solution too long.

cutting occurs when the chemicals start to etch away the copper under the protective coating. This, in turn, will cause an open conductor (will not carry current). In some cases, the conductor will crack easily and open the path to the component.

When you have a good printed circuit board, you can solder the parts into your circuit. The big advantage of the PC board system is that it gets rid of a lot of wires. It also decreases the size of your circuit, so that it will fit in a smaller package or chassis box. Fig. 18-44 shows a few finished circuits that have been mounted on PC boards.

OTHER PROJECTS

Now that you have covered the basics of electricity and electronics, consider advanced studies in this field. The projects presented throughout this book have been kept simple and directly related to the subject area of

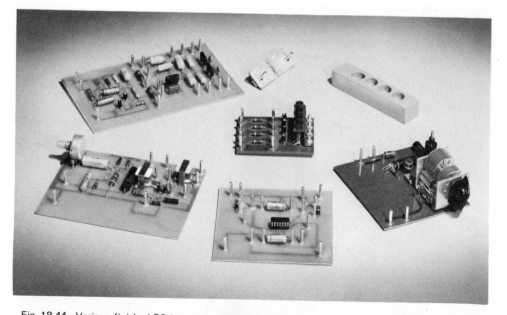

Fig. 18-44. Various finished PC boards are shown with components mounted in place and connected to printed circuits underneath.

Electricity and Basic Electronics

the chapter in which they appeared.

If you are interested in building more complicated projects, page through the various electricity and electronics magazines and paperback project books that are available. They contain helpful information and excellent projects to further your knowledge and improve your circuit-building ability.

Fig. 18-45, for example, shows a typical, advanced IC project. The photo enlarger pictured is controlled automatically by a shop-built timing controller unit and a sensor that picks up the reflected light of the enlarger. The sensor then signals the controller, and a built-in timer IC times the exposure.

The Automatic Photo Enlarger Controller project will allow you to select the proper exposure time and cut down on wasted photo paper. This particular project is just one of hundreds described and illustrated in magazines and books devoted to the field of electricity and electronics.

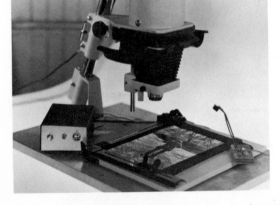

Fig. 18-45. Automatic photo enlarger controller unit (at left) and sensor (at right) work together to properly time the exposure of the enlarged print. (Project by Bruce Swail)

TEST YOUR KNOWLEDGE

1. Draw the symbol for the "AND."
2. Show the TRUTH TABLE for the "AND."
3. What are the two basic steps in making a printed circuit board?
4. An integrated circuit can contain at least five transistors as part of its circuitry. True or False?
5. A thick film circuit is made of deposited resistors. How are the diodes and transistors added to the circuit?
6. What is an LED? *Light Emitting Diode*
7. What do the letters MOSFET stand for? *Metal Oxide Field Effect XSTR*
8. What must be done to turn on an SCR?
9. What must be done to turn off an SCR?

10. Draw the symbol for the "OR GATE."
11. Make a TRUTH TABLE for the "OR GATE."
12. The copper on a printed circuit board may be _____ if it is left in the etching solution too long.
13. The input circuit for a JFET is from _____ to the _____.
14. The output circuit for a JFET is from _____ to the _____.

THINGS TO DO

1. How is the copper clad attached to the printed circuit board material? How can it be put on one side only, or on both sides?
2. Make a list of the different types of solid state devices. Show the symbols for each one. How many can you find?
3. How many different sizes of SCRs are there? What is the minimum amount of current they will carry? What is the maximum current?
4. What are some of the uses for light emitting diodes?
5. If you were given a number on an IC, how can you tell what it will do? Does each manufacturer label them in a different manner? How do you know where to start counting for each lead?
6. Solid state "LOGIC" symbols are different for different companies. How do you know which one to follow and use? Are some of the symbols the same for different companies?
7. Make a list of at least 10 different "LOGIC" symbols. Does each have a "TRUTH TABLE?"
8. Make a printed circuit board with your own personalized symbol or name. How do you tell which company made a printed circuit board? Do they use special symbols?
9. You sometimes hear the words "consumer products" being used in the electrical field. What does this mean?
10. What are majority and minority carriers?
11. Where do you find unijunction transistors being used?
12. If you were to increase the temperature surrounding a transistor, how would it affect operation? Does the heat found in foundries or other industrial plants affect the operation of the transistor? What affects does space travel have on the transistor?
13. What affect does frequency have on the operation of a transistor? Can you use just any transistor for a switching or an amplification application? Why?
14. How is it possible for deep space probes to send signals back to the earth?

Chapter 19

POWER SUPPLIES AND THE OSCILLOSCOPE

In the last three chapters, we have studied solid state devices and the basic principles of electronics. In this chapter, we will cover two related and very important areas of electricity and electronics.

First, consider that every circuit you have worked with had some type of power supply to provide for the flow of electrons. Therefore, in the first half of this chapter, we will discuss some of the various power supplies that are available to you.

Second, many of your circuit tests were made with meters. However, some of the values you tried to measure would last for only a fraction of a second on the dial of a conventional meter. With this in mind, the second half of this chapter is devoted to methods of checking these circuits with a high speed meter called the "oscilloscope."

TYPES OF POWER SUPPLIES

There are many different types of power supplies used in the electricity and electronics field. The name "power supply" tells you that we must have some way of supplying power to many types of circuits. Since there are so many different means of supplying power, we will cover only the more common types you will find in beginning circuit work.

BATTERIES

The simple battery is one type of power source that people often forget. Yet, it has many advantages that some expensive power supplies do not have:
1. The battery is easy to hook up in most simple circuits. There are many different types of clips available for holding it in place on the back of a circuit or on the chassis.
2. There is no need to take along adapters and plugs when you want to connect the battery.
3. You are not forced to depend on having a nearby electrical outlet if you want to use your circuit.

4. When the life of the battery is used up, you can buy a new one for a low cost.

Think of some other advantages of batteries. They are an important type of power supply.

There are times when a battery is not what is needed for your circuit. When this happens, you have to find some other type of power supply. This means you will use ac or dc produced by rectifying an ac signal — at increased cost. The more you want a power supply to do — or the greater the number of different voltages you want it to produce — the more it will cost.

TRYING OUT A NEW POWER SUPPLY

You can get various types of power supplies in two ways. You can buy one already assembled or build it yourself. If you decide to build it yourself, be very careful when installing the various types of solid state devices. Watch the polarity. If you hook up a diode or rectifier the wrong way, certain things could happen — including failure of the device. The same is true for transistors, electrolytic capacitors and many other components.

I.C. SAYS:

"Here is a warning worth repeating. Make sure you check and recheck the polarity of your circuit when building any power supply. Otherwise, wrong polarity can ruin some of these solid state devices."

After the power supply has been assembled and wired, test it by applying the voltage in easy steps. *Do not bring it up to full voltage right away and hope that everything is wired right.* If you do, you create big problems. By increasing the voltage in easy steps, you can make a series of checks to be sure that each area being tested is satisfactory. By doing it this way, you can correct any problems that come up without ruining other parts of the circuit.

Again, remember to use heat sinks when soldering. If

261

you use rectifiers in the circuits, mount them on a heat sink if needed. Also, use silicone grease between the high power transistor and its insulating washer.

Finally, install a fuse in the power supply circuit. Fusing the circuit has saved experts a lot of expense just because they used common sense ahead of time. Mistakes are possible no matter how many times you check your circuit. While a fuse will not always protect your circuit from being damaged, it will help when you get careless.

THE 120V AC POWER SOURCE

Looking at common types of power supplies, batteries rank first. The second most common type of power supply is the 120 volt ac found at your wall outlets. This ac voltage is used to drive motors, light lamps and operate heaters.

As with batteries, many people overlook the 120 volt ac "convenience outlet" as a source of power. Actually, batteries and ac outlets are called "power sources" rather than power supplies.

Power supplies, on the other hand, usually are classified into four different types (with examples):
1. dc in, dc out (converters).
2. dc in, ac out (inverters).
3. ac in, ac out (step up or step down transformers).
4. ac in, dc out (regulated or unregulated power supply).

CONVERTERS

The first type of power supply on our list has a dc input and gives a dc output. It is known as a "converter" because it converts dc from one level to another. With a simple converter circuit, you can convert the 6 volt dc of a battery to 125 volts.

Most converter circuits use two transistors, Fig. 19-1. This is known as an "oscillator circuit." The two transistors are said to oscillate or switch back and forth.

When a converter is connected to the primary of a transformer, it will produce what the transformer "sees" as ac. If you connect the secondary of the transformer across a full wave bridge, it will produce dc output. At this point, however, all you need to know is that a converter will change one level of dc to another level of dc. Shortly, you will see that many things can be done to the output.

INVERTERS

The second type of power supply on our list has dc for an input and gives an ac output. It is known as an "inverter" and works much the same way as the converter. However, when dc is changed to ac in the inverter, it is not necessary to rectify the output of the transformer, Fig. 19-2. With the inverter, it is possible to get a square wave out of the ouput of the transformer. The voltages and currents you are working with will determine the size of the resistors, capacitors and other components.

LINE REGULATORS

The third type of power supply has an ac input and an ac output. It is called a "line regulator" or "voltage changer." Its main purpose is to change ac from the input to either a higher or lower voltage on

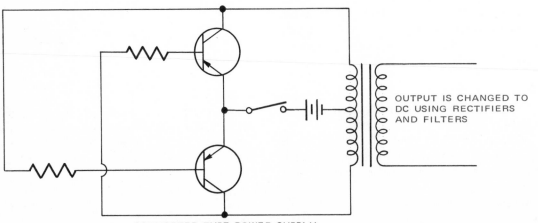

OUTPUT IS CHANGED TO DC USING RECTIFIERS AND FILTERS

CONVERTER TYPE POWER SUPPLY

Fig. 19-1. A converter circuit uses two transistors that switch back and forth to create ac at the transformer input. Output is changed back to dc at a different level than the power source.

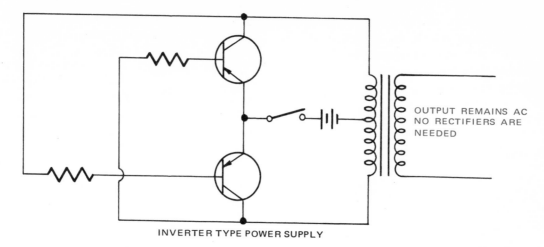

INVERTER TYPE POWER SUPPLY

OUTPUT REMAINS AC
NO RECTIFIERS ARE
NEEDED

Fig. 19-2. An inverter circuit is similar to the converter circuit except that the ac created by the two transistors remains ac on the output.

the output. You recall from the chapter on transformers that this was done with a step up or step down transformer. If you want a higher voltage, use a step up transformer. If you want a lower voltage, use a step down transformer.

AC TO DC

The fourth type of power supply has an ac input with a dc output. Because ac to dc is the most common type, we will cover it in more detail than any of the others.

The ac to dc type of power supply can be either:
1. Regulated.
2. Unregulated.

REGULATED POWER SUPPLIES

A regulated power supply, Fig. 19-3, is one which will maintain a constant output voltage or current. The output is constant, even though there may be a change in the line voltage, output load or temperature of the air around the device (ambient temperature). An unregulated power supply on the other hand, will not keep the voltage or current constant.

The regulated type of power supply would seem to have a big advantage over the unregulated type. You would think that everyone would use the regulated type. However, the regulated type of power supply costs much more than the unregulated type.

UNREGULATED POWER SUPPLIES – HALF WAVE

The easiest type of power supply to understand is one which will give us a half wave rectified output, Fig. 19-4. If we connect one diode or rectifier in a circuit, it will give us pulsating dc since only one alternation can get to our load. The other alternation is blocked because it is the wrong polarity.

You recall that a diode must be biased with the

Fig. 19-3. A high voltage regulated power supply will produce a constant output voltage regardless of load changes. (Heath Co.)

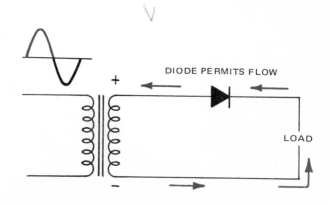

DIODE PERMITS FLOW
LOAD

Fig. 19-4. This is a sine wave used as the input to an unregulated power supply. On the first half cycle, electrons flow through the diode because it has correct bias voltage.

correct polarity before it will allow electrons to flow. In Fig. 19-4, you can see that the first half cycle will pass the electrons. On the second half cycle, Fig. 19-5, they are blocked. This results in a wave that is dc but pulsing on and off every half cycle, Fig. 19-6.

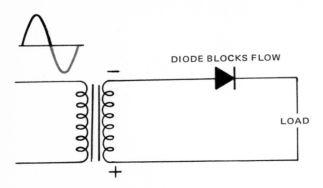

Fig. 19-5. On the second half cycle of the sine wave of the unregulated power supply, the diode is not biased correctly and will block the flow.

Fig. 19-6. Pulsating dc comes from the power supply because the negative half cycle of the sine wave is blocked.

POWER SUPPLY FILTER

There are ways to help reduce this pulsing action or "ripple." One way is to wire a capacitor in parallel with the load, where it serves as a filter (see Chapter 15). In this way, the capacitor will charge during the first half cycle, Fig. 19-7. The first pulse of dc will also pass through the load. The second half cycle is when the diode will not conduct electrons. At this time, the capacitor will go to work. Instead of falling to zero, the

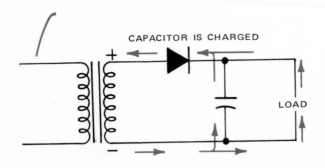

Fig. 19-7. With a capacitor added to the circuit shown in Fig. 19-4, we get flow through the diode, and the capacitor is charged on the first half cycle.

capacitor will discharge. See Fig. 19-8.

During the second half cycle, the load "sees" a continued flow of electrons in the same direction as during the first half cycle. And, the flow of electrons from the capacitor will not run out before the next half cycle begins. Remember that the time needed for a capacitor to discharge is about five RC time constants. As it starts to discharge, another charging action will follow, keeping the capacitor almost fully charged.

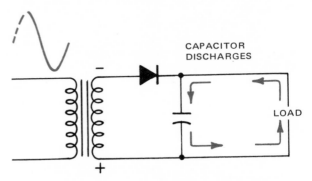

Fig. 19-8. As the sine wave falls, the capacitor will discharge through the load until the diode is again biased to conduct.

The charging and discharging actions of the capacitor will result in a wave similar to the one shown in Fig. 19-9. Note that the output of this power supply is much higher than the one without the capacitor. The capacitor takes away some of the pulsing that results from using only one diode. The "ripple" will be less, but you cannot eliminate it with this type of power supply.

PULSATING DC WITH REDUCED RIPPLE RESULTS FROM USING CAPACITOR

Fig. 19-9. By using a capacitor in our power supply circuit with one diode, we will eliminate part of the sine wave, reduce ripple and raise the output.

CENTER TAP TRANSFORMER

One way to reduce the power supply output ripple is to install a transformer using a center tap. Then, if you add a second diode, the first half cycle will take the path shown in Fig. 19-10. The load will "see" the first half cycle as a pulse. The second half cycle will also pass a pulse. See Fig. 19-11.

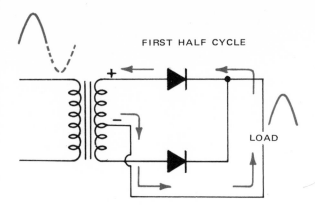

FIRST HALF CYCLE

LOAD

Fig. 19-10. By using a center tap transformer with two diodes, we will get the path of flow shown in this schematic for the first half cycle of the sine wave.

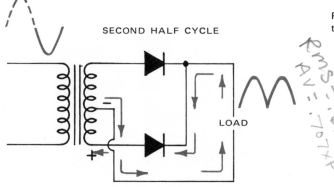

SECOND HALF CYCLE

LOAD

Fig. 19-11. Working with the circuit set up for Fig. 19-10, this is the second half cycle of the wave. Note that we have used only one half of the transformer voltage for each half cycle.

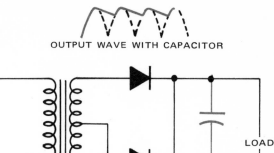

OUTPUT WAVE WITH CAPACITOR

LOAD

OUTPUT WITHOUT CAPACITOR

Fig. 19-12. Placing a capacitor in the circuit with a center tap transformer with two diodes will give us a power supply with very little ripple.

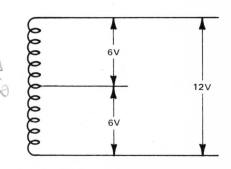

6V

6V

12V

Fig. 19-13. Output voltage from a center tap transformer is one half of the total output.

Using two diodes will give you a real advantage. Now, it is possible to get both half cycles through the load. This means there will be much less "ripple." There no longer will be a large gap to be filled by a discharging capacitor.

If a capacitor is used in this circuit, Fig. 19-12, the output ripple will be very small. The capacitor will discharge for a shorter period of time. The output wave would be much smoother if you saw it on an oscilloscope.

You might think that this center tap transformer type of power supply would be the best one to make. However, it has problems too. The biggest problem can be seen by looking at the circuit. Note in Figs. 19-10 and 19-11 that each alternation used only one-half of the transformer. This means that only half of the voltage is useful at any one time. If 6 volts is needed, the transformer must give 6 volts from each side to center, Fig. 19-13. The full 12 volts could not be used.

FULL WAVE RECTIFIERS

There *is* a way that the full output of the transformer can be used. First, look at just one half cycle, Fig. 19-14. Here, we are using two diodes to pass electrons through the load. Pay close attention to the direction of electron flow through the load. On the second half cycle, Fig. 19-15, we can use two other diodes. The direction through the load is not changed. Full transformer output can be used.

If we have a 12 volt transformer, the full 12 volts will be supplied to the load. This is an advantage, but it means that we must use four diodes. You recall that we worked with four diodes in an earlier chapter. The finished power supply, Fig. 19-16, is our full wave bridge rectifier. You can buy this bridge rectifier as an

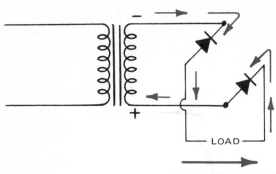

Fig. 19-14. With two diodes connected to the power supply circuit, note how the first half cycle of flow passes through the load. Also note the direction of flow.

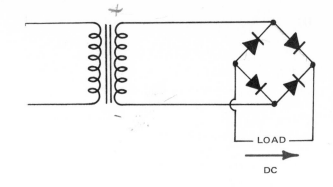

Fig. 19-16. Putting all four diodes or a full wave bridge rectifier into the Fig. 19-14 circuit will result in full transformer dc output through the load.

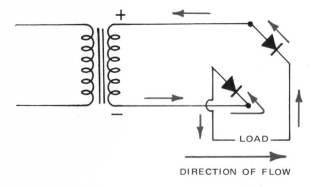

Fig. 19-15. By using two other diodes in the Fig. 19-14 circuit, flow through the load is maintained in the same direction on the second half cycle of flow.

Fig. 19-17. This small dc power supply will deliver up to 30 volts. (Acme Electric Corp.)

assembly rather than buying four separate diodes. This power supply can be filtered to further smooth out and almost eliminate the ripple.

POWER SUPPLY SIZE

Power supplies are available in quite a range of sizes. The small dc type shown in Fig. 19-17 can produce about 30 volts. Plants in industry use much larger power supplies. Industrial units must provide the power needed for:

1. High frequency heating.
2. Annealing furnaces used to soften metals.
3. Large industrial controls for giant spot welders.
4. Hardening the surfaces of metals and gears.
5. Huge presses for stamping out automobile sheet metal parts.

There are many other places where large power supplies are needed. Many of those used in industry are operated around 460 volts, Fig. 19-18. Most important, however, is that the voltage must be controlled.

Fig. 19-18. Large industrial power supplies often are mounted on I beams. This one operates at 460 volts. (Pillar Corp.)

INDUSTRIAL POWER SUPPLIES

With large changes on the output, unregulated power supplies lose control. Often, therefore, large and small power supplies are regulated. In some cases, the voltage is controlled within plus or minus one percent. Others must do even better.

Fig. 19-19 shows the inside of a 300 kW, 3 kHz power supply. Over a dozen semiconductor devices are built into it. These large units are water cooled to aid in getting rid of the heat that is created during loading. In this way, they are able to run much longer without overheating. Water cooling is also very important when it comes to regulation. Both large and small power supplies can be regulated to control the output within close tolerances.

FEEDBACK

There are several ways to regulate the output of a power supply. Most regulation works on the idea of "feedback." Feedback is a simple idea. For example, we may ask "How tall are you?" Your answer would be feedback. You would be giving information back to us. *Feedback is information being fed back to the source.*

Fig. 19-19. Arrow points to one of 15 water-cooled SCRs used in this power supply unit to keep the heat down.

Suppose that transformer output is 2 amps. If we added 1 more amp to the load, it would climb to 3 amps. If this information could be relayed to the input, it would be known as feedback.

Feedback is used in regulated power supplies. This means that more circuitry must be added to the power supply. As a result, costs will go up.

Among other things, feedback is used to tell the input when the output load increases or if more heat is being generated at the output. This is important because one of the main reasons for using a regulated power supply is to control the output. In many cases, controlled output is needed because a change in 1 or 2 volts can turn on a transistor or SCR. You recall that an SCR carrying 200 amps can be affected by a change of only a few volts.

There are many other uses for regulated power supplies. However, there is one main point to remember: *regulated power supplies control the output by using feedback to the input.* In this way, we can maintain close control of the output voltage.

POWER SUPPLY PROJECT

A 12 volt power supply, Fig. 19-20, seems to be the one most needed for use in the home. This need came about when car tape players became popular. Many people wanted to remove them at night for use in playing tapes at home. Therefore, a good 12 volt power supply was needed.

Here are plans for a power supply that can be built at a reasonable cost. The amount of filtering could be improved, but it is not really necessary. No voltage regulator is needed. Tape players are regulated for constant speed.

The circuit has been laid out for you, showing diode type of construction, Fig. 19-21. Keep the leads short and compact, so that the assembly can be placed in a

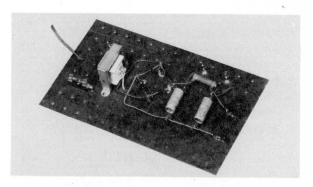

Fig. 19-20. This power supply unit can be used to power car tape players in the home. (Project by Bill Martinez)

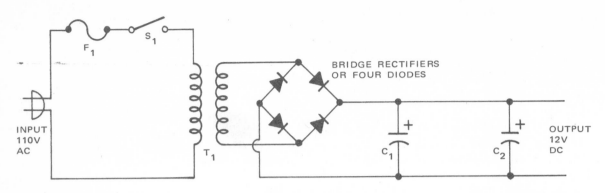

Fig. 19-21. Schematic shows hookup for power supply for home use. $C_{1,2}$—Electrolytic capacitors. F_1—Fuse. S_1—Switch. T_1—110/12V transformer.

small case. Buy a bridge rectifier or use four separate diodes. Use whatever capacitors you can find to make up the capacitance to around 6000 μF. Note that they are electrolytic capacitors.

Obtain the following parts and assemble them according to the construction procedure given.

No.	Item
1	110/12 volt transformer.
1	Bridge rectifier or four diodes.
2-3-4	Capacitors to total 6000 μF at 25 WVDC.
1	SPST switch.
1	Pilot light.
1	Fuse.
1	Line cord and plug.
1	Case.
2	Output terminals.

CONSTRUCTION PROCEDURE

Assemble the parts and hook up the circuit as shown in Fig. 19-21.
1. Lay out parts and determine size of case.
2. Mark positions of switch, light, output terminals and other attachements on case.
3. Drill all holes in case.
4. Bend case to shape. Paint it or cover it with contact paper.
5. Mount parts in case.
6. Wire circuit according to Fig. 19-21.
7. Install fuse in series with transformer primary.
8. Add decals to front of case.

THE OSCILLOSCOPE

The oscilloscope, Fig. 19-22, is a testing instrument designed to give a technician a lot of information about circuits. Most people just call it a "scope."

There are many types of scopes. A common type used by those beginning the study of electronics is shown in Fig. 19-22. Most scopes, however, have the same basic parts and will read the same basic values in a circuit.

Fig. 19-22. Typical oscilloscope used for taking measurements in electronic circuits. (Hickok Teaching Systems, Inc.)

CATHODE RAY TUBE

The heart of all scopes is the cathode ray tube (CRT), Fig. 19-23. The CRT makes it possible for you to see what the circuit is doing.

The inside face (screen) of the CRT is coated with phosphor. Since phosphor will give off a glow when an electron beam strikes it, this glow will be seen on the screen. Your eyes will see the glow for a short time, even after the electron beam has moved. This is called "persistence of vision." It is the same thing that lets you see a picture on a TV screen instead of a bunch of dots and lines.

Persistence of vision also takes the flicker out of movies. As you know, a movie is really a group of still pictures being run through a projector. It is this

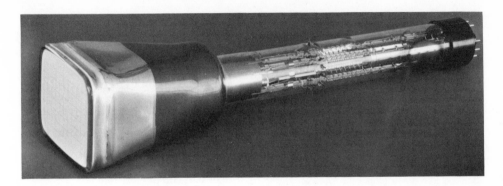

Fig. 19-23. This cathode ray tube (CRT) is ready for installation in a scope. Note its many parts. (Hewlett-Packard)

persistence of vision that makes it possible for the electron beam to move across the screen and show up as a line to your eyes.

The scope is nothing more than a voltmeter. It will however, measure voltage changes during a fraction of a second. This makes it very valuable for use in checking circuits that have a pulse lasting just a fraction of a second. See Fig. 19-24.

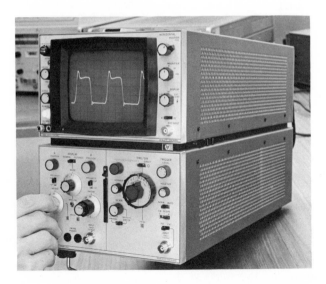

Fig. 19-24. Technician uses an oscilloscope to check a fast pulse circuit. (Hewlett-Packard)

Take a pencil and move it quickly across the surface of a pool of water. Note that you will create a line which quickly disappears. If you start at point 1 and draw a line to point 2, this will take a certain amount of time. If you draw a line to point 3, a little further on, it will take a little longer. This is the same way that a scope works, Fig. 19-25. It takes a certain amount of time to move a beam from the left to the right side of the screen. Therefore, the horizontal axis of the scope is used to show the amount of time.

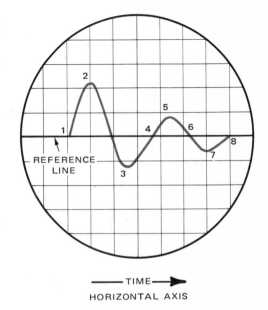

Fig. 19-25. The points along the line being traced out by a scope can have many different shapes.

Often you hear someone ask how tall a mountain is, or how deep a lake is. What they really want to know is "how high" or "how deep" it is with respect to the ground at sea level. In other words, they are using the ground as a reference axis. A scope does the same thing. Earlier you saw traces of sine waves and many of them also had a reference line. This reference line gives us the point from which we judge the value of our voltage.

GRATICULE

The screen of the scope, then, shows us the time on the horizontal axis and the voltage on the vertical axis, Fig. 19-26. The horizontal axis is also called the X-axis. The vertical axis is also called the Y-axis. These graphs are very similar to police lie detectors, sales graphs showing profits or losses, or medical records of patients

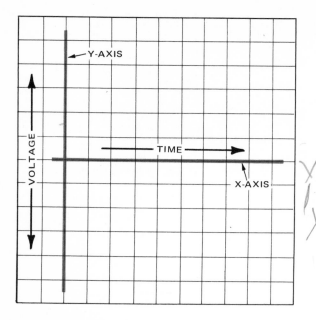

Fig. 19-26. The horizontal axis on a scope screen shows the time. The vertical axis shows the voltage.

with heart trouble.

To help make measurements on a scope, a graticule is mounted on the front of the CRT. See Fig. 19-27. The graticule has a series of lines on the order of graph

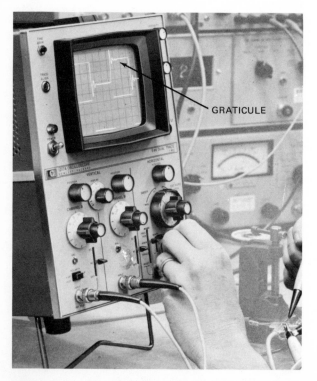

Fig. 19-27. A graticule is placed in front of a CRT to make it easier to read scope traces. (Hewlett-Packard)

paper. The horizontal spaces help us measure the amount of time needed (time interval). The vertical space gives us information about the voltage.

DEFLECTION PLATES

Inside the CRT are groups of deflection plates, Fig. 19-28. *Deflection plates make the trace move on the screen.* You recall that an electron has a negative charge. Therefore, the electron beam is nothing more than a stream of electrons.

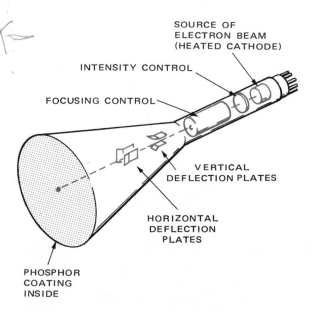

Fig. 19-28. Location of various controls and deflection plates inside a cathode ray tube.

The electrons come from the electron gun, which is in the small end of the tube. They are forced to travel down the length of the tube toward the screen. The tube is a big glass envelope with a vacuum inside.

As the beam travels down the tube, it will pass between two sets of deflection plates. *Remember, unlike charges attract.* If the negatively charged electron beam passes between a set of plates, it will be attracted to one of the plates if it has a positive charge. Because of the high speed of the beam, it will move toward or be deflected toward the positive plate. The beam can be bent up or down. See Fig. 19-29. The amount of deflection depends directly on the amount of voltage put on the plates. If you had a small voltage, the deflection will be small. A large voltage will cause a large deflection.

The first set of plates are the vertical deflection plates. They will move the beam up and down on the screen. The second set of plates are the horizontal deflection plates. They cause the beam to move sideways on the screen. If you turn the horizontal or vertical centering dials on the scope, you will see how they move the beam.

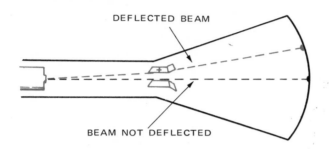

Fig. 19-29. The electron beam can be deflected by changing the charge on the deflection plates. This attracts the beam and moves it on the screen of the CRT.

INTENSITY CONTROL

By adjusting a control, you can stop the beam from appearing on the face of the CRT. The scope shown in Fig. 19-22 has an intensity control. You can make the beam very bright by turning this control one way, or you can make the beam disappear by turning it in the opposite direction.

To see how this intensity control works, first place a wire screen (grid) in front of the beam so you can control it. Then, put a small negative charge on the screen. Some of the beam will be blocked and will, therefore, be less bright (less intense). If you put a strong negative charge on the wire screen, you can block the beam completely. See Fig. 19-30. This is true because like charges repel. The grid also helps to focus the beam when we are not blocking it.

I.C. WARNS:

"If you make the beam very bright, you can damage the face of the CRT. The beam must be moving across the face of the screen. If not, you will burn the phosphor. This will show up as a dark, burned spot on the face of the CRT."

Burned phosphor will not disappear; the CRT is ruined. To avoid this problem, turn down the intensity of the beam or take the beam out of focus when it is not moving on the screen.

FOCUS

The beam is focused by another dial on the front of the scope. Focusing the scope is similar to using a magnifying glass to focus sunlight down to a pinpoint

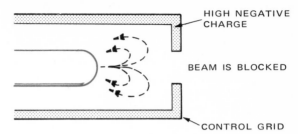

Fig. 19-30. By putting a high negative charge on the control grid, you can block the electron beam.

size. If you were to focus the spot of sunlight on your hand, you would burn yourself. This is the same thing that happens to the phosphor of the scope with a finely focused beam.

Again, when using the scope, move the beam across the face of the CRT so that it will not build up enough heat to burn the phosphor. Also, when the scope is on but not being used, turn down the intensity and/or put the beam out of focus.

PRECAUTIONS

Scopes usually have two types of power supply. One carries medium amounts of current at about 300 volts dc. The other carries low currents at over 1000 volts dc. With this in mind, do not take the protective metal case off the scope. Only those who know how to handle and repair scopes know where these voltages are present. See Fig. 19-31.

Fig. 19-31. Scopes and other electronic equipment have some complicated circuits. Do not work on this equipment until you know where to look for voltage levels that may be dangerous. (Hewlett-Packard)

Another caution concerns the CRT, which has a vacuum inside its glass housing. Any CRT or picture tube on a television receivier should be handled with extreme care. Either of these tubes can cause severe cuts and injury if dropped or broken. The chemical coating inside the CRT adds to this hazard. It can cause problems of infection to cuts from the broken glass. Technicians working with these tubes use a special technique to release the vacuum very slowly. Do not try it unless you are given proper instructions on how to do it safely.

SCOPE HOOKUPS

There are a number of buttons, dials and jacks on the front of a scope. Each is used to make certain types of measurements. The jacks are used for plugging in test leads, Fig. 19-32. Then, by connecting these test leads to the circuit, the technician is able to operate the scope and see the results under test.

Some scopes even have two beams so that you can compare two different signals at one time. Each time new features are added to a scope, the more the test results can tell you. This is why scopes come in so many different sizes and shapes.

Fig. 19-32. This technician is using a test lead to pick up a signal from a circuit and take a measurement with a scope. (Autonetics Div. of North American Rockwell)

STORAGE SCOPES AND CAMERAS

Some scopes even have the ability to read one cycle, then store the trace on the CRT. These are called "storage scopes." By using a storage scope, the technician is able to study very high speed signals. It also will let him decide which specific pulse he would like to study. These scopes cost much more than standard models.

Some companies make cameras, Fig. 19-33, so that you can take a picture of the CRT. This is not the same as a storage scope. However, it does provide a way in which you can compare different traces at different times. With it, you can make pictures to show other people if the need arises.

Fig. 19-33. Camera used to take pictures of traces that appear on the CRT. (Hewlett-Packard)

PROBES

Besides the test leads, probes can be used with a scope, Fig. 19-34. There are several types of probes, so the one that you use will depend on the type of circuit being tested. Commonly used probes include:
1. Low capacitance probe.
2. RF probe.
3. Demodulator probe.
4. High impedance probe.

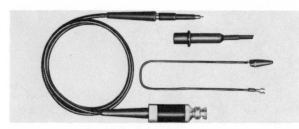

Fig. 19-34. Typical probe used to make checks on circuity with an oscilloscope. (Heath Co.)

Basically, all probes attach to jacks on the front of the scope. Then, they can be held to the circuit or attached to it. When the probes are attached, Fig.

19-35, the technician is free to make adjustments by means of controls on the scope. There are times, however, when you are only looking for voltages or certain waveshapes. Then, you will just touch the probe to the circuit and go on to the next test point.

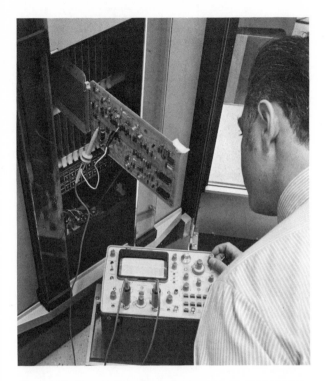

Fig. 19-35. By attaching the probe to a circuit, the technician is free to make adjustments on the scope. (Hewlett-Packard)

MEASURING UNKNOWN FREQUENCIES

The number of things you can measure by using a scope is almost endless. For example, a scope can be used for the measurement of unknown frequencies. To do this, you will need to use a sinewave generator in addition to the scope.

The sinewave generator is an instrument marked off with frequencies which you can set. This check is made by using the known frequency of the sinewave generator to find the value of an unknown frequency, Fig. 19-36. It is often used when working on television receivers.

LISSAJOUS FIGURES

To check for an unknown frequency, you must know about "Lissajous figures." These are a group of figures of various shapes named in honor of the French scientist Lissajous. He discovered that these figures

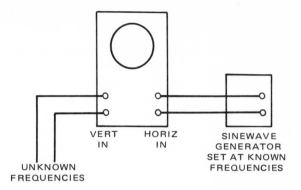

Fig. 19-36. Using the scope and sinewave generator to find unknown frequencies. Lissajous figures will help give the results.

could be developed in several ways. However, we are interested in using them to measure frequencies.

We can demonstrate how Lissajous figures help us find unknown frequencies. First, hook up the scope to the sinewave generator and to the unknown frequency as shown in Fig. 19-36. Then, adjust the sinewave generator until a figure appears on the scope.

Suppose the figure on the scope is a circle. Next, draw a right angle and place it by the figure on the scope, Fig. 19-37. Place it so that the left side of the figure touches the vertical line. Next, set it so that the horizontal line touches the bottom of the figure. Actually, you do not need to draw the right angle because of the graticule on the scope. Finally, count the places where the figure touches each line.

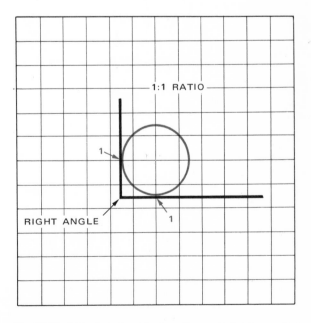

Fig. 19-37. This circle is a Lissajous figure which tells you that the two frequencies being measured are equal. The figure touches each line in one place.

RATIOS FOR LISSAJOUS FIGURES

Our circle in Fig. 19-37 touches the horizontal line in one place, and it touches the vertical line in one place. Therefore, it has a 1 to 1 ratio. This means that the unknown frequency is equal to the frequency of the sinewave generator.

Suppose that our Lissajous figure touched the horizontal line in two places, Fig. 19-38. Now, we have a 2 to 1 ratio. Therefore, the vertical input to the scope would be twice the frequency of the horizontal input signal. This means that our unknown frequency is twice the frequency of the sinewave generator. If the generator had been set on 300 Hz, the unknown frequency would be 600 Hz:

$$\text{Unknown frequency} \times \frac{\text{Points touching vertical line}}{} = \text{Known frequency} \times \frac{\text{Points touching horizontal line}}{}$$

Therefore: Unknown frequency x 1 = 300 Hz x 2
Unknown frequency = 600 Hz

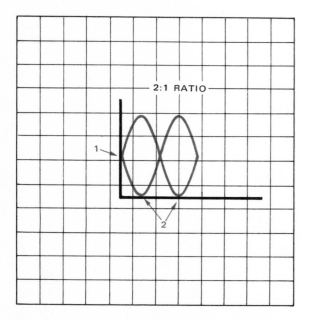

Fig. 19-38. A 2 to 1 ratio Lissajous figure touches horizontal reference line in two places and vertical line in one place.

If we had a 3 to 1 ratio, Fig. 19-39, the unknown frequency would be three times as large as the known frequency. If our sinewave generator had been set at 400 Hz, our unknown frequency would be equal to 1200 Hz:

Unknown frequency x 1 = Known frequency x 3
Unknown frequency   = 400 Hz x 3
Unknown frequency   = 1200 Hz

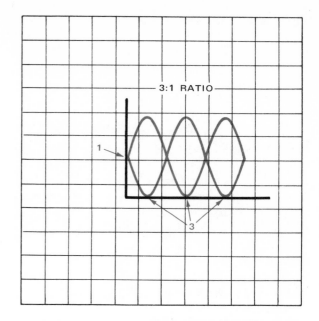

Fig. 19-39. In a 3 to 1 ratio Lissajous figure, the figure touches the horizontal line in three places and vertical line in one place.

All the Lissajous figures do not end up with high ratios on the horizontal axis. Some will have the greater number touching the vertical axis, Fig. 19-40. In this case, the ratio is 1 to 3. From this, it is easy to figure out what the unknown frequency would be.

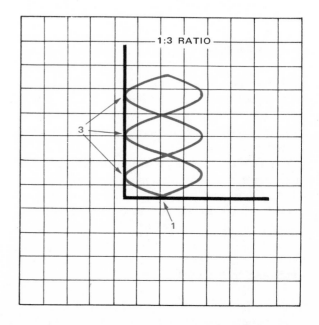

Fig. 19-40. In a 1 to 3 ratio Lissajous figure, the figure touches the horizontal line in one place and vertical line in three places.

GETTING READABLE RATIOS

In many cases, you will get figures that touch the horizontal axis and vertical axis several times and end up with ratios of 3 to 5, 4 to 6 or some other combination. At this point, you will have trouble counting the number of spots where the figures touch the horizontal and vertical lines. Then, your best bet is to adjust the sinewave generator to different frequencies until you can get a ratio that is easy to read.

In some cases, the figures will tend to turn slowly. When this happens, the sinewave is not adjusted so that the ratio reads in whole numbers. The Lissajous figure may turn in either direction. It still means the same thing. Again, the best solution is to change the known frequency until you get a figure that is easy to read. If this is not possible, you will have to try some other method used to check for unknown frequencies.

PRACTICE USING SCOPE

The use of Lissajous figures is just one example of what a scope can do for you. There are many other uses possible. However, it is best to check on the manufacturer's suggested hookup for any check you wish to make.

Each company makes their scope a little different from the others. The only way to really learn how to hook them up and take readings is through practice. Too often people see the scope and are afraid to use it. Some companies have scopes in their electrical shop just collecting dust.

Once you learn how to read a scope, you will have one of the best tools available for checking circuits and correcting faults. A scope can save many hours of troubleshooting when used by people who know how to read them.

SIGNAL TRACER PROJECT

Do you need a small signal tracer to help find trouble spots in audio systems? You can build this project, Fig. 19-41, quickly and easily. Basically, it amplifies minute signals that can be heard on the earphone. As you probe the circuit, the point where you lose the sound is where your problem is.

Obtain the following parts and assemble them according to the construction procedure given.

No.	Item
1	9 volt battery.
1	0.0022 μF capacitor (C_1).
2	2.0 μF electrolytic 15 volt capacitors ($C_{2,3}$).

1	IN38B diode.
1	2N464 transistor.
1	1.2 meg ohm resistor.
1	2 meg ohm potentiometer with SPST switch, or substitute an SPST slide switch.
1	Probe and alligator clip.
1	Battery clip hookup.
1	Earphone and earphone jack.
1	Chassis box and perfboard.
	Miscellaneous hardware and wiring.

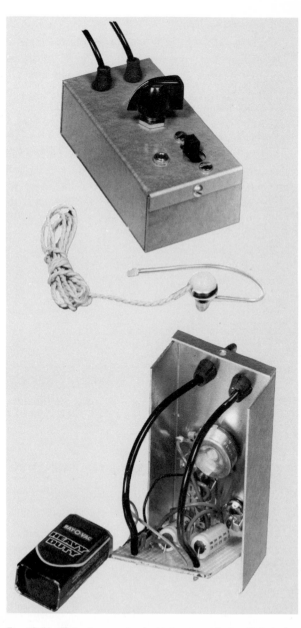

Fig. 19-41. This shop-made signal tracer project will help in troubleshooting audio systems. Top—Finished project. Bottom— Assembled parts. (Project by Mike O'Berski)

CONSTRUCTION PROCEDURE

To build the signal tracer:

1. Lay out the probe lead and battery lead, earphone jack and potentiometer or slide switch to determine correct hole size and hole locations in cover of chassis box.
2. Drill holes and mount parts in cover.
3. Mount parts on perfboard as shown in Fig. 19-41.
4. Connect all parts in circuit, Fig. 19-42. Observe correct polarity and make mechanically sound connections.
5. Solder all connections.
6. Coat all soldered connections with a heavy layer of nonconductive material. The circuit must be insulated from the metal box when the cover is installed.

To use the signal tracer, plug in the earphone lead and turn on the switch. Clip the input ground lead to the chassis of the audio system to be tested. Probe the circuit with the "hot" lead, starting at the input audio stage. Adjust the potentiometer, or slide switch, until you hear a very small audio signal. Then, using the probe, work your way toward the speaker until the problem area is pinpointed by the complete loss of sound.

Do not leave the power "on" when the unit is not in use. It may damage the transistor.

TEST YOUR KNOWLEDGE

1. Which of the following power supplies is also known as a converter:
   a. dc in, dc out. *CONVERTER*
   b. dc in, ac out. *INVERTER*
   c. ac in, ac out. *STEP UP OR STEP DOWN*
   d. ac in, dc out. *UNREGULATED & REGUL*
2. Which of the following is also known as an inverter:
   a. dc in, dc out.
   b. dc in, ac out. *INVERTER*
   c. ac in, ac out.
   d. ac in, dc out.
3. What is a regulated power supply? *Voltage Remains C*
4. A half wave rectifier will produce a *HALF* wave.
5. One way to reduce the ripple in a wave is through the use of *CAP.* or *CENTER TAP XFMR*
6. How many diodes are used in a full wave bridge rectifier circuit? *4*
7. How many diodes are needed in a full wave rectifier that uses a transformer with a center tapped secondary winding? *2*
8. Explain the advantage of a full wave rectifier. *NO WAS VOLTAG*
9. What is a graticule?
10. An unknown frequency can be found on a scope with the use of a sinewave generator and Mossimbique figures. True or False? *FALSE*

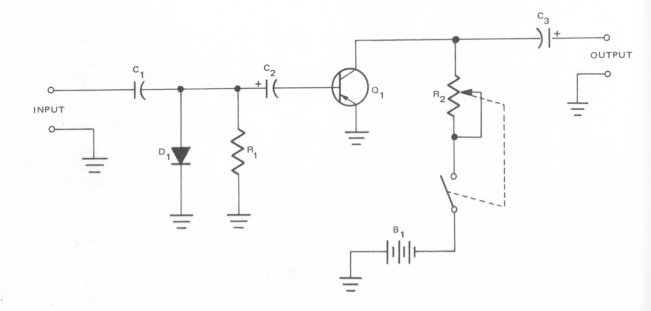

Fig. 19-42. Schematic for signal tracer includes: B_1—Battery. C_1—Capacitor. $C_{2,3}$—Electrolytic capacitors. D_1—Diode. Q_1—Transistor. R_1—Resistor. R_2—Potentiometer or slide switch.

11. As the electron beam moves down the length of the CRT, it is moved up and down or sideways by DEFlection PIATES

12. In order to make some measurements possible on an oscilloscope, it is necessary to connect _____ to the input leads.

THINGS TO DO

1. Find out the best and safest method to release the vacuum from a CRT or picture tube from a TV receiver.

2. How is the electron beam on a scope generated? Why does it move toward the front of the CRT? How fast does the beam move?

3. You may have heard the word "astigmatism" when dealing with scopes. What is it? How does it affect the scope?

4. Make a list of the different types of probes used with a scope. Tell the advantages of each.

5. What types of power supplies do you work with? Are they regulated or unregulated power supplies? Why do you use that particular type? What does a good regulated power supply cost? What does an unregulated power supply cost?

6. How did Lissajous make his discovery? How did he use it?

7. How many different types of jobs can be found which require electrical or electronics experience? What do these jobs pay? Does an electrical engineer know how to use all electrical equipment? What does this person do that is different from what a technician does?

8. What kind of background does a person that wires homes have? Must the work be inspected? By whom? Why?

9. What is the difference between a dual trace and a dual beam scope?

10. How is it possible for a scope to retain the image on the CRT after only one sweep? How does a camera mounted on a scope catch the speed of the trace on scopes other than storage scopes?

11. What is the price range for various types of scopes? What do some of the accessories cost?

12. Why are some electronic calculators so much more expensive than others?

13. How does the memory on a calculator work? How is it able to add and subtract?

14. How does a digital clock work? What makes it not count over 12 hours or 24 hours for the various types?

15. What is a Schmitt trigger? How is it used?

Special scopes can be built into the machines they are meant to check. The equipment shown is in a television studio.

Chapter 20
CAREER OPPORTUNITIES

Some people think that job opportunities in the electricity/electronics field are changing too fast. Consider, however, that "changes" are the main reason why so many jobs are available in this field. Changes in electricity/electronics have resulted from new demands of the world around us. For example, the need to store large amounts of information for later recall led to the development of computers.

In another example, people in countries around the world wanted to be warned about storms to prevent disasters. They no longer wanted to be at the mercy of the weather. If they knew in advance about the coming of a hurricane or tornado, many lives could be saved. Weather satellites now in use provide us with that early warning information.

When you stop and think about it, you can see that this same need created many other jobs in research, development and communications.

FIVE AREAS OF OPPORTUNITY

Jobs in electricity/electronics generally can be broken down into five areas of career opportunities: construction workers, maintenance workers, technicians, engineers and supervisors. To get any of these jobs, you need to start preparing early, Fig. 20-1. The more you know about electricity and electronics, the better and higher paying jobs you can get.

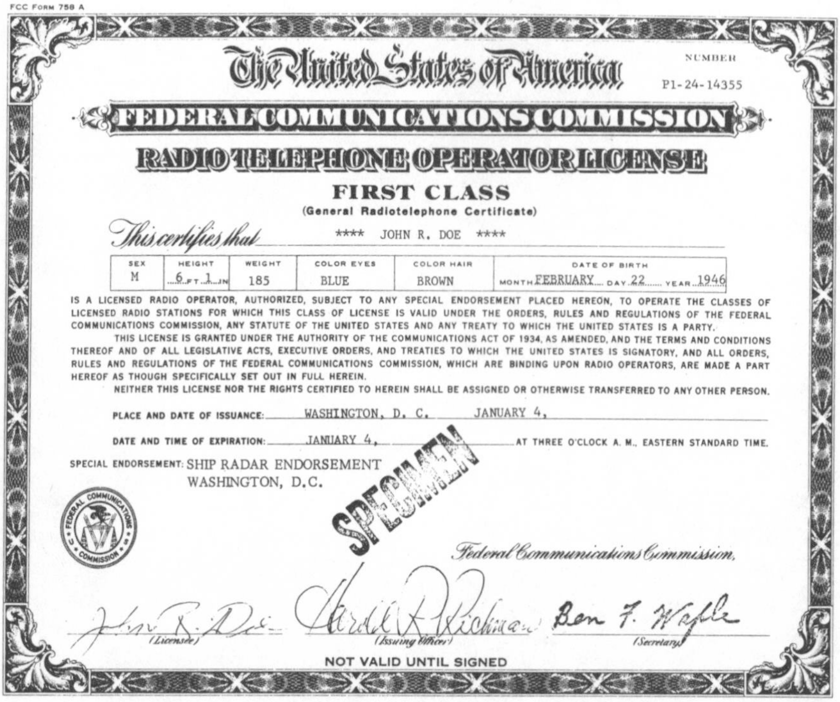
FCC FORM 758 A

The United States of America

NUMBER P1-24-14355

FEDERAL COMMUNICATIONS COMMISSION

RADIO TELEPHONE OPERATOR LICENSE

FIRST CLASS

(General Radiotelephone Certificate)

This certifies that **** JOHN R. DOE ****

SEX	HEIGHT	WEIGHT	COLOR EYES	COLOR HAIR	DATE OF BIRTH
M	6 FT 1 IN	185	BLUE	BROWN	MONTH FEBRUARY DAY 22 YEAR 1946

IS A LICENSED RADIO OPERATOR, AUTHORIZED, SUBJECT TO ANY SPECIAL ENDORSEMENT PLACED HEREON, TO OPERATE THE CLASSES OF LICENSED RADIO STATIONS FOR WHICH THIS CLASS OF LICENSE IS VALID UNDER THE ORDERS, RULES AND REGULATIONS OF THE FEDERAL COMMUNICATIONS COMMISSION, ANY STATUTE OF THE UNITED STATES AND ANY TREATY TO WHICH THE UNITED STATES IS A PARTY.

THIS LICENSE IS GRANTED UNDER THE AUTHORITY OF THE COMMUNICATIONS ACT OF 1934, AS AMENDED, AND THE TERMS AND CONDITIONS THEREOF AND OF ALL LEGISLATIVE ACTS, EXECUTIVE ORDERS, AND TREATIES TO WHICH THE UNITED STATES IS SIGNATORY, AND ALL ORDERS, RULES AND REGULATIONS OF THE FEDERAL COMMUNICATIONS COMMISSION, WHICH ARE BINDING UPON RADIO OPERATORS, ARE MADE A PART HEREOF AS THOUGH SPECIFICALLY SET OUT IN FULL HEREIN.

NEITHER THIS LICENSE NOR THE RIGHTS CERTIFIED TO HEREIN SHALL BE ASSIGNED OR OTHERWISE TRANSFERRED TO ANY OTHER PERSON.

PLACE AND DATE OF ISSUANCE: WASHINGTON, D. C. JANUARY 4,

DATE AND TIME OF EXPIRATION: JANUARY 4, AT THREE O'CLOCK A. M., EASTERN STANDARD TIME.

SPECIAL ENDORSEMENT: SHIP RADAR ENDORSEMENT WASHINGTON, D.C.

SPECIMEN

Federal Communications Commission,

(Licensee) (Issuing Officer) Ben F. Waple (Secretary)

NOT VALID UNTIL SIGNED

Fig. 20-1. Some jobs in electricity/electronics require that you earn a special license. (Cleveland Institute of Electronics, Inc.)

CONSTRUCTION WORKERS

The first area of opportunity we will look at is construction, Fig. 20-2. This includes workers who build hundreds of the items we use everyday. They do the wiring of our homes, factories and schools. Their work gives us the electricity we have grown to depend on in our everyday living.

CONSTRUCTION WORKERS must have the skills to be sure that all wiring is safely installed. To help insure this, cities enforce local safety codes and the

Fig. 20-2. These tools and parts are needed for constructing automotive electrical and electronic circuits.

National Electrical Code. All building wiring is done according to these codes. CHIEF ELECTRICAL INSPECTORS are sent out by the city to approve the wiring before it is put in service.

The construction area of electricity/electronics also includes jobs for ASSEMBLERS. These workers wire and solder various components for television sets, stereo sets, printed circuit boards, computers, industrial controls, citizen band (CB) radios, airplanes and almost any other electronic device.

ASSEMBLERS must be able to identify many electronic components and install them into circuits. Once the parts are installed, QUALITY CONTROL

INSPECTORS check out the finished work. Inspectors, Fig. 20-3, put the wired circuits through a series of tests. Some circuits are so small that they have to be inspected under a microscope, Fig. 20-4.

Fig. 20-3. A quality control inspector tests a radar system. (Autonetics Div. of North American Rockwell, Inc.)

MAINTENANCE WORKERS

Once the products have been built, we must have a group of qualified MAINTENANCE WORKERS to maintain them. Those doing maintenance work must have skill and knowledge, Fig. 20-5. They are the ones who keep so many of our home appliances in good operating condition. Maintenance workers are the service experts who fix our dishwashers, ranges, typewriters, radios and television sets.

Another group of maintenance workers are needed to keep TV and radio stations broadcasting. They keep the transmitters working day after day. Other groups in communications fix CB radios, aircraft communication systems and all the telephone and microwave equipment which we rarely see.

Trained persons are needed to keep our large and small manufacturing firms running. Most schools cannot do the necessary training. Each company has its own specialized equipment. Since it is almost impossible to hire people with the needed skills, the companies do their own training.

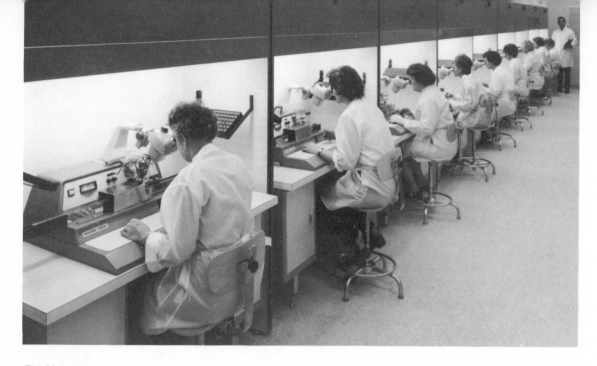

Fig. 20-4. Inspectors put integrated circuits through tests to be sure they will work before being installed in a circuit. (Sprague Electric Co.)

Fig. 20-5. A journeyman maintenance worker uses some of his skill and knowledge to check an aircraft component. (Autonetics Div. of North American Rockwell)

Fig. 20-6. This part of on-the-job training is performed at the direction of the instructor. Next, the employee will practice the task just learned. (Packard Electric Div., General Motors Corp.)

Many companies set up training programs with the cooperation of the employees and the Federal government. An APPRENTICE has to go through an On-the-Job Training (OJT) program, Fig. 20-6. Usually, this is combined with special classes or self-study courses at home. It takes about 8000 hours (four years) to complete the program.

Upon finishing the training, the person is granted JOURNEYMAN ELECTRICIAN status. With the journeyman card, an ELECTRICIAN has an advantage when job hunting is necessary. Every company knows that this job hunter has had at least four years of planned electrical training.

A number of companies have set up another kind of television transmission. See Fig. 20-7. Called close circuit television (CCTV), Fig. 20-7. They use CCTV to help them troubleshoot the equipment in their plants. First, they make video tapes, then play them back for viewing. When troubleshooting high-speed equipment, the tapes are played back in slow motion. Skilled maintenance workers are needed to keep this type of equipment operating.

Fig. 20-7. Using a closed circuit television (CCTV) setup to troubleshoot high-speed machinery. The tape is played back later to look for faults with machine operation. (Guide Div., General Motors Corp.)

TECHNICIANS

A TECHNICIAN, Fig. 20-8, is specially trained to keep the manufacturer's equipment running. Generally, the basic training is done by a vocational technical school or community college. In addition, some companies send their technicians to programs set up by their equipment suppliers.

Once trained, the job of the technician is to work with existing equipment. But, the job involves more than that. A technician also works with engineers to help design and build even better and better machines. To do this, a technician must be able to work with hands and mind, Fig. 20-9. If you do become a technician, you must continue to keep up to date. Each new component discovered and built by the electronic companies must be understood. You might be able to use them in new machine designs.

New components may make it possible for you to design faster and more accurate machines. Millions of dollars are saved by companies whose technicians can make their production operations faster and more accurate.

Technicians may be found working in the following places:
1. Electronically controlled steel mills.
2. Automotive electronics design departments.
3. Laboratories.
4. Electronic companies such as Hewlett-Packard, Texas Instruments, Bell Laboratories, General Electric and Westinghouse.

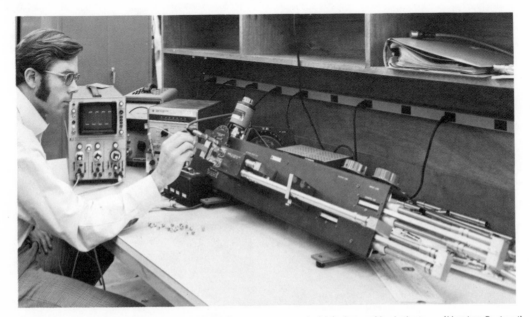

Fig. 20-8 Technicians with the ability to read scope traces are in high demand by industry. (Hewlett-Packard)

Fig. 20-9. Job opportunities are growing for people having electricity/electronics training. Here, an engineer and a technician are running operational tests on an electronic device. (Autonetics Div. of North American Rockwell)

These are only a few of the hundreds of places that need the help of skilled technicians. Some companies need between three and ten technicians for every engineer they have on their payroll.

ENGINEERS

Many of the recent discoveries in the electronic field have come about because of the work of ENGINEERS, Fig. 20-10. Engineers usually have gone through a four year college training program. Since they must have this training, engineers get more responsibility and higher pay than the first three job opportunities mentioned.

Engineers design and follow the building of new equipment until it is finally installed. They must make sure that it will run with the least amount of problems. Once the equipment is in operation, the engineers usually assign technicians to keep it running.

However, if a major problem should occur after installation, the engineers are called back. Then, they must redesign or find a better way of doing that operation.

People who go into engineering must have a good math and science background. This is necessary because so much of the work of an engineer is abstract, Fig. 20-11. They have to be able to visualize things before putting them on paper.

Engineers must have a good understanding of all the principles of physics, electronics and English. You can imagine what would happen if an engineer was not able to write a report on a new discovery so that people could understand it. Even the smartest engineer would be of little value to the company if no one understood what he said or wrote.

SUPERVISORS

Probably the highest paying job in the field of electricity/electronics is that of a SUPERVISOR, Fig. 20-12. A supervisor must have knowledge and skills

Fig. 20-10. Engineering designers check for possible errors and omissions in electrical drawings for a new industrial machine. (Packard Electric Div., General Motors Corp.)

Fig. 20-11. Two engineers energize a typical industrial control. (Packard Electric Div., General Motors Corp.)

Fig. 20-12. Management and supervisory people with electronic knowledge are needed in research and development, training and operations.

Fig. 20-13. Will you have the skills needed to handle a key job in electricity/electronics? This technician is checking the time, temperature and force needed by an injection molding machine. (Guide Div., General Motors Corp.)

related to construction, maintenance, technical and engineering responsibilities.

The supervisor has the added responsibility of making sure that others do their jobs right. Most supervisors have handled a great variety of work assignments. Only in this way could you gain the skill and knowledge needed to supervise efficiently.

Supervisors also gain the skill needed through a lot of study. They read technical and management reports and keep up to date on new discoveries. They attend conferences, national meetings and seminars. Supervisors have to write hundreds of reports each year and they must be skilled in handling people.

Once new equipment is built and installed, more job opportunities are present, Fig. 20-13. People having the skills needed are the ones who will get the best jobs. These skills and choice jobs will not be handed to you. You must work and work hard to earn them.

Fig. 20-14. Terminal operator studies CRT screen.

Typical jobs in electricity and electronics include:

Construction workers
    Assembler
    Checker
    Sorter
    Inventory Clerk

Maintenance Workers
    Adjuster
    Repair specialist
    Aircraft electrician
    Home appliance service specialist
    Industrial electrician

Technicians
    Office machine service specialist
    Sound technician
    Radio operator
    Field technician
    Computer repair specialist
    CCTV maintenance specialist

Through 20 chapters, ELECTRICITY AND BASIC ELECTRONICS has introduced you to a fast growing field full of career opportunities. If you want a place in the electricity/electronics field in the future, start now to improve your skills. Study the Reference Section that follows. Learn the words and definitions in the Dictionary of Terms. Refer to the Index for subject areas you may want to review.

REFERENCE SECTION

$AV = .637 \times P$

$Rms = .707 \times P$

This chapter contains reference information to aid you in your study of electricity and electronics. Included are formulas in common use, dozens of symbols, scientific notation, color codes for resistors and diodes, frequency bands, metric prefixes, twist drill sizes and decimal equivalents.

FORMULAS IN COMMON USE

Ohm's Law for dc circuits

$I = \dfrac{E}{R}$

$E = IR$

$R = \dfrac{E}{I}$

$P = IE$

$I = \dfrac{P}{E}$

$E = \dfrac{P}{I}$

$R = \dfrac{E^2}{P}$

$P = I^2 R$

$I = \sqrt{\dfrac{P}{R}}$

$E = \sqrt{PR}$

$R = \dfrac{P}{I^2}$

$P = \dfrac{E^2}{R}$

Resistance

In Series $\quad R_T = R_1 + R_2 + R_3$

In Parallel $\quad \dfrac{1}{R_T} = \dfrac{1}{R_1} + \dfrac{1}{R_2} + \dfrac{1}{R_3}$

Two Resistors in Parallel

$$R_T = \dfrac{R_1 R_2}{R_1 + R_2}$$

Equal Resistors in Parallel

$$R_T = \dfrac{R}{N}$$

R = any one resistor
N = number of resistors in parallel

Ohm's Law for ac circuits

$I = \dfrac{E}{Z}$

$E = IZ$

$Z = \dfrac{E}{I}$

$P = EI \cos \theta$

$I = \dfrac{P}{E \cos \theta}$

$E = \dfrac{P}{I \cos \theta}$

$Z = \dfrac{E^2 \cos \theta}{P}$

$P = I^2 Z \cos \theta$

$I = \sqrt{\dfrac{P}{Z \cos \theta}}$

$E = \sqrt{\dfrac{PZ}{\cos \theta}}$

$Z = \dfrac{P}{I^2 \cos \theta}$

$P = \dfrac{E^2 \cos \theta}{Z}$

$\cos \theta$ = power factor

Capacitance

In Parallel $\quad C_T = C_1 + C_2 + C_3$

In Series $\quad \dfrac{1}{C_T} = \dfrac{1}{C_1} + \dfrac{1}{C_2} + \dfrac{1}{C_3}$

Two Capacitors in Series

$$C_T = \dfrac{C_1 C_2}{C_1 + C_2}$$

$Q = CE$

Q = coulombs
C = farads
E = volts

Inductance

Inductors in Series

$$L_T = L_1 + L_2 + L_3$$

Inductors in Parallel

$$\frac{1}{L_T} = \frac{1}{L_1} + \frac{1}{L_2} + \frac{1}{L_3}$$

Two inductors in Parallel

$$L_T = \frac{L_1 L_2}{L_1 + L_2}$$

Coupled Inductance (Fields Aiding)

Series $\quad L_T = L_1 + L_2 + 2M$

Parallel $\dfrac{1}{L_T} = \dfrac{1}{L_1 + M} + \dfrac{1}{L_2 + M}$

Coupled Inductance (Fields Opposing)

Series $\quad L_T = L_1 + L_2 - 2M$

Parallel $\dfrac{1}{L_T} = \dfrac{1}{L_1 - M} + \dfrac{1}{L_2 - M}$

M = Mutual Inductance

Mutual Inductance

$$M = K \sqrt{L_1 L_2}$$

K = coefficient of coupling

Reactance

$$X_L = 2 \pi fL \qquad X_C = \frac{1}{2 \pi fC}$$

Resonance

$$f_r = \frac{1}{2 \pi \sqrt{LC}} \quad \text{or} \quad \frac{.159}{\sqrt{LC}}$$

$$L = \frac{1}{4\pi^2 f_r^2 C}$$

$$C = \frac{1}{4\pi^2 f_r^2 L}$$

$$2\pi = 6.28 \qquad 4\pi^2 = 39.5$$

Impedance

$$Z = \sqrt{R^2 + X^2}$$

where $X = X_L - X_C$

Power Factor

$$P_{true} = EI \cos \theta \qquad P_{apparent} = EI$$

$$PF = \text{Power Factor} = \cos \theta$$

Time Constants

R (in ohms) x C (farads) = t (in seconds) RC

$$t \text{ (secs)} = \frac{L \text{ (in henrys)}}{R \text{ (in ohms)}} \qquad \frac{L}{R}$$

Peak, RMS and Average ac Values

$$E_{rms} = E_{peak} \times .707 \qquad E_{peak} = E_{AV} \times 1.57$$

$$E_{peak} = E_{rms} \times 1.41 \qquad E_{rms} = E_{AV} \times 1.11$$

$$E_{AV} = E_{peak} \times .637$$

Wavelength

$$\lambda = \frac{3 \times 10^8}{f} \text{ meters} \qquad f = \frac{3 \times 10^8}{\lambda} \text{ cycles}$$

Conductance

$$G = \frac{1}{R} \qquad R = \frac{1}{G}$$

G (in siemens)

Figure of Merit Q

$$Q = \frac{X_L}{R_L}$$

Gain

$$db = 10 \log \frac{P_{out}}{P_{in}} \qquad \frac{10\ W}{1\ W}$$

$$db = 20 \log \frac{E_{out}}{E_{in}} \qquad \frac{1\ V}{2\ V}$$

$$db = 20 \log \frac{I_{out}}{I_{in}} \qquad \frac{1\ mA}{2\ mA}$$

Electricity and Basic Electronics

ELECTRICAL AND ELECTRONIC SYMBOLS

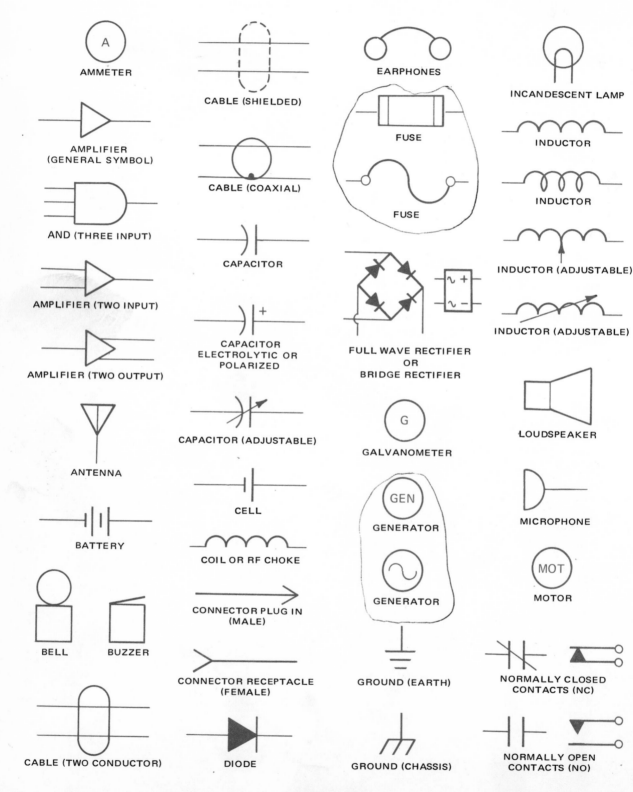

AMMETER

AMPLIFIER (GENERAL SYMBOL)

AND (THREE INPUT)

AMPLIFIER (TWO INPUT)

AMPLIFIER (TWO OUTPUT)

ANTENNA

BATTERY

BELL   BUZZER

CABLE (TWO CONDUCTOR)

CABLE (SHIELDED)

CABLE (COAXIAL)

CAPACITOR

CAPACITOR ELECTROLYTIC OR POLARIZED

CAPACITOR (ADJUSTABLE)

CELL

COIL OR RF CHOKE

CONNECTOR PLUG IN (MALE)

CONNECTOR RECEPTACLE (FEMALE)

DIODE

EARPHONES

FUSE

FUSE

FULL WAVE RECTIFIER OR BRIDGE RECTIFIER

GALVANOMETER

GENERATOR

GENERATOR

GROUND (EARTH)

GROUND (CHASSIS)

INCANDESCENT LAMP

INDUCTOR

INDUCTOR

INDUCTOR (ADJUSTABLE)

INDUCTOR (ADJUSTABLE)

LOUDSPEAKER

MICROPHONE

MOTOR

NORMALLY CLOSED CONTACTS (NC)

NORMALLY OPEN CONTACTS (NO)

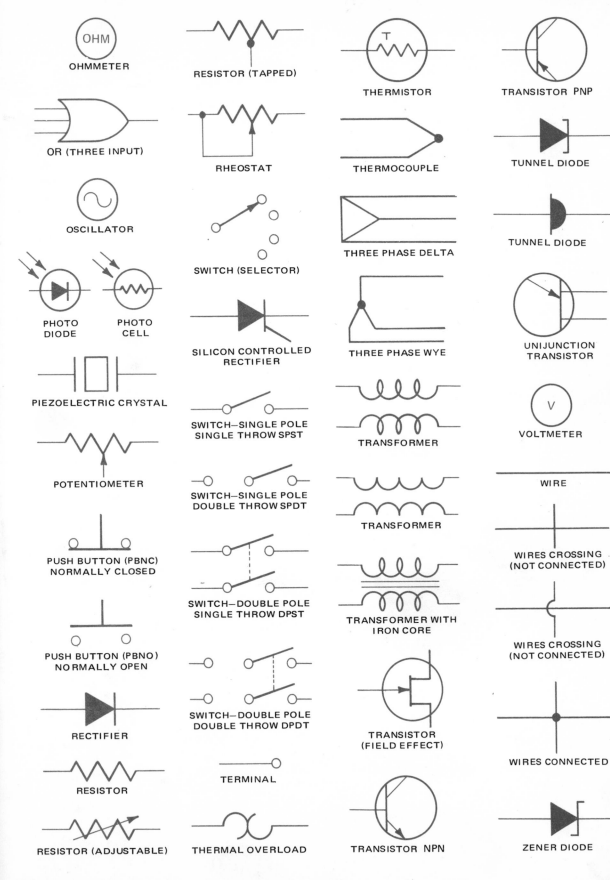

OHMMETER

OR (THREE INPUT)

OSCILLATOR

PHOTO DIODE

PHOTO CELL

PIEZOELECTRIC CRYSTAL

POTENTIOMETER

PUSH BUTTON (PBNC) NORMALLY CLOSED

PUSH BUTTON (PBNO) NORMALLY OPEN

RECTIFIER

RESISTOR

RESISTOR (ADJUSTABLE)

RESISTOR (TAPPED)

RHEOSTAT

SWITCH (SELECTOR)

SILICON CONTROLLED RECTIFIER

SWITCH—SINGLE POLE SINGLE THROW SPST

SWITCH—SINGLE POLE DOUBLE THROW SPDT

SWITCH—DOUBLE POLE SINGLE THROW DPST

SWITCH—DOUBLE POLE DOUBLE THROW DPDT

TERMINAL

THERMAL OVERLOAD

THERMISTOR

THERMOCOUPLE

THREE PHASE DELTA

THREE PHASE WYE

TRANSFORMER

TRANSFORMER

TRANSFORMER WITH IRON CORE

TRANSISTOR (FIELD EFFECT)

TRANSISTOR NPN

TRANSISTOR PNP

TUNNEL DIODE

TUNNEL DIODE

UNIJUNCTION TRANSISTOR

VOLTMETER

WIRE

WIRES CROSSING (NOT CONNECTED)

WIRES CROSSING (NOT CONNECTED)

WIRES CONNECTED

ZENER DIODE

SCIENTIFIC NOTATION

You probably noticed that some numbers used in formulas are very small and some are very large. When you have both small numbers and large numbers in the same formula, that formula can be very difficult to use in solving problems. To make these numbers easier to multiply and divide, a method called "scientific notation," or "engineer's shorthand," has been devised. It is a fast way of working out problems, based on being able to move the decimal point either to the left or to the right. Here is how it works:

10^2 is read as 10 squared and is the same as 10 x 10 (10 times 10), and it is equal to 100. Therefore, 10^2 = 100.

10^3 is read 10 cubed or 10 to the third power. It is the same as 10 x 10 x 10. Therefore, 10^3 = 1000.

10^4 = 10 x 10 x 10 x 10 or 10,000.

10^5 = 10 x 10 x 10 x 10 x 10 or 100,000.

EXPONENTS

An "exponent" is a mathematical symbol that indicates the power of a given number. In our first example of scientific notation, the exponent is 2. This means that 10^2 is 10 to the second power, and that 10 is to be multiplied by itself.

In our second example, 10^3 is 10 to the third power and number 10 is to be multiplied by itself twice. Likewise, 10^4 is 10 to the fourth power, and 10^5 is 10 to the fifth power. Note, too, that the exponent tells you the number of zeros after the 10 has been multiplied.

10^2 = 1 followed by 2 zeros or 100.
10^3 = 1 followed by 3 zeros or 1000.
10^4 = 1 followed by 4 zeros or 10,000.
10^5 = 1 followed by 5 zeros or 100,000.

POSITIVE EXPONENTS

If you want to change a large number by scientific notation, you can do it by moving the decimal point. For positive exponents (10^2, 10^3, etc.), you would move the decimal point to the left. Moving it to the left will increase the exponent by one for every digit the decimal point is moved. Here, for example, is how you would change a resistor rated at 120,000 ohms by using scientific notation:

120,000 ohms = 1.2×10^5 ohms

We have moved the decimal point five places to the left as follows:

1,20,000 with the decimal point moved one place to the left would be

120,000 = 12,000 x 10

120,000 with the decimal point moved two places to the left would be

120,000 = 1200 x 10^2

120,000 with the decimal point moved three places to the left would be

120,000 = 120 x 10^3

120,000 with the decimal point moved four places to the left would be

120,000 = 12 x 10^4

Finally, 120,000 with the decimal point moved five places to the left would be

120,000 = 1.2×10^5

Note, too, that each time we moved the decimal point, the exponent increased by one digit.

Try another example. Change 6700 ohms by using scientific notation:

6700 ohms = 6.7×10^3 ohms

Note that we moved the decimal point three places to the left, and the exponent over the 10 is 3.

NEGATIVE EXPONENTS

If you want to change a small number by scientific notation, you can do it by moving the decimal point to the right. The only difference in this method is that you will go to negative exponents (10^{-2}, 10^{-3}, etc.).

For example, change .00013 amp by using scientific notation:

.00013 amp = 1.3×10^{-4} amp

We have moved the decimal point four places to the right as follows:

.00013 with the decimal point moved one place to the right would be

$$.00013 = .0013 \times 10^{-1}$$

.00013 with the decimal point moved two places to the right would be

$$.00013 = .013 \times 10^{-2}$$

.00013 with the decimal point moved three places to the right would be

$$.00013 = .13 \times 10^{-3}$$

Finally, .00013 with the decimal point moved four places to the right would be

$$.00013 = 1.3 \times 10^{-4}$$

Try another example. Change .000006 amp by scientific notation:

$$.000006 \text{ amp} = 6 \times 10^{-6} \text{ amp}$$

Since we moved the decimal point six places to the right, the exponent is -6. The following table will help you see both methods of moving the decimal:

$$
\begin{aligned}
1,000,000,000 &= 1 \times 10^9 \\
100,000,000 &= 1 \times 10^8 \\
10,000,000 &= 1 \times 10^7 \\
1,000,000 &= 1 \times 10^6 \\
100,000 &= 1 \times 10^5 \\
10,000 &= 1 \times 10^4 \\
1000 &= 1 \times 10^3 \\
100 &= 1 \times 10^2 \\
10 &= 1 \times 10 \\
1 &= 1 \\
1 \times 10^{-1} &= .1 \\
1 \times 10^{-2} &= .01 \\
1 \times 10^{-3} &= .001 \\
1 \times 10^{-4} &= .0001 \\
1 \times 10^{-5} &= .00001 \\
1 \times 10^{-6} &= .000001 \\
1 \times 10^{-7} &= .0000001 \\
1 \times 10^{-8} &= .00000001 \\
1 \times 10^{-9} &= .000000001
\end{aligned}
$$

MULTIPLYING BY SCIENTIFIC NOTATION

Once you are able to convert with exponents, multiplying and dividing numbers by scientific notation is much easier. *To multiply, you add exponents. To divide, you subtract exponents.*

Study the following examples that show how to multiply numbers by scientific notation:

$$6000 \times 60,000$$

First, convert to scientific notation.

$$(6 \times 10^3) \times (6 \times 10^4)$$

Adding the exponents gives

$$(6 \times 6) \times 10^7$$

because 3 plus 4 equals 7, and this is equal to

$$36 \times 10^7 \text{ or}$$

moving decimal point one place to left equals

$$3.6 \times 10^8 \text{ or}$$

moving decimal point eight places to right equals

$$360,000,000.$$

If you want to multiply numbers by scientific notation and one exponent is plus while the other is minus, you still add the components. For example:

$$6000 \times .0006$$

First, convert to scientific notation.

$$(6 \times 10^3) \times (6 \times 10^{-4})$$

Adding the exponents gives

$$(6 \times 6) \times 10^{-1}$$

because +3 and -4 equals -1, and this is equal to

$$36 \times 10^{-1} \text{ or}$$

moving decimal point one place to left equals

$$3.6$$

If you want to multiply numbers by scientific notation and both exponents are minus, you still add the exponents. For example:

.0006 x .006

First, convert to scientific notation.

$(6 \times 10^{-4}) \times (6 \times 10^{-3})$

Adding the exponents gives

$(6 \times 6) \times 10^{-7}$

because -4 plus -3 equals -7, and this is equal to

36×10^{-7} or

moving decimal point seven places to left equals

.0000036

DIVIDING BY SCIENTIFIC NOTATION

When dividing numbers by scientific notation, you subtract exponents. Be careful when the signs are different. They may give you a problem if you are not paying attention. Try these examples:

$120{,}000 \div 400$

First, convert to scientific notation.

$$\frac{120{,}000}{400} = \frac{12 \times 10^{4}}{4 \times 10^{2}}$$

Since it is easier to divide 12 by 4, you would not change the 120,000 to 1.2×10^{5}. By subtracting the exponent in the denominator (numbers below line) from the exponent in the numerator (numbers above line) you get

$$\frac{12 \times 10^{2}}{4}$$

because 10^{4} minus 10^{2} equals 10^{2}. Remember, you are subtracting exponents. Then, 12 is divided by 4 to get

3×10^{2} or

moving the decimal point two places to right equals

300.

If you want to use scientific notation to divide numbers with negative exponents, you still subtract the exponents. For example:

$.00012 \div .004$

First, convert to scientific notation.

$$\frac{12 \times 10^{-5}}{4 \times 10^{-3}}$$

When you subtract a negative number, it becomes

$$\frac{12 \times 10^{-2}}{4}$$

because the negative exponent (-3) in the denominator becomes positive, and this is equal to

3×10^{-2} or

moving decimal point two places to left equals

.03

From these various examples, you can see that scientific notation, gives you a fast way of multiplying or dividing very large numbers or very small numbers.

RESISTOR COLOR CODE

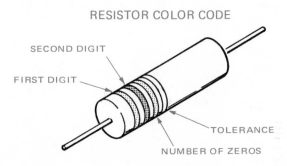

COLOR	DIGIT
BLACK	0
BROWN	1
RED	2
ORANGE	3
YELLOW	4
GREEN	5
BLUE	6
VIOLET	7
GRAY	8
WHITE	9

GOLD	5%
SILVER	10%
NONE	20%

} TOLERANCE

THIS EXAMPLE:

RED	VIOLET	RED	SILVER
2	7	00	± 10%

2700 Ω ± 10%
2.7K Ω ± 10%

TWIST DRILL SIZES AND DECIMAL EQUIVALENTS

DRILL	EQUIVALENT	DRILL	EQUIVALENT	DRILL	EQUIVALENT	DRILL	EQUIVALENT
.1 mm	.003937	38	.101500	B	.238000	33/64	.515625
.25 mm	.009842	37	.104000	C	.242000	17/32	.531250
80	.013500	36	.106500	D	.246000	13 1/2 mm	.531495
79	.014500	7/64	.109375	E 1/4	.250000	35/64	.546875
78	.016000	35	.110000	6.5 mm	.255905	14 mm	.551180
1/64	.015625	34	.111000	F	.257000	9/16	.562500
77	.018000	33	.113000	G	.261000	14 1/2 mm	.570865
.5 mm	.019685	32	.116000	17/64	.265625	37/64	.578125
76	.020000	3. mm	.118110	H	.266000	15 mm	.590550
75	.021000	31	.120000	I	.272000	19/32	.593750
74	.022500	1/8	.125000	7. mm	.275590	39/64	.609375
73	.024000	30	.128500	J	.277000	15 1/2 mm	.610235
72	.025000	29	.136000	K	.281000	5/8	.625000
71	.026000	3.5 mm	.137795	9/32	.281250	16 mm	.629920
70	.028000	28	.140500	L	.290000	41/64	.640625
69	.029200	9/64	.140625	M	.295000	16 1/2 mm	.649605
68	.031000	27	.144000	7.5 mm	.295275	21/32	.656250
67	.032000	26	.147000	19/64	.296875	17 mm	.669290
1/32	.031250	25	.149500	N	.302000	43/64	.671875
66	.033000	24	.152000	5/16	.312500	11/16	.687500
65	.035000	23	.154000	8. mm	.314960	17 1/2 mm	.688975
64	.036000	5/32	.156250	O	.316000	45/64	.703125
63	.037000	22	.157000	P	.323000	18 mm	.708660
62	.038000	4. mm	.157480	21/64	.328125	23/32	.718750
61	.039000	21	.159000	Q	.332000	18 1/2 mm	.728345
1. mm	.039370	20	.161000	8.5 mm	.334645	47/64	.734375
60	.040000	19	.166000	R	.339000	19 mm	.748030
59	.041000	18	.169500	11/32	.343750	3/4	.750000
58	.042000	11/64	.171875	S	.348000	49/64	.765625
57	.043000	17	.173000	9. mm	.354330	19 1/2 mm	.767715
56	.046500	16	.177000	T	.358000	25/32	.781250
3/64	.046875	4.5 mm	.177165	23/64	.359375	20 mm	.787400
55	.052000	15	.180000	U	.368000	51/64	.796875
54	.055000	14	.182000	9.5 mm	.374015	20 1/2 mm	.807085
1.5 mm	.059055	13	.185000	3/8	.375000	13/16	.812500
53	.059500	3/16	.187500	V	.377000	21 mm	.826770
1/16	.062500	12	.189000	W	.386000	53/64	.828125
52	.063500	11	.191000	25/64	.390625	27/32	.843750
51	.067000	10	.193500	10 mm	.393700	21 1/2 mm	.846455
50	.070000	9	.196000	X	.397000	55/64	.859375
49	.073000	5. mm	.196850	Y	.404000	22 mm	.866140
48	.076000	8	.199000	13/32	.406250	7/8	.875000
5/64	.078125	7	.201000	Z	.413000	22 1/2 mm	.885825
47	.078500	13/64	.203125	10 1/2 mm	.413385	57/64	.890625
2. mm	.078740	6	.204000	27/64	.421875	23 mm	.905510
46	.081000	5	.205500	11 mm	.433070	29/32	.906250
45	.082000	4	.209000	7/16	.437500	59/64	.921875
44	.086000	3	.213000	11 1/2 mm	.452755	23 1/2 mm	.925195
43	.089000	5.5 mm	.216535	29/64	.453125	15/16	.937500
42	.093500	7/32	.218750	15/32	.468750	24 mm	.944880
3/32	.093750	2	.221000	12 mm	.472440	61/64	.953125
41	.096000	1	.228000	31/64	.484375	24 1/2 mm	.964565
40	.098000	A	.234000	12 1/2 mm	.492125	31/32	.968750
2.5 mm	.098425	15/64	.234375	1/2	.500000	25 mm	.984250
39	.099500	6. mm	.236220	13 mm	.511810	63/64	.984375
						1	1.000000

DIODE COLOR CODE

Crystal diodes can be marked according to the EIA color code, which is the same code used with resistors. The color bands start from the cathode end with the numbers following the 1N prefix. For example, the diode in the drawing is a 1N462.

There are a number of other ways of coding diodes. Some companies will use other letters and numbers. If you want to be sure of the code, use the reference catalog of the company that made the diode in question.

BLACK	0
BROWN	1
RED	2
ORANGE	3
YELLOW	4
GREEN	5
BLUE	6
VIOLET	7
GRAY	8
WHITE	9

THIS EXAMPLE:
YELLOW 4 BLUE 6 RED 2
1N462

FREQUENCY BANDS

ABBREVIATION	DESCRIPTION	FREQUENCY
VLF	VERY LOW FREQUENCY	LESS THAN 30 kHz
LF	LOW FREQUENCY	30 TO 300 kHz
MF	MEDIUM FREQUENCY	300 TO 3000 kHz
HF	HIGH FREQUENCY	3 TO 30 MHz
VHF	VERY HIGH FREQUENCY	30 TO 300 MHz
UHF	ULTRA HIGH FREQUENCY	300 TO 3000 MHz
SHF	SUPER HIGH FREQUENCY	3 TO 30 GHz
EHF	EXTREMELY HIGH FREQUENCY	30 TO 300 GHz

METRIC PREFIXES

PREFIX	ABBREVIATION	MULTIPLE	
TERA	T	10^{12}	1,000,000,000,000
GIGA	G	10^{9}	1,000,000,000
MEGA	M	10^{6}	1,000,000
KILO	k	10^{3}	1000
HECTO	h	10^{2}	100
DEKA	da	10	10
DECI	d	10^{-1}	0.1
CENTI	c	10^{-2}	.01
MILLI	m	10^{-3}	.001
MICRO	μ	10^{-6}	.000001
NANO	n	10^{-9}	.000000001
PICO	p	10^{-12}	.000000000001

DICTIONARY OF TERMS

ABSOLUTE MAXIMUM SUPPLY VOLTAGE: The most voltage that can be applied to a circuit without the danger of causing permanent change to the characteristics of that circuit.

ABSOLUTE MINIMUM RESISTANCE: The resistance between the wiper and the end terminal of a potentiometer when the wiper is moved to the point where you would expect to get the lowest resistance.

ABSOLUTE TEMPERATURE: Temperature measured from a point called absolute zero, usually defined as 0 K (kelvin) or -273 C (celsius), 0 R (rankine) or -460 deg. F (fahrenheit).

ABSOLUTE ZERO: (Same as above.)

AC (ac): Abbreviation for alternating current.

ACCELERATING ELECTRODE: A conductor element in a cathode ray tube used to increase the speed of electrons or ions toward the anode.

ACCELERATOR: A device used to give a high velocity to charged particles such as electrons or protons. It is used in scientific research to study the structure of the atom.

ACCEPTOR: An impurity used in semiconductors. It accepts electrons from other atoms and creates what is called a "hole" in the structure of the crystal.

ACCEPTOR CIRCUIT: A circuit tuned to a single frequency or said to be "at resonance." Therefore, it will accept signals at the resonant frequency.

AC CIRCUIT BREAKER: A device used in an ac circuit to allow current flow and open the circuit when a fault occurs or an overload is put on the circuit.

AC COUPLING: Connecting one ac circuit to another ac circuit by using a capacitor. This allows the ac passage but blocks the dc signals.

AC/DC (ac/dc): Equipment made so that it can be operated from either an ac or a dc power source.

AC GENERATOR: A device called an "alternator," used to change mechanical energy to alternating current.

ACHROMATIC: A term used to mean black and white television.

AC POWER SUPPLY: One which gives an ac output voltage. Some may give a whole range of ac output voltages from which you can choose.

AC TRANSDUCER: A device which is activated by an ac signal or produces an ac signal. In the case of an ac input, the output could be some other form of energy such as pressure or light. These same things, pressure or light, could be used as inputs to produce the ac signal output.

ADAPTER: A device used to change a receptacle, such as a plug, so that it can be used in another way. For example: an adapter which allows a sander with two prongs and a ground to be plugged into an outlet which does not have a ground outlet.

ADDRESS: A way of telling the specific location on a tape or drum of a computer so that a piece of information can be located in the storage memory.

ADJACENT CHANNEL: The frequency band just above or below the one to which you refer.

ADJUSTABLE RESISTOR: One which has the resistance wire partly exposed so that the user can move the lug from one position to another when a different resistance value is needed.

ADP: Abbreviation for automatic data processing.

AF: Abbreviation for audio frequency.

AGING: Applying a voltage to devices such as capacitors, diodes, etc., until their characteristics level off to a constant value.

AIEE: Abbreviation for American Institute of Electrical Engineers.

AIR CAPACITOR: One in which air is the insulating material between its plates.

ALGOL: Short for ALGOrithmic Language in computer language.

ALIGN: To adjust tuned circuits so that you get the maximum signal response.

ALKALINE CELL: One that uses an electrolyte that gives the cell more capacity than the standard zinc-carbon cell. Electrolyte is potassium hydroxide.

ALLIGATOR CLIP: A spring-loaded metal clip used to make temporary electrical connections.

ALNICO: An alloy made from ALuminum, NIckel, CObalt and iron, used in making magnets.

ALPHA: The emitter-to-collector current gain of a transistor. It is taken from the Greek letter α.

ALPHA CUTOFF FREQUENCY: Frequency at which the current gain of the transistor drops to 0.707 of its maximum gain.

ALPHA PARTICLE: An electrically charged particle given off by many radioactive materials.

ALTERNATING CURRENT (ac): A flow of electrons which regularly reverses its direction of flow.

ALTERNATOR: A device used to convert mechanical energy into electrical energy in the form of an alternating current.

ALUMINA: A ceramic used as substrates (supporting material) in thin film circuits. It is used because it can withstand high temperatures and has little effect on the circuit.

AM: Abbreviation for amplitude modulation.

ASA: Abbreviation for American Standards Association.

AWG: Abbreviation for American Wire Gage, and it is the term used for giving the size of solid wire.

AMMETER: An instrument used to measure current in ac or dc circuits.

AMMETER SHUNT: A low resistance conductor placed in parallel with the meter movement to avoid possible damage to meter when measuring high currents.

AMP: Abbreviation for ampere.

AMPERE (A): The unit of measure for current based on the flow of one coulomb per second past a given point. Also, one volt across one ohm of resistance causes a flow of one ampere.

AMPERE-HOUR: A current of one ampere flowing for one hour. It is a term used to tell the amount of energy a battery can give before it has to be recharged.

AMPERE-TURN: Unit used to describe the magnetomotive force produced by a coil. It is found by multiplying the number of turns of wire in the coil by the current flowing in the coil.

AMPLIFICATION: A term used to describe the increase in current or voltage from one side to the other side of a device. A transistor, for example, can amplify (increase) the current from base to collector.

AMPLIFIER: A device that takes a small input signal and converts it to a much larger output signal.

AMPLITUDE: The amount of change in a quantity, such as a sine wave.

AMPLITUDE MODULATION (AM): A way of changing the waveform of a signal by impressing it on a carrier frequency. This is done so that it is possible to have different radio stations broadcast at different frequencies and not have them all found at one point on a dial.

AM RECEIVER: A device used to change amplitude modulate signals into audio frequencies.

ANALOG COMPUTER: An automatic electronic calculating machine where both the input and output signals are constantly varying, such as measuring the amount of rotation, temperature change, speed change, etc. (as opposed to a digital computer).

AND CIRCUIT: An electronic circuit where all the inputs must be present to get an output.

ANGLE OF DEFLECTION: The amount that an electron beam in a cathode ray tube is bent with respect to its normal position.

ANGLE OF LAG: The angle between the voltage and current vectors of an ac circuit.

ANGSTROM: Unit of measurement for the wavelength of light. It is equal to one ten thousandth of a micron.

ANNEALING: Process of heating material, such as copper, and cooling it so that it becomes softer.

ANODE: The positive electrode or terminal.

ANODE TERMINAL: The terminal of a diode that is positive with respect to the other terminal (cathode) when it is forward biased.

ANTENNA: Part of a radio receiver used to "catch" the radio waves from the transmitter or from space. In the case of the transmitter, it is that part which is used to radiate the signal into space.

ANTENNA CORES: Ferrite cores used for radio antennas.

ANTIMAGNETIC: Material that will not remain in the magnetic state, such as copper, brass and other nonferrous alloys.

APPARENT POWER: The product of the current times the voltage in a circuit where the two reach their peaks at different times.

APPLIANCE: Equipment used in the home, such as radios, television sets, toasters and dryers.

ARC: An electrical discharge between two points having different electrical potentials.

ARMATURE: The moving part of a motor or generator, and the vibrating part of a buzzer or relay.

ARMATURE WIRE: Soft wire used for winding coils.

ARMORED CABLE: Two or more insulated wires inside a flexible metal covering, protecting them from damage.

ASA: Abbreviation for American Standards Association.

ASBESTOS: An insulating material.

ASSEMBLY: The complete unit made up of various individual parts.

ASTABLE: Not stable. A device that has two states

and constantly alternates between them, usually at some specific rate.

ASTABLE MULTIVIBRATOR: A circuit that switches back and forth between two stable states.

ATMOSPHERIC NOISE: The crackle heard in radio reception caused by atmospheric interference.

ATOM: The smallest particle that makes up an element.

ATOMIC FISSION: Splitting the nucleus of an atom of material such as uranium.

ATOMIC NUMBER: The number of protons in the nucleus of an atom.

ATOMIC WEIGHT: The weight of the protons and neutrons in the nucleus of an atom.

ATTENUATION: The drop in amplitude of a signal from one point to another during its transmission.

ATTENUATOR: A network of resistors that reduces the amplitude of a signal without too much change in the frequency.

AUDIBLE: Sound that can be heard by the human ear.

AUDIO: Frequencies from 20 to 20,000 hertz, which is roughly the range that can be heard by the human ear.

AUDIO FREQUENCY (AF): The number of complete cycles per second within the range that can be heard by the human ear.

AUDIO FREQUENCY TRANSFORMER: An iron core electrical device used to transfer signals from one circuit to another or for impedance matching.

AUTOMATIC CIRCUIT BREAKER: A device used to open a circuit automatically when an overload occurs. It can be reset manually after the problem is fixed.

AUTOMATIC FREQUENCY CONTROL (AFC): A circuit that automatically maintains the frequency of an oscillator within certain limits.

AUTOMATIC GAIN CONTROL (AGC): A circuit which maintains the output volume of a receiver even though the strength of the signal it receives is always changing.

AUTOMATION: A method of automatically controlling machines in the manufacturing process, without having people involved.

AUTOTRANSFORMER: An electrical device with only one winding that acts as both the primary and secondary circuit to step up or step down the voltage.

AUXILLIARY CIRCUIT: A circuit which is not the main circuit.

AVALANCHE: A condition where the current in the reverse direction builds up very quickly as electrons flow past a junction.

AVALANCHE BREAKDOWN: The rapid passing of electrons through a diode in the reverse biased direction, but usually without ruining the diode.

AVALANCHE DIODE: A diode that has the ability to break down in the reverse direction when a specified point is reached. For this reason, it is often used to regulate voltage. A zener diode.

AVERAGE VOLTAGE: Value of alternating current or voltage of a sine wave found by dividing the area from 0 to 180 deg. of one cycle by the distance along the X axis. It is equal to 0.637 times E_{max}.

AVIONICS: Short way of saying AVIation electrONICS.

AWG: Abbreviation for American Wire Gage which is a way of measuring the diameter of wire.

AXIAL LEADS: Leads coming out of the ends of a resistor or capacitor rather than out from the side of the device.

B: The symbol for the base of a transistor.

BACK TO BACK: When two semiconductors are connected in opposite directions to control current without rectifying it. One semiconductor conducts for one half cycle, the second conducts for the other half cycle.

BALANCE: A way to control the volume from each speaker in a stereo system.

BALANCED BRIDGE: A circuit that is adjusted so that its output voltage is zero.

BALANCED CIRCUIT: One which has both sides electrically balanced to some reference point.

BALLAST RESISTOR: A resistor in series with the load used to keep the current constant even though the line voltage is changing.

BANANA JACK: A socket which will take a special plug shaped like a banana yet make a good electrical connection.

BANANA PLUG: A male connector that fits into a banana jack.

BAND: A part or range of frequencies which falls between two defined limits.

BAND ELIMINATION FILTER: A filter that passes frequencies on either side of the band, but not the one which it is designed to stop.

BAND PASS FILTER: A filter which will pass one band yet decreases inputs of a higher or lower frequency.

BAND REJECT FILTER: See BAND ELIMINATION FILTER.

BANDWIDTH: The frequency limits of a given waveband for transmitting a modulated signal.

BARE CONDUCTOR: A wire not covered with any insulation.

BAR MAGNET: A piece of metal in the shape of a bar which holds its magnetism.

BARRIER: Something which acts as a insulator between circuits or prevents the flow of electrons. In some cases, it will act as a block to the passage of electrons until they reach a certain level.

BARRIER VOLTAGE: The minimum potential difference necessary to cause current flow through a PN junction.

BASE: The control region of a transistor, which blocks or allows flow between the emitter and collector.

BASS: The low audio frequency sound range.

BATCH PROCESS: A way of making groups of resistors, capacitors, etc., at the same time.

BAT HANDLE: The lever on a toggle switch, which is shaped like a baseball bat.

BATTERY: Two or more cells connected to deliver a dc voltage.

BATTERY ACID: The electrolyte in a storage battery.

BATTERY CHARGER: Device used to take the ac from an outlet and convert it to dc for increasing the charge of a storage battery.

BEAM: Electromagnetic radiation concentrated so that it is focused in one direction.

BEAM CURRENT: That part of the electron stream that is focused on the screen of a cathode ray tube.

BEAT: The rhythmic increase and decrease of amplitude which result from the combination of two waves at different frequencies.

BEAT FREQUENCY: A frequency produced when two different frequencies are combined. There are two beat frequencies. One is equal to difference between them. The other is the sum of the two frequencies.

BELL TRANSFORMER: A small iron core transformer used to step down the primary voltage to operate doorbells and chimes.

BENCH TEST: A test conducted in a shop or laboratory rather than in the home.

BETA: The Greek letter β, used to specify the current gain from base to collector in a transistor.

BETA PARTICLE: An electrical charge given off by radioactive material.

BIAS: Voltage applied to a transistor or diode to block passage or allow the flow of electrons in the device.

BIAS CURRENT: The current through the base-emitter junction of a transistor used to set its operating point.

BIDIRECTIONAL ANTENNA: One which has a maximum response in only two directions.

BIMETALLIC STRIP: A strip formed by welding two dissimilar metals together. It is often used in temperature sensing.

BINARY: A numbering system that uses only the numbers 0 and 1.

BINARY CODE: System using 0 and 1 or ON and OFF as a code.

BINARY DIGIT (BIT): In the binary numbering system, a character which is either a 0 or a 1.

BINDING POST: A device used for making electrical connections.

BISTABLE MULTIVIBRATOR (FLIP FLOP): A circuit that has two stable states and can be switched back and forth.

BIT: Abbreviation for binary digit.

BLACK BOX: A method used to describe a circuit where you only want to know about its inputs and outputs but are not concerned about how and what makes up the individual parts of the circuit.

BLEEDER RESISTOR: A resistor used to discharge a capacitor after its circuit has been deenergized, to stabilize voltages or to draw a fixed amount of current.

BLOCK DIAGRAM: A way of showing how a circuit acts by representing its parts in blocks and explaining them, instead of using the components of the circuit.

BLOCKING: Putting a high voltage bias on a transistor to stop current from flowing.

BLOW: The act of opening a circuit, usually because of too high a current.

BRAIDED WIRE: A wire made up of many strands that are woven together.

BREADBOARD: To put the components together on a temporary circuit so that it can be checked before soldering.

BREAKDOWN VOLTAGE: The voltage that causes a big change in the characteristics of a diode.

BRIDGE RECTIFIER: Using four rectifiers to convert ac to full wave rectified dc.

BROADCAST BAND: The band of frequencies from 535 to 1605 kilohertz from which AM radio stations broadcast.

BRUSH: A piece of graphite or carbon used to make the electrical connection on the commutator of a motor.

BULK ERASER: A piece of equipment used to degauss or erase the magnetic fields which are found on a piece of audio tape or video tape.

BUSBAR: A heavy copper bar used to carry high current to three or more circuits, usually in factories.

BUTTON: A container that holds the granules in a carbon microphone.

BX CABLE: A flexible metal tube which houses two or more pieces of insulated wire.

BYPASS CAPACITOR: One which provides a low impedance ac path around some part of the circuit.

BYPASS FILTER: One which provides a low attenuation path around some part of the circuit.

C: The symbol for a capacitor, carbon or coulomb.

CABLE: Two or more conductors insulated from one another, but in the same covering.

CALIBRATE: To set up a piece of equipment by comparing it to some standard (something with a known fixed value).

CALL LETTERS: A fixed set of letters assigned to transmitting stations by the government.

CANNIBALIZATION: Taking parts off a similar piece of equipment so that at least one piece can be kept in operation.

CAPACITANCE: The electrical characteristic which permits the storage of electrical energy in the electrostatic field between two conductors. The amount of charge that can be stored by a capacitor as measured in farads.

CAPACITIVE REACTANCE: Opposition to any change in voltage between two conductors having capacitance. It is measured by the formula X_c equals $\frac{1}{2 \pi fc}$.

CAPACITOR: Two conductors separated by an insulator or dielectric used to store an electrical charge.

CAPACITOR START MOTOR: An induction motor that has a capacitor wired in series with its windings for easier starting.

CAPSTAN: A part of a tape recorder used to pull the tape at a constant speed so the recording and playback speeds are always constant.

CARBON: A black material used in resistors and brushes because of its special ability to conduct electricity.

CARRIER WAVE: A single frequency wave which when it is modulated by a transmitting station is used to send information over the air.

CARTRIDGE: The removable part of a phonograph pickup.

CARTRIDGE FUSE: A glass covered fuse with metallic ends for connecting it in the circuit you wish to protect.

CATHODE: The general name for any negative electrode such as the negative side of a diode or electrical cell.

CATHODE RAY TUBE (CRT): A tube used in television sets and oscilloscopes where the electron beam is capable of being deflected to produce a visible pattern.

CATV: Abbreviation for community antenna television.

CAT WHISKER: A small pointed wire used to make contact on the surface of a semiconductor.

CCW: Abbreviation for counterclockwise.

CDI: Abbreviation for capacitive discharge ignition.

CELL: A single power unit, such as those used in flashlights, which will produce electrical energy to supply power to operate a circuit.

CELSIUS SCALE: Temperature scale based on water freezing at 0 C and boiling at 100 C. It is also called the "centigrade scale."

CENTER TAP: An electrical connection in the center of a resistor or winding of a transformer.

CERAMIC: An insulating material capable of withstanding high temperatures.

CHARACTERISTIC CURVE: A graph showing how the values change for some component or circuit.

CHARGE: The amount of electrical energy stored in a capacitor or battery.

CHARGED PARTICLE: An extremely small piece of matter which has a negative or positive electrical charge such as an ion, electron or proton.

CHARGER: A device used to change ac to dc for the purpose of increasing the electrical charge of a battery.

CHASSIS: The framework, such as a sheet metal box, upon which components are mounted or attached in building a electrical circuit.

CHASSIS GROUND: An electrical connection to the metal frame or chassis that holds the circuit.

CHATTER: The unwanted and repeated opening and closing of electrical contacts.

CHIP: A single piece of material on which all the parts of an integrated circuit are found.

CHOKE: An inductor.

CHOKE COIL: An inductor used to limit the flow of alternating current.

CHOPPER: A device used to interrupt a beam of light at regular intervals.

CIRCLE CUTTER: A tool used to cut holes in a panel or chassis.

CIRCUIT: An electrical path having a power source, load and conductors to carry current.

CIRCUIT BREAKER: A device that automatically opens a circuit when its current carrying ability has been exceeded.

CIRCULAR MIL: The cross-sectional area of a round wire having a diameter of 0.001 in.

CLOSED LOOP: A circuit where the output is being fed back to the input. In this way, the output can be constantly compared and checked so that changes can be made automatically.

COAXIAL CABLE: A transmission wire completely surrounded by a second wire, which is insulated from the first. This is done so that the inner wire is protected from external magnetic fields.

COIL: Wrapping a number of turns of wire on a coil form or an iron core.

COIL FORM: An insulating support for a coil, made from plastic, paper or other insulating material.

COLD: Another way of saying that the circuit is

disconnected or that the wire is not carrying current.

COLLECTOR: One of the leads on a transistor.

COLLECTOR JUNCTION: The point between the collector and base of a transistor.

COLOR CODE: A way to identify the value of components by reading the color patches on the body of the component in a specified order.

COMMON BASE AMPLIFIER: A transistor amplifier where the base is common to both the input and output circuits.

COMMON COLLECTOR AMPLIFIER: A transistor amplifier where the collector is common to both the input and output circuits.

COMMON EMITTER AMPLIFIER: A transistor amplifier where the emitter is common to both the input and output circuits.

COMMUTATION: A method of converting ac to dc in a generator by using a pick off system that will reverse itself every half cycle.

COMMUTATOR: A group of bars providing connections between armature coils and brushes.

COMPARISON BRIDGE: A special circuit used to balance voltages by means of a four-part circuit arrangement that will give a zero error signal when balance exists.

COMPONENT: A single part used in a circuit or a series of parts that are considered to work together as a piece of the complete circuit.

COMPONENT LAYOUT: Method of placing all parts to be used in the circuit on a chassis to make them fit and work the best possible way.

COMPUTER: An automatic electronic machine that performs calculations.

CONDENSER: The term used by automotive mechanics to label a capacitor.

CONDUCTOR: Any item or material which is used to carry current; a wire.

CONDUIT: Specially built tubing used to hold wire and protect it from damage.

CONNECTION: The point where two or more wires or components are tied together electrically.

CONNECTOR: A device used to make a mechanical or electrical connection.

CONSTANT: A fixed value that does not change.

CONSTANT VOLTAGE POWER SUPPLY: A power supply which is regulated so that the voltage does not change when it is loaded.

CONTACT: A current-carrying part of a switch or relay that is capable of opening and closing to allow current flow.

CONTACT BOUNCE: Repeated opening and closing of relay contacts when they should be closed. This is caused by springs and is similar to the bouncing of a rubber ball on a floor.

CONTINUITY: An unbroken path for electron flow.

CONTINUITY TEST: An electrical check to see if a part of the circuit has a completed path for electron flow.

CONVERGENCE: Focusing the electron beams of a cathode ray tube to a single point.

COPPER LOSSES: Heat losses or I^2R losses in motors, generators and transformers that result from the resistance of the copper wire.

CORE: The magnetic material inside a relay, coil or transformer used to increase the strength of the magnetic field. Also magnetic material inside coils of an electromagnet.

COULOMB: The quantity of electricity that will flow past a point in a circuit as measured in one second when the current is maintained at one ampere. It is also represented by 6.24×10^{18} electrons.

CPS: Abbreviation for cycles per second, now known as hertz.

CRIMP: To crush a connector around a wire to make an electrical connection.

CROSSOVER: The point on a print where two conductors insulated from one another pass each other.

CRT: Abbreviation for cathode ray tube.

CRYSTAL: A piece of material that will vibrate at one special frequency determined by the way it is cut. It is used in certain types of radio work.

CRYSTAL DIODE: A diode formed by a small semiconductor crystal and cat whisker. They form the PN junction of the diode.

CURRENT: The flow of electrons through a conductor, measured in amperes.

CURRENT AMPLIFICATION: The ratio of the current at the output to current at the input. Amplification takes place when the ratio is greater than 1, which tells you that the output is higher.

CURRENT DENSITY: The amount of current that passes through a given cross-sectional area.

CURRENT LIMITING RESISTOR: A resistor put in a circuit to limit the current to some specified value.

CUTOFF: The point where the emitter base junction of a transistor is reverse biased or has zero bias, and there is no collector current.

CYCLE: The change in an alternating wave from zero to a positive peak to zero to a negative peak and back to zero. This is the sine wave normally seen for an ac cycle.

DAMPEN: To progressively lower the amplitude.

DAMPING: A method of making the pointer on a meter come to rest instead of overshooting the value of the reading and constantly going back and forth until it finally stops.

D'ARSONVAL MOVEMENT: A meter where the movement is supported on jeweled bearings between the poles of a permanent magnet. When current is passed through the movement, a pointer attached to a small spring will rotate to indicate the amount being measured.

DB (dB): Abbreviation for decibel.

DC TRANSDUCER: A device that will convert a mechanical signal into a dc output.

DCWV: Abbreviation for direct current working voltage. It is the rating found on capacitors which tells the maximum continuous voltage that can be applied.

DEAD: Said of a part of a circuit that is not carrying current. This term applies to a circuit that usually carries current.

DEAD SHORT: A path for current which bypasses the load. Therefore has almost no resistance and very high current.

DEBUG: To troubleshoot and correct problems in a piece of equipment when it is first being set up.

DECADE BOX: A box that contains a group of precision resistors, capacitors or coils wired to a series of switches for use in substituting many values in circuits. It is used for temporary connections into circuits.

DECAY: A slow drop in the value of a quantity.

DECI: A prefix which means one-tenth.

DECIBEL: A unit used to give the relative increase or decrease in power.

DECIMAL CODE: A code which uses a scale of numbers from 0 to 9.

DECIMAL TO BINARY: Changing a decimal code number written in a scale from 0 to 9 to a binary code number which uses a scale of 0 to 1.

DECOUPLING NETWORK: A group of chokes, resistors and/or capacitors placed in some circuits to stop them from harmful coupling between one another.

DE-ENERGIZE: To disconnect a device from its power supply.

DEFLECTION: To bend the electron beam in a cathode ray tube so that it gives a pattern on the face of the tube.

DEFLECTION PLATES: Two pairs of parallel electrodes inside a cathode ray tube used to change the electron beam in either a horizontal or vertical direction.

DEGAUSSER: A device used to demagnetize taped signals. Also, another device found in television sets to reduce magnetic fields which are not wanted because of the effect they have on the picture tube.

DEGRADATION: The gradual lowering of quality.

DELTA CONNECTION: The terminal connections in a three phase electrical system which are in the shape of a triangle similar to the Greek letter delta.

DEMODULATOR: A device used to separate the original wave which was joined to a carrier wave for purposes of transmitting.

DEPLETION REGION: The region in a reverse biased semiconductor where there is a lack of electrons and holes at its junction.

DEPOLARIZATION: A way of removing the hydrogen gas bubbles that try to form inside a primary cell.

DERATE: To reduce the rating of the voltage, current or power of a device so that its reliability will be improved when subjected to higher temperatures.

DETECTOR PROBE: A device used with a vacuum tube voltmeter or oscilloscope to analyze the signals of certain waves.

DETUNE: To change the value of a tuned capacitor or inductor that was at the resonant frequency.

DIAC: A two-lead alternating current semiconductor that will conduct when its breakdown voltage is reached. It has properties of both diodes and transistors.

DIAGRAM: A schematic or print used to show the function or sequence of a circuit.

DICING: The process used to saw a formed crystal wafer into blanks.

DIELECTRIC: The insulating material between the plates of a capacitor.

DIFFERENCE IN POTENTIAL: The voltage difference between two points in a circuit.

DIFFERENTIAL AMPLIFIER: A special circuit that will work when there is a difference between the two voltages or currents at the inputs to two transistors.

DIFFUSED JUNCTION: A junction formed in a semiconductor by adding impurities at a high temperature.

DIFFUSION: A way of adding impurities such as boron and arsenic to a semiconductor material by using heat and special containers so that it can be controlled to very small values.

DIGITAL COMPUTER: A device which operates on, and records in terms of, numbers which are represented by a series of pulses.

DIGITAL VOLTMETER: A meter which gives voltage readouts in numbers instead of by means of a pointer moving on a scale.

DIODE: A semiconductor device that has a PN junction. Usually, it is thought of as allowing electron flow in one direction and blocking flow from the other direction. However, there are some special types which do not work that way.

DIODE TRANSISTOR LOGIC (DTL): A type of circuit that uses diodes as inputs to control or activate the transistor outputs.

DIP: Abbreviation for dual in-line package, a type of

integrated circuit.

DIP SOLDERING: A way of soldering the bottom of a printed circuit by floating it on a molten pool of solder. This method is used when the componets have been mounted on the top.

DIRECT CURRENT (dc): The flow of electrons in only one direction.

DIRECT CURRENT GENERATOR: A device that converts mechanical energy into dc electrical energy. Brushes pick up energy from a rotating armature.

DISCHARGE: The release of energy stored in a battery or a capacitor.

DISCONNECT: To unplug or break the power to an electrical circuit.

DISCRETE COMPONENT: Parts such as resistors, capacitors, etc., which are made up prior to being installed into circuits or on printed circuit boards.

DISPLAY: The visual part of a signal seen on the cathode ray tube.

DISSIPATION: Losing electrical energy in the form of heat.

DISTORTION: The unwanted change in a signal caused by some outside source.

DONOR: The impurities added to semiconductors because they give up electrons.

DOPE: To add impurities to semiconductors in a controlled way.

DOUBLE POLE DOUBLE THROW SWITCH (DPDT): A switch with two terminals that can be connected to one of two other pairs of terminals, depending on how it is thrown.

DPDT: Abbreviation for double pole, double throw switch.

DOUBLE POLE SINGLE THROW SWITCH (DPST): A switch with two terminals that can be connected to or disconnected from another pair of terminals, depending on how it is thrown.

DPST: Abbreviation for double pole, single throw switch.

DRAIN: The part of a field effect transistor which corresponds to the collector of a transistor.

DRIFT: The unwanted change in the voltage of a power supply over a period of time.

DRY CELL: A cell which produces electricity through chemical means. It is commonly called a "battery."

DTL: Abbreviation for diode transistor logic.

DUPLEX: Two-in-one, such as two wires in one cable, two receptacles in one outlet, or transmitting and receiving at the same time.

DUTY CYCLE: The amount of time a device is "on," as opposed to the time it is "off" or in idle condition.

DV/DT: Actually: $\Delta V/\Delta t$; the change in voltage with respect to time.

DYNAMOMETER: A device used to measure the power output of a rotating machine.

E: The symbol for electromotive force. Also, the symbol for the emitter of a transistor.

E CORE: The shape of the laminations used in building some types of transformers.

EDDY CURRENTS: Currents that are induced in a body. In transformers, these currents usually circulate in the iron and copper windings and produce unwanted heat.

EDISON BASE: The screw base used for electric lamps.

EFFECTIVE VALUE: In ac sine wave voltages, it is 0.707 times the peak value and is the heat equivalent of the same amount of dc. Also see RMS.

EFFICIENCY: The ratio of the output of a device to its input.

EHF: Abbreviation for extremely high frequency.

EIA: Abbreviation for Electronic Industries Association.

ELECTRIC EYE: Another way of saying photoelectric cell.

ELECTRIC MOTOR: A device used to convert electrical energy into rotating mechanical energy.

ELECTRICAL DISCHARGE MACHINING: A way of removing metal from a part with a controlled spark in a dielectric.

ELECTRICAL LOAD: A device which is the resistive part of a circuit.

ELECTRICAL NOISE: Unwanted electrical energy in a circuit, usually caused by some other source.

ELECTRODE: A part of a battery. Also, a piece of metal used in electroplating. A terminal of an electrical component that gets an electrical signal into or out of that component.

ELECTROLYTE: The paste in a dry cell. Also, the fluid in a storage battery or material found in an electrolytic capacitor.

ELECTROLYTIC CAPACITOR: A capacitor made with an aluminum positive plate, electrolyte as a negative plate and a thin oxide layer on the aluminum as the dielectric. Its great advantage is larger capacitance from a smaller package.

ELECTROMAGNET: A temporary magnet which has magnetism only when current is passing through a coil of wire.

ELECTROMAGNETIC INDUCTION: Emf is generated in a conductor as an expanding and collapsing magnetic field causes electrons to flow.

ELECTROMOTIVE FORCE (emf): The force that causes electricity to flow because a difference in potential exists between two points.

ELECTRON: A negatively charged particle which orbits the nucleus of an atom.

ELECTRON BEAM: A narrow stream of electrons forced to move in a certain direction.

ELECTRONIC FLASH: A source of controlled light used by photographers to take pictures. Commonly called a "strobe."

ELECTRONIC INDUSTRIES ASSOCIATION (EIA): A trade association in the electronics industry.

ELECTRON MICROSCOPE: An electrical instrument used to magnify a sample of a material over 300,000 times larger so that you can see how it is made up.

ELECTROPLATE: A method used to deposit a thin layer of metal on the surface of an object.

ELECTROSCOPE: An apparatus with gold leafs used to demonstrate charges.

ELECTROSTATIC CHARGE: The electric charge stored by a capacitor.

ELEMENT: A substance that cannot be divided into simpler substances by chemical means. It is composed of protons, electrons and neutrons.

EMF: Abbreviation for electromotive force.

EMITTER (E): The lead of a transistor shown with the arrowhead on it.

EMPIRICAL: Information that can be found by actual measurement or observation.

ENCAPSULATION: The protective coating of plastic put around fragile electrical and electronic components.

ENERGIZED: Connected to a source of voltage. Also, called "hot" or "live" by many people.

ENERGY: That which is capable of producing work.

ENERGY CONVERSION: To change energy from one form to another.

ENTRANCE CABLE: The large power line which brings electricity into a building from an outside power line.

EPITAXIAL PROCESS: A method by which semiconductor devices are made by depositing a thin layer of single crystal material on a substrate (supporting material). This is done by using a vapor and controlling both the resistance and thickness of the material being deposited.

EQUIVALENT CIRCUIT: A way of taking a complicated circuit and finding out what it is like in a simpler, more easy to understand version that is the same electrically.

ETCHED PRINTED CIRCUIT: A way of making printed circuits by using a chemical to dissolve the copper from a specially made insulating board.

EXCITATION CURRENT: The current caused by applying voltage to the field windings of a motor.

EXOTHERMIC: Heat being produced, usually the result of a chemical reaction.

EXPONENT: A mathematical symbol which indicates the power of a given number.

EXTREMELY HIGH FREQUENCY (ehf): The range of frequencies from 30 to 300 gigahertz (10^9).

F: Symbol for a fuse. Also, abbreviation for fahrenheit and farad.

FACE: The front part of a meter that shows the readings.

FADE: The very slow lowering of a signal's amplitude.

FAHNESTOCK CLIP: A spring device used for making temporary connections in electrical circuits.

FAILURE RATE: The average number of parts that will fail in a given period of time.

FARAD (F): The unit of capacitance that is the result of a charge of one coulomb producing a difference in potential of one volt.

FAULT: A defect or break in a line. Also, a problem with an electrical circuit.

FCC: Abbreviation for Federal Communications Commission, a body which regulates all electrical communication systems in the United States.

FEEDBACK: Transferring voltage from the output of a circuit back to its input. Also, information being fed back to the source.

FEMALE: The recessed part of a device into which another part (male) fits.

FERRITE: An iron type material used for cores because it is easy to magnetize and it demagnetizes very rapidly.

FERRITE ANTENNA: An antenna found in radios. Its coil is wound around a ferrite core.

FET: Abbreviation for field effect transistor.

FIELD EFFECT TRANSISTOR (FET): A semiconductor device that conducts from one terminal called the "source" to another terminal called the "drain" when a voltage is applied to a third terminal called a "gate."

FIELD MAGNETS: Electromagnets used to make up the magnetic field of motors, generators and other electrical devices.

FILM RESISTOR: A resistor made from a very thin layer (film) of material deposited on an insulated form.

FILTERS: Circuit networks made from resistors, capacitors and inductors. They are used to block or reduce the amplitude of specific bands of frequencies.

FINITE: Having a fixed limit.

FISHPAPER: An insulating paper used to wrap coils and transformer windings or other conductors which are close together to protect against a possible short.

FISSION: Splitting the nucleus of an atom.

FLAT PACK: A rectangular integrated circuit in a flat package.

FLIP FLOP: A multivibrator circuit having two stable states and the signal is switched back and forth between them.

FLOATING: A condition when a device is not grounded.

FLUX: A material used in soldering to clean the surface of any oxide so that a good connection will result.

FLUX DENSITY: Number of lines of flux per cross-sectional area of a magnetic circuit.

FM BAND: The band of frequencies from 88 to 108 MHz.

FORWARD BIAS: The voltage applied to a PN junction so that it conducts.

FREE ELECTRONS: Electrons that circulate among the atoms of a substance without being tied to any particular atom.

FREQUENCY: The number of complete cycles per second as measured in hertz.

FREQUENCY MODULATION (FM): A system of radio transmission where the amplitude of the carrier remains constant but the frequency is varied.

FUSE: An electrical device put in a circuit to protect against overloading. Current above the rating of the fuse will melt the fusible link and open the circuit.

FUSESTAT: Trade name for a time delay fuse.

FUSETRON: A fuse which allows short periodic overloads without blowing.

GAIN: The ratio of the output ac voltage to the input ac voltage. Common examples also include current gain or power gain.

GALVANOMETER: A meter used to measure very small amounts of current.

GANGED CAPACITORS: Two or more variable capacitors mounted on the same shaft and varied by turning one knob.

GATE: One of the leads on a FET or SCR. Normally used to control current through the device when it is properly biased.

GAUSS: Unit of measure for magnetic flux density equal to one line of magnetic flux per square centimetre.

GENERATOR: A device used to convert rotating mechanical energy into electrical energy.

GERMANIUM DIODE: A two terminal semiconductor device made from germanium (a brittle, grayish-white metallic element).

GIGAHERTZ (GHz): A term used to represent 10^9 cycles per second.

GRATICULE: A calibrated screen placed at the face of a cathode ray tube of an oscilloscope for measuring purposes.

GRID DIP METER: A test instrument for use in measuring resonant frequencies, detecting harmonics and checking relative field strength of signals.

GROMMET: A rubber washer used to prevent a wire from touching exposed metal on a chassis.

GROUND: The voltage reference point in a circuit.

GROWLER: A device used to locate short-circuited coils by setting up a magnetic field. It makes a growling noise when a short is present.

H: Symbol for Henry.

HALF CYCLE: The interval required for the travel of 180 deg. on a sine wave.

HALF WAVE RECTIFICATION: A pulsating dc produced by passing only one half of a sine wave.

HAM: An amateur radio operator.

HARMONIC FREQUENCY: A frequency which is a multiple of a fundamental frequency.

HARNESS: Wires and cables which are tied together for easier handling.

HEAT LOSS: The loss of electrical energy in the form of heat.

HEAT SHRINK: A type of tubing placed over bare wire for insulation. It will shrink in diameter when heat is applied to it.

HEAT SINK: A metal base on which semiconductor devices are mounted to help dissipate heat. Also, a device used when soldering components to stop them from overheating. It usually is placed between the component and the joint being soldered.

HENRY (H): The unit of measure for inductance. A coil has one henry of inductance if an EMF of one volt is induced when the current through the inductor is changing at the rate of one ampere per second.

HERTZ (Hz): The unit of measure for frequency which is equal to one cycle per second.

HETERODYNE: The process of combining two signals of different frequencies to obtain two other frequencies: the sum frequency and the difference frequency.

HIGH FIDELITY (HI-FI): The sound output of an amplifier and loudspeaker combination, which is a resonably true reproduction of the quality of the original sound.

HIGH FREQUENCY (hf): The frequency band from 3 to 30 megahertz.

HIGH TENSION: Wires carrying voltages as high as several thousand volts.

HIGH VOLTAGE PROBE: A probe used for measuring high voltages. Therefore, it has a high internal resistance.

HIPOT: A device which has high voltages and is used for testing insulation breakdown or leakage of transformers.

HOLE FLOW: The idea that a positive charge carrier in

a P type material will attract electrons which, when combined, will form new holes. It is similar to the idea that a flow of positive charges will exist in a semiconductor.

HOLLERITH CODE: The code used when holes are punched into an 80 column card. The holes represent numbers and letters for use in keypunch machines.

HOLOGRAM: A way of recording three dimensional images on a two dimensional film plate through the use of lasers.

HOOD: The shield placed over the front of an oscilloscope to make it easier to see the image on the CRT.

HORSEPOWER (hp): The unit of power equal to 746 watts: 33,000 ft. lb. per minute or 550 ft lb. per second.

HOT: Connected to a source of power; energized.

HP (hp): Abbreviation for horsepower.

HYBRID: A circuit made up of thin film and discrete integrated circuits.

HYDROELECTRIC: Producing electricity by water power.

HYDROMETER: A device used to measure the specific gravity of the electrolyte in a storage battery.

HYSTERESIS LOOP: A curve used to show flux density when the magnetic force is increased and decreased.

I: Symbol for current. Also, abbreviation for indicator.

IC: Abbreviation for integrated circuit.

IEEE: Abbreviation for Institute of Electrical and Electronic Engineers.

IF: Abbreviation for intermediate frequency.

IGFET: Abbreviation for insulated gate field effect transistor.

IMPEDANCE: The total opposition which a circuit offers to the flow of alternating current at a given frequency. It is the combination of resistance and reactance.

IMPREGNATED COILS: Coils of wire completely covered with electrical varnish or epoxy to protect them from vibration and moisture.

IMPURITY: Material such as boron or arsenic which is added to semiconductor material in order to make P and N material.

INCREMENT: A small change in the value of something.

INDUCED CURRENT: Current which flows in a conductor because of a changing magnetic field.

INDUCTANCE: That property of a circuit or device which opposes any change in current through it. Its symbol is L.

INDUCTIVE REACTANCE: The opposition of an inductor to a changing current. Its symbol is X_L.

INFINITY: A number larger than any other number which can possibly exist. Its symbol is ∞ .

INPUT: The current, voltage or power applied to a circuit.

INSULATED GATE FIELD EFFECT TRANSISTOR (IGFET): A specially made diffused transistor which has very high input impedance and an insulated gate.

INSULATED WIRE: A conductor covered with any nonconducting material.

INTEGRATED CIRCUIT (IC): A circuit made from transistors, resistors, etc., all placed in a package smaller than the traditional transistor and sometimes called a "chip."

INTERFERENCE: Any disturbance, natural or otherwise, that interfers with the reception of signals.

INTERLOCK CIRCUIT: A circuit which will not activate until after one or more other actions have taken place. Often, these actions are made to occur in a certain sequence by means of relays. This prevents accidental activation of the circuit.

INTERMEDIATE FREQUENCY: It is the difference frequency obtained by beating the incoming signal against the local oscillator frequency.

INTERMITTENT: Something that occurs at intervals and not in a regular pattern.

INTERNAL RESISTANCE: The resistance within a source of voltage such as that within a transformer primary.

INTERNATIONAL MORSE CODE: A system that uses dots and dashes for signals.

INTERPOLATION: The method used to find a value between two known values.

ION: An electrically charged atom.

IR DROP: The voltage drop across a resistor when current is flowing through it.

IRON CONSTANTAN: A combination of two metals used in thermocouples.

ISOLATION TRANSFORMER: An electrical device used to link, by magnetic means, one or more parts of isolated circuits.

JACK: A socket for plugging in a wire.

JFET: Abbreviation for junction field effect transistor.

JOGGING: A way of inching a motor by quickly pushing the start and stop buttons in rapid succession.

JOINT: The connection of two wires.

JOULE: The work done by a force of 1 newton acting through a distance of 1 meter.

JUMPER: A short piece of wire used to eliminate or bypass portions of a circuit temporarily.

JUNCTION DIODE: A PN junction diode which will conduct more easily in one direction than it will in the opposite direction.

JUTE: A fiber used to cover an electrical cable.

K (k): Abbreviation for kilo.

KEEPER: A piece of metal put over the ends of a magnet to keep it from being demagnetized.

KELVIN SCALE: A temperature scale using absolute zero at -273 C.

KEYPUNCH: A machine used for punching holes into cards or paper tape as information.

KHZ (kHz): Abbreviation for kilohertz.

KILO (k): Prefix used to represent 1000.

KILOWATT HOUR: The amount of energy supplied by 1000 watts for one hour.

KIRCHHOFF'S LAWS: 1 — At any junction in a circuit, the algebraic sum of the currents around any closed loop is zero. 2 — In a simple circuit, the algebraic sum of the voltages around the circuit is zero.

KNOCKOUT: The removable slug in the side of an electrical box.

KRAFT PAPER: An electrical insulating material.

KV (kV): Abbreviation for kilovolt.

KVA (kVA): Abbreviation for kilovolt ampere.

KW (kW): Abbreviation for kilowatt.

L: Symbol for a coil or inductance.

LAG: The property of an ac circuit where the current reaches its peak value later than the voltage. In a purely inductive circuit, it does this 90 deg. later than voltage and is said to "lag" the voltage by 90 deg.

LAMINATED: Made in layers or stacked together, such as cores in a transformer.

LAMP: A device which produces light.

LASCR: A light activated silicon controlled rectifier (SCR) which has its gate activated by a light source.

LASER: Light amplification by stimulated emission of radiation, which is a device that takes incoherent light and produces a very narrow, highly intense beam of coherent light.

LATTICE: The way the atoms are arranged in semiconductor material.

LAW OF MAGNETISM: Like poles repel, unlike poles attract.

LAYOUT: The way the parts are positioned on a chassis as shown on a diagram.

LEFT HAND RULE: Grasp a conductor carrying current with your left thumb pointing in the direction of current flow and your fingers will show the direction of the magnetic field.

LENZ'S LAW: Induced emf in any circuit is always in the opposite direction of the effect that produced it.

LEYDEN JAR: A special kind of construction for experimentation, using a jar and metal foil. It was the original device used to study capacitors.

LIGHT EMITTING DIODE (LED): A PN junction diode that emits light when it is biased in the forward direction.

LIMIT SWITCH: A device used to make and break electrical circuits by tripping a switch that is part of the body of this device.

LINE CORD: A cord which is plugged into an outlet to provide power to a device.

LINE DROP: The voltage loss in a power transmission line as measured between two points.

LINE FREQUENCY: The frequency which is supplied by the power source.

LINES OF FORCE: The flux lines given off by an electric field. Also, the magnetic field surrounding a current-carrying device.

LISSAJOUS FIGURES: The patterns produced on the screen of an oscilloscope by changing the amplitude and phases of sine waves and feeding those to the horizontal and vertical inputs of the scope.

LIVE: A current-carrying member of a circuit, electrically hot.

LOAD: The device that is being driven by the source of power. It absorbs the power from the supply voltage and converts it to heat, light, etc. It is the resistance connected across a circuit that determines the current flow and energy used.

LODESTONE: A substance which is naturally magnetic as it occurs in nature.

LOGIC: A method of using the symbols AND, OR and NOT to represent complex circuit functions.

LOOP: A closed electrical circuit or path.

LOSS: The dissipation of energy without any useful work.

LOW CAPACITANCE PROBE: One type of test probe used with an oscilloscope.

LOW PASS FILTER: A filter that allows all frequencies below a predetermined frequency to pass frequencies higher than the predetermined frequency.

LUG: Terminal that is soldered or crimped on the end of a wire so that it can be attached to some part of a circuit.

M: Abbreviation for meg. Also symbol for mutual inductance.

MA (mA): Abbreviation for milliampere.

MAGNET: A substance, such as iron or steel, which attracts magnetic material. In its natural state, it is called "lodestone."

MAGNET KEEPER: A bar of iron that is put across the poles of a horseshoe magnet when not in use to prevent the magnet from being demagnetized.

MAGNETIC CIRCUIT: The complete path the lines of flux follow from the north pole to the south pole and through the magnet to the north pole.

MAGNETIC DEFLECTION: Using a magnetic field to bend the electron beam inside an oscilloscope.

MAGNETIC FIELD: The area around a magnet in

which the effects of magnetism are felt.

MAGNETIC FLUX: All of the magnetic lines surrounding a magnet. The symbol is phi (Φ).

MAGNETIC PICKUP: A phono cartridge that uses magnetism to produce an electrical output. It is connected to the stylus.

MAGNETIC SATURATION: A condition that exists in a magnetic material when a further increase in magnetizing force produces very little increase in flux density.

MAGNETISM: The invisible force that attracts certain materials such as steel.

MAGNETOMOTIVE FORCE (mmf): The force that produces the flux in a magnetic circuit.

MAGNITUDE: The size or numerical value assigned to a quantity.

MAINTENANCE: The process of keeping equipment repaired and/or in good working order.

MAJORITY CARRIERS: The principal charge carriers in a doped semiconductor. These are the holes in a P type material and the electrons in an N type material.

MANUAL: Hand operated. Also, the book of instructions that explains how to operate or repair a piece of equipment.

MASER: Amplification by stimulated emission of radiation; microwave.

MASK: A thin piece of material used in the process of making semiconductor devices.

MASONITE: A trade name for compressed material used in the construction of electrical panels.

MATRIX: A coding system for putting signals in a grid-shaped network so they can be found when needed by a computer. Similar to the way dates are put on a calendar.

MEG: Abbreviation for megohm.

MEGA: Short way of saying "one million."

MEGAHERTZ (MHz): One million hertz.

MEGGER: An ohmmeter used for making very high resistance measurements.

MEGOHM (MO): One million ohms.

MELT: The molten semiconductor material from which the basic crystal ingots are formed.

MERCURY BATTERY: A special battery which gives an almost constant output for its entire life.

MERCURY SWITCH: A special switch that has mercury in it. As the switch is opened or closed, the mercury moves so that it breaks or makes the circuit.

MESA TRANSISTOR: A transistor made by etching away the chip in a special way. The etching leaves mounds or hills that look like mesas.

METAL FILM RESISTOR: A resistor made by depositing a thin layer of metal on a substrate.

METAL OXIDE SEMICONDUCTOR (MOS): A device formed by diffusing a thin layer of metal oxide and separating or insulating the various parts as required.

METER: An electrical measuring instrument.

MF OR MFD (μ F): Abbreviation for microfarad.

MH (mH): Abbreviation for millihenry.

MHO: Previously used as the unit of conductance.

MICA: A transparent insulating material which is used because of its heat resistance.

MICRO: A prefix meaning one millionth (1/1,000,000).

MICROELECTRONIC: Reducing the size of electronic parts and circuits so that they are extremely small in size.

MICROFARAD (mF, mfd, or μ F): One millionth of a farad.

MICROVOLT: One millionth of a volt.

MICROWAVE: Short wavelength radio frequencies used in certain types of communication systems.

MIL: One thousandth of an inch. Also, a term used for wire measurement.

MILLI (m): A prefix meaning one thousandth (1/1000).

MILLIAMMETER: A meter used for measuring small current values.

MILLIAMPERE (mA): One thousandth of an amp.

MINIATURIZATION: The process of making something smaller in size.

MINORITY CARRIERS: The normally undesirable current carriers in a doped semiconductor. These are electrons in P material and holes in N material. They flow opposite the direction of majority carriers.

MODULATION: The process of varying the amplitude, frequency or phase of a carrier wave, depending on how you want to transmit the signal.

MOLECULE: The smallest division of matter that has the characteristics of that substance.

MONOCHROMATIC: Something that consists of a single color.

MONOSTABLE: A circuit having only one stable state.

MORSE CODE: System of dots and dashes used in telegraph systems.

MOS: Abbreviation for metal oxide semiconductor.

MOSFET: Abbreviation for metal oxide semiconductor field effect transistor.

MOTOR: A device used to convert electrical energy into rotating mechanical energy.

MU (μ): The Greek letter used as an abbreviation for micro.

MULTIMETER: A meter that can measure different values such as voltage, current and resistance.

MULTIPLEX: A method used in radio and telephone communications where several messages are transmitted in one or both directions over a single path.

MULTIVIBRATOR: A circuit using two transistors

that oscillate. The output of one is coupled to the input of the second. The output is a series of pulses.

MUTUAL INDUCTANCE: The condition that exists when the magnetic field of one conductor is linked to the magnetic field of a second conductor. In this way, each field will have some affect on the emf of the other conductor.

MYLAR: Trade name of a transparent plastic used as dielectric material in capacitors.

N: Symbol used for the north seeking pole of a magnet.

NAND GATE: The combination of an AND and a NOT used in logic circuits.

NANO: A prefix meaning one billionth.

NATIONAL ELECTRICAL MANUFACTURERS ASSOCIATION: Abbreviation NEMA. It is an organization of companies that manufacture electrical products.

NATIONAL ELECTRICAL CODE: A set of regulations that deal with electrical wiring of buildings and construction requirements to make sure they are safe and free of hazards.

NATURAL MAGNET: A substance which when found in its natural state has the properties of a magnet.

NC: Abbreviation for normally closed. It is a term that deals with the contacts of a relay.

NEGATIVE: The terminal having an excess of electrons which flow toward the positive terminal.

NEMA: Abbreviation for the National Electrical Manufacturers Association.

NEON BULB: A glass bulb that will not light until the ionization point of the neon gas inside has been reached.

NETWORK: The combination of two or more electrical circuits.

NEUTRAL: Neither positive nor negative.

NEUTRON: One of the three parts of an atom. It is the one having no electrical charge and is located in the nucleus of the atom.

NICHROME: A high resistance wire used for heating elements.

NICKEL CADMIUM CELL: An electrical cell which is popular because it has the ability to be recharged.

NO: Abbreviation for normally open. It is a term that deals with the contacts of a relay.

NOISE: Unwanted electrical disturbances often found in communications and circuits. Many times this disturbance results from a source outside the circuits.

NOMOGRAPH: A chart with three values in which a straightedge is placed across two of the known values, and the third is read directly off the chart.

NONFERROUS: Not made of iron.

NONFLAMMABLE: Material that will not burn very easily when exposed to flame or high temperatures.

NONMAGNETIC: Material which will not be attracted to a magnet.

NORMAL CONDITION: A relay that is de-energized.

NORMALLY CLOSED: Contacts that are closed when the relay is in the de-energized state.

NORMALLY OPEN: Contacts that are open when the relay is in the de-energized state.

NPN TRANSISTOR: A transistor that has N type material for both the emitter and collector and P type material for its base.

N TYPE MATERIAL: Semiconductor material which has free electrons for its majority carriers.

NUCLEUS: The core of the atom.

NUMERICAL CONTROL: A method of controlling the motion of machines by using a coded message (either tape, cards or drum stored), which is read by a computer.

OHM: The unit of measure for resistance. It is symbolized by the Greek letter omega (Ω). There is one ohm of resistance when an EMF of one volt causes a current of one amp.

OHMMETER: A meter used to measure resistance.

OHM'S LAW: A relationship between the values for voltage (E), current (I) and resistance (R). Discovered by a scientist named OHM, it states $E = I \times R$.

OHMS PER VOLT: A term used to describe the sensitivity rating of a meter.

OMNIDIRECTIONAL MICROPHONE: A microphone that picks up sound waves from almost any direction.

ON DELAY: A circuit that will produce an output signal after a time lapse. Often found in relay circuits and logic circuits.

OP AMP: Abbreviation for operational amplifier.

OPEN: To break the path for current in a circuit.

OPERATING LIFE: The minimum length of time a device is normally expected to operate without failing.

OPERATIONAL AMPLIFIER (OP AMP): A special circuit device used to do mathematical operations such as add, subtract, average, etc.

OPPOSITION: Two forces which act against one another.

ORBITAL ELECTRONS: The electrons that orbit around the nucleus of an atom.

ORDINATE: The vertical line on a graph.

OR GATE: A logic device that produces a signal when one of the inputs is present.

ORIFICE: A small opening in the path of flow.

OSCILLATOR: An electronic device that generates ac voltages.

OSCILLOSCOPE: An electronic test instrument that uses a cathode ray tube for visually displaying signals.

OUTLET: The point where current can be taken from an electrical circuit.

OUT OF PHASE: A term used to describe alternating current or voltage wave forms that do not pass through the minimum and maximum values at the same time.

OUTPUT: The energy delivered by a device or circuit.

OVERLOAD: To apply excessive load to a device, which results in a distorted output and, in some cases, damage to the equipment.

OXIDE: The thin layer of ferric oxide put on magnetic tape for the purpose of being able to magnetize them to produce a sound.

OZONE: A type of oxygen produced during an electrical discharge.

P: Abbreviation for power.

PADDER: A small variable capacitor usually placed in series with a tuning capacitor to adjust for the lower operating frequencies of the tuning capacitor.

PANEL: A material on which electrical parts or controls are mounted.

PANEL METER: A measuring instrument built so that it is easy to mount on a panel.

PAPER CAPACITOR: A capacitor which uses a special paper for its dielectric.

PARALLEL: A method of connecting electrical components so that there are two or more paths of flow.

PARALLEL CIRCUIT: A circuit which contains two or more paths for electrons to flow from a common voltage source.

PASSBAND: The frequencies that will pass through a filter network without reducing the amplitude.

PASTE: A thick, moist material that acts as the electrolyte in a dry cell.

PATCH: To connect a circuit by using a temporary wiring hookup.

PEAK: The highest point on a sine wave.

PEAK INVERSE VOLTAGE (PIV): The value of the voltage applied across a diode in the reverse direction.

PEAK TO PEAK: The value of a sine wave from the positive peak to the negative peak.

PERIOD: The time required to complete one cycle.

PERMANENT MAGNET: Bar of steel or other material that has been magnetized and retains its magnetism for a long time (years).

PERMEABILITY: The relative ability of a substance to conduct magnetic lines of force as compared to air. It is symbolized by the Greek letter mu (μ).

PF (pF): Abbreviation for picofarad.

PHASE: The relationship between the current and voltage in an ac circuit with respect to their angular displacement.

PHASE SHIFT: The change in the relationship between two alternating current voltages with the same frequency.

PHONE JACK: A receptacle designed for use with phone plugs.

PHONE PLUG: A special plug used with audio equipment such as headphones.

PHONO JACK: A receptacle designed for use with phono plugs.

PHONO PLUG: A special plug smaller in size than a phono plug. It is used on shielded wire for feeding audio signals to speakers and audio preamplifiers. Sometimes called an RCA plug.

PHOSPHOR: The layer of material coating the inside of cathode ray tubes that produces light when hit by an electron beam.

PHOTOCELL: Another way of saying "photoelectric cell."

PHOTODIODE: A PN junction device that will conduct electrons when exposed to light.

PHOTOELECTRIC CELL: A cell which is affected by light. It will generate a small voltage or become conductive, depending on the type of cell in use.

PI (π): Another way of giving a value to a ratio used when working with circles. It is approximately equal to 3.1416.

PICO (p): A prefix meaning 10^{-12}.

PICOFARAD (pF): A way of labeling a capacitor equal to one millionth of a microfarad (one millionth of one millionth farad).

PIEZOELECTRIC: The ability of some types of crystals to produce a voltage when put under certain types of stress.

PILE: A nuclear reactor.

PILOT LAMP: A light used to indicate whether a circuit is energized or operating in the proper sequence.

PI NETWORK: A three branch network of components hooked up in the shape of the symbol π .

PIV: Abbreviation for peak inverse voltage.

PLATE: A term used to refer to a single electrode found in a capacitor or battery.

PLUG: The male part of a connector.

PLUGGING: A way of slowing down a motor by hitting the reverse button while the motor is still running in the forward direction.

PN JUNCTION: The line of separation between P type and N type semiconductor materials.

PNP TRANSISTOR: A transistor which has P type material for its emitter and collector and N type material for its base.

POLARITY: The property of a device or circuit to have positive and negative poles.

POLARIZATION: The property found in an electrolytic cell to increase its resistance during electrolysis, thereby, shortening its life. Also, a way of making plugs and receptacles so they can be mated together only one way.

POLE: One of the ways used to name the ends of magnets. Also, one electrode found on a battery.

POLYGRAPH: A recorder that can measure small changes in blood pressure and body resistance. Often called a lie detector.

POLYPHASE: Having more than one phase.

POSITIVE: A point that attracts electrons (as opposed to negative, which is the source of electron flow). Also, any number that is numerically greater than zero.

POSITIVE ELECTRODE: The positive terminal of a battery or electrolytic capacitor.

POSITIVE ION: An atom that has lost electrons. Therefore, it is positively charged.

POSITIVE PLATE: The plate in a capacitor or battery that has the fewer number of electrons.

POT: A short way of saying "potentiometer." Also, the process of completely covering and sealing a component · or circuit with a hot liquid insulating material which turns solid as it cools.

POTENTIAL: The difference in the voltage between two points of a circuit.

POTENTIOMETER: A variable resistor used as a voltage divider. Also, an electrical device used to give an electrical output signal which is proportional to some rotary movement.

POTTED CIRCUIT: A circuit that has been completely covered and sealed with an insulating material.

POWER (P): The rate of doing work.

POWER AMPLIFIER: A special amplifier used for providing a lot of current to a load, such as a motor or speaker.

POWER DISSIPATION: The ability of a device to get rid of the heat it generates.

POWER FACTOR: The ratio of the true power to the apparent power of a circuit.

POWER FACTOR CORRECTION: The addition of capacitors to an inductive circuit to increase the power factor by putting the current more closely in phase with the voltage.

POWER GAIN: The amount that a device will amplify a signal. It is the ratio of power output to the power absorbed by the input.

POWER LINE: The conductors used to transport electricity from one point to another.

POWER LOSS: The loss in a circuit caused by resistance, heat, etc. Sometimes called the I^2R loss.

POWER TRANSFORMER: A transformer used to supply output voltages from an ac power source.

POWER TRANSISTOR: A transistor that can deliver high current and power without overheating.

PREAMPLIFIER: An amplifier between the signal source (usually a low level output) and the main amplifier.

PRECISION: The ability to make the same measurements a number of times and have them as close together as possible.

PRESSURE: The force per unit area.

PRESSURE SWITCH: A switch that is activated by a change in the pressure of some liquid. It is also used with gases.

PRIMARY: The input winding on a transformer.

PRIMARY CELL: A cell that is not designed to be recharged.

PRIMARY CURRENT: Current which flows through the input side of a winding of a transformer.

PRINTED CIRCUIT: A circuit made on a special insulated board on which thin strips of copper are the wires.

PRINTED CIRCUIT BOARD: The insulated board on which the printed circuit is formed.

PRINTED WIRING: A method of making the conductors by printing very thin ribbons of material on an insulated board.

PROBE: A test lead with a pointed metal end used to contact the part of a circuit to be measured by a vacuum tube voltmeter or observed on a scope.

PROTON: The positively charged particle in the nucleus of an atom.

PROTOTYPE: A hand-assembled model of the finished product used for evaluation of design, construction and operation before mass production begins.

PROXIMITY SWITCH: A switch designed to open or close when an object, such as a piece of metal, is brought very near.

P TYPE MATERIAL: A semiconductor material which has holes for its majority current carriers.

PULL IN VOLTAGE: The minimum voltage which can be used to energize a relay.

PULSATING CURRENT: Current that will rise and fall, but generally not fall below the zero reference point.

PULSE: A momentary voltage or current.

PULSE GENERATOR: A device used for generating a controlled series of electrical pulses.

PUNCHED CARD: A card slightly larger than 3 in. x 7 in. used for coding information in the form of punched holes. Normally, each card has 80 columns with 12 locations in each column where a hole can be punched.

PUSH PULL CIRCUIT: A circuit having two components hooked up 180 deg. out of phase with one another so that each one will operate on one half cycle.

PYROMETER: A device used to measure high temperatures, such as the temperature of molten iron.

Q: Figure of merit. It is the ratio of the reactance of a capacitor or inductor to its resistance.

QUAD: A combination of four items used in a circuit. Also, the short way of saying "quadruple."

QUADRANT: One fourth of a circle or 90 deg.

QUALITY CONTROL: To control the appearance and operation of a finished product by inspecting it during manufacturing.

QUANTITY: The amount or number of an item.

QUARTZ: A mineral (silicon dioxide) used in some electrical devices because of its piezoelectric properties.

QUICK DISCONNECT: A device used to quickly lock and/or remove two connecting halves.

QUIESCENCE: The condition in which a circuit has power supplied to it but does not have a signal supplied to it.

R: The symbol for resistance.

RACEWAY: Channel that holds electrical conductors (wires, bars and cables).

RADAR: Radio Detecting and Ranging is a term used for a system of measuring the speed and location of an object by measuring the amount of time required for a signal to bounce off that object and return.

RADIAL LEAD: The lead coming out of the side of a component rather than from its end.

RADIAN: A system for measuring angles in which π is the unit of measure rather than degrees. A circle contains 2π radians.

RADIATE: To give off rays.

RADIATION: To give off electromagnetic waves into space.

RADIO: Electronic equipment used to transmit or receive electromagnetic waves without the use of sending or receiving wires.

RADIOACTIVITY: The emission of alpha particles, gamma rays, etc., from an unstable nucleus as it decays to a more stable state.

RADIO FIX: A method used to find the position of a source of radio waves. Also, a method used by ships and planes to determine their own location by using radio signals from stations whose locations are known.

RADIO FREQUENCY CHOKE: An inductor that has a high impedance to RF currents.

RADIOISOTOPE: A radioactive particle used as a tracer in the medical and science fields because it emits rays that can be traced with instruments.

RANGE: The maximum distance a radio transmitter can effectively send signals.

RASTER: The lighted area on a television picture tube produced by the scanning lines moving across the screen when there is no picture on the screen.

RATIO: The value obtained by dividing one number by another.

RC: Symbol for resistance capacitance.

RC TIME CONSTANT: The time period needed for the voltage across a capacitor in an RC circuit to increase to 63.2 percent of the source voltage or decrease to 36.7 percent of the voltage across the capacitor. The t (in seconds) equals R (in ohms) x C (in farads).

REACTANCE: The opposition to the flow of alternating current in a circuit because of its inductance or capacitance. It is expressed by the symbol X.

REACTIVE LOAD: A load caused by capacitance or inductance (as opposed to a resistive load).

REACTOR: A device used to split the atom under controlled means.

RECEIVER: That part of a communications system which accepts radio waves and converts them into audible form.

RECEPTACLE: A contact device installed at an outlet so that a plug can be connected.

RECIPROCAL: A number found by dividing it into 1. Therefore, it is 1 over the number.

RECTIFICATION: The conversion of ac into dc.

RECTIFIER: The component used to convert ac into pulsating dc.

REED RELAY: A relay that is made with two magnetic strips inside a coil. They are attracted to each other each time the coil is energized, closing the contacts that are mounted on the strips.

REED SWITCH: Two overlapping magnetic strips, mounted inside a glass housing, that are closed by a magnetic field.

REEL: A cylindrical device used for holding wire for storage purposes.

REFERENCE LEVEL: The starting point from which other measurements are made, such as 0 deg. on a thermometer or 0 deg. on a sine wave.

REL: The unit of measurement for reluctance.

RELAY: An electrical device that will open or close contacts through which another circuit can be opened or closed. In this way, the relay can be used in conjunction with one circuit to control a second circuit.

RELEASE: To disconnect.

RELUCTANCE: Opposition to the flow of magnetic flux.

REMOTE CONTROL: A method of controlling another device from some distance away by the use of wires, light or other means.

RENEWABLE FUSE: A fuse that can be repaired by

changing its link, then it is reused.

REPEL: To force away or apart, as like magnetic poles do.

REPULSION INDUCTION MOTOR: A squirrel cage type motor having two sets of windings, one for repulsion and one for induction.

REPULSION START MOTOR: A motor that develops starting torque by the interaction of rotor currents and a single phase stator field.

RESIDUAL MAGNETISM: The amount of magnetism left after the magnetic force is removed or the source of electromagnetism has its circuit opened.

RESIST: The material placed on printed circuit boards to prevent the copper from being etched away by chemicals used in the process.

RESISTANCE: Opposition to the flow of current.

RESISTANCE SUBSTITUTION BOX: An assembly of resistors connected by switches to a pair of test leads. Any one of the resistors may be connected into a circuit for checking purposes. This device is not left in a circuit permanently.

RESISTANCE WIRE: A wire such as Nichrome that has high resistance to the flow of current. Therefore, it is used in heating elements and wirewound resistors.

RESISTOR: A device that will oppose the flow of current in an electrical circuit.

RESISTOR COLOR CODE: A means of identifying the various values of resistors by coding them with colored rings around the body. Each color represents a specific number or multiplying value.

RESOLUTION: The degree to which nearly equal values can be distinguished from one another.

RESONANCE: The condition that exists when the inductive reactance equals the capacitive reactance in the circuit.

RESONANCE CURVE: A graph that shows how a tuned circuit will react at various frequencies.

RESONANT FREQUENCY (f_r): The frequency at which inductive reactance equals capacitive reactance in a tuned circuit.

RETENTIVITY: The ability of a material to retain its magnetism after the magnetizing force is moved.

REVERSE BIAS: The voltage applied to a PN junction so that it does not conduct or conducts only a very small amount.

REVERSE BREAKDOWN VOLTAGE: The point at which a further increase in voltage in the reverse direction on a semiconductor causes a sharp increase in current without much more voltage.

REWIND: The process of repairing a motor by removing and replacing its wound coils with new coils.

RF (rf): Abbreviation for radio frequency.

RF CHOKE: A coil that has a high inductive reactance at radio frequencies.

RF SIGNAL GENERATOR: A test instrument that will generate many bands of radio frequencies. It is used in the repair and alignment of radio equipment.

RHEOSTAT: A variable resistor placed in series with the load to control current. It has one fixed terminal and a movable contact.

RIGHT HAND RULE: A method used with motors to show the direction of the field (first finger), direction of current (second finger) and motion of conductor (thumb).

RIPPLE: A series of peaks found after ac has been rectified to dc.

RMS: Abbreviation for root mean square.

ROOT MEAN SQUARE: The value of ac that would produce the same heating effect as an equal amount of dc. In ac, the value is 0.707 times the peak value, and it is the same as the effective value.

ROSIN CORE SOLDER: A material used to join electrical components. It is a mixture of tin and lead, and it has a fluxing agent in the hollow center.

ROTATING FIELD: The magnetic field found in electrical motors where the poles keep shifting to cause rotation.

ROTOR: The rotating part of an electric motor.

RPM (rpm): Abbreviation for revolutions per minute.

RTL: Abbreviation for resistor transistor logic.

S: Symbol for siemen. (s) Symbol for second.

SAFETY FACTOR: A way of rating a component so that its stated limit may be exceeded temporarily without causing failure.

SAL AMMONIAC: Ammonium chloride. It is a material found in electric cells as part of its electrolyte.

SAPPHIRE: A gem used for bearings in electrical instruments and in needles of phonograph players.

SATURATED: A point where a transistor will not give an increase in current if the base current is increased. Also, a permanent magnet or the core of an electromagnet which is conducting as many lines of magnetic flux as it can be made to conduct under normal conditions.

SAWTOOTH GENERATOR: An oscillator which produces an alternating waveform that has the shape of teeth on a saw (series of $\wedge\wedge$s).

SCHEMATIC: An electrical diagram that shows components as symbols and electrical connections of components to each other.

SCOPE: Short way of saying "oscilloscope."

SCR: Abbreviation for silicon controlled rectifier.

SCREEN: The surface of an oscilloscope that shows the visible pattern traced out by the electron beam.

SEC: Abbreviation for the secondary winding of a transformer. Also, abbreviation for second.

SECONDARY: The output winding of a transformer.

SECONDARY CELL: A cell which, after it has been discharged is meant to be recharged.

SEED: A single crystal from which semiconductor crystal material is grown.

SELENIUM RECTIFIER: A rectifier made by depositing selenium on aluminum plates and coating it with a conductive metal.

SELF INDUCTION: That property which causes a counter emf to be generated to oppose a change in current.

SEMICONDUCTOR: An electronic device that has been made from material somewhere between the range of conductors and insulators.

SEMICONDUCTOR CHIP: A small piece of material from which semiconductor devices are made.

SEMICONDUCTOR DIODE: A device made from P and N type material forming a PN junction that passes current easily in one direction but opposes the flow of current in the other direction.

SEQUENCE: The order in which operations occur.

SERIES CIRCUIT: A circuit in which components are connected from end to end, and the current must flow through each one in sequence to complete its path.

SERIES PARALLEL NETWORK: A circuit which has only one path for current in some places but more than one path for current in other places.

SERIES RESISTOR: A resistor that is connected in series to some device to control voltage or current.

SERVO: A device that controls and delivers power to move something. Short for servomotor.

SHELF LIFE: The length of time a component may be stored and still be useful.

SHELL: The various energy levels of electrons around an atom.

SHIELDED CABLE: A cable that has an outer shield or casing which is intended to decrease the interference from sources outside the cable.

SHOCK: Current flow through a living organism, especially a person.

SHORT CIRCUIT: A usually undesired path for current which bypasses a desired path.

SHORT OUT: A method used to bypass some component by connecting a wire between its terminals.

SHUNT: A low value resistor used in parallel with a meter to increase the amount of current it can measure. Also, a component or circuit connected in parallel with another component or circuit.

SIDEBANDS: Frequencies above and below carrier frequencies as a result of modulation.

SIGNAL: The part of an electrical voltage or current which is used to carry intelligence or information.

SILICON: A material used to make semiconductors.

SILICON CONTROLLED RECTIFIER (SCR): A semiconductor device having an anode, cathode and gate used for current values that are usually higher than those carried by transistors.

SILICONE GREASE: A material used to aid in heat dissipation of semiconductors.

SINE WAVE: A wave form of a single frequency alternating current whose displacement is the sine of an angle proportional to time or distance.

SINGLE POLE DOUBLE THROW (SPDT): A switch used to connect one terminal to one of two other terminals.

SINGLE POLE SINGLE THROW (SPST): A switch used to open or close one circuit.

SINGLE SHOT MULTIVIBRATOR: A multivibrator designed to provide a single pulse on demand.

SINK: A device used to drain off excess energy, such as a heat sink drawing off excess heat.

SLICE: A wafer cut from a silicon ingot and used for making integrated circuits.

SLIDER: The movable contact in rheostats.

SLIP: The difference between the actual speed of a motor and its synchronous speed.

SLUG: An adjustable metallic piece found inside a coil used for tuning circuits.

SOLAR CELL: A cell which converts the energy of light into electrical energy. A photovoltaic cell.

SOLDER: An alloy of tin and lead used for providing a low resistance electrical connection which is also mechanically strong.

SOLDERING IRON: One kind of device used to provide the heat to melt solder.

SOLDERLESS CONNECTOR: A device used to connect wires to terminals or other wires by means of crimping, etc.

SOLENOID: An electromagnet that has an air core in which a small magnetic shaft is free to move and operate a valve or other device.

SOLID STATE: Electronic circuits using semiconductor devices such as transistors, diodes and silicon controlled rectifiers.

SOURCE: One of the leads of a field effect transistor. Also, a device or part of a circuit that supplies electrical energy to another device or circuit.

SOUTH POLE: The pole of a magnet where the lines of force enter after having left the north pole.

SPAGHETTI: An insulating tubing used in electrical work.

SPDT: Abbreviation for single pole, double throw.

SPEAKER: A device used to convert electrical signals into sound.

SPEED OF LIGHT: The distance light will travel in

one second, which is approximately 300,000,000 metres per second or 186,000 miles per second.

SPEED OF SOUND: The distance sound will travel in one second, which is about 1080 feet per second.

SPIKE: A usually undesirable very narrow pulse of electricity with so great an amplitude that it interferes with or deteriorates normal circuit operation.

SPKR: Abbreviation for speaker.

SPLICE: Joining two or more conductors.

SPST: Abbreviation for single pole, single throw.

SPURIOUS PULSE: A pulse or spike which was not generated on purpose.

SQUARE WAVE: An electrical wave which alternates abruptly between high and low values of approximately equal duration.

SQUIRREL CAGE INDUCTION MOTOR: An induction motor having a rotor that looks similar to a squirrel cage in construction.

STATIC ELECTRICITY: Electricity at rest. Also, electricity caused by friction.

STATOR: The stationary (nonrotating) part which contains the primary windings of a motor.

STATOR PLATES: The fixed plates of a variable capacitor.

STEADY STATE: The condition of a circuit after all the transients have settled down.

STEP DOWN TRANSFORMER: A transformer used to reduce the voltage.

STEREOPHONIC: Sound which is broken into two or more channels to give the effect of depth because different signals can be received by each ear.

STORAGE BATTERY: A common name for the lead acid battery used in automobiles. It is made from storage cells connected in series.

STORAGE OSCILLOSCOPE: A scope that has the ability to retain the image on its screen for a period of time so it can be observed or studied.

STRANDED WIRE: A conductor which is made from fine strands of wire twisted together.

STRIP: To remove insulation from wire.

STROBOSCOPE: A bright light that can be adjusted to flash a specific number of times to coincide with some repeating motion.

SUBMINIATURIZATION: A method of reducing the size of a circuit by using different construction techniques.

SUBSTATION: A place where electrical power is transformed for a large number of users.

SUBSTRATE: An insulating material on which circuits are formed.

SURGE: A sudden change in current or voltage.

SWEEP: The crossing rate for the electron beam on the CRT of an oscilloscope. The horizontal line caused by

the deflection of an electron beam across the face of the CRT from left to right.

SWITCH: A device used to make and break the path for current in a circuit.

SYMBOL: A method used to represent an electrical or electronic component by using letters, numbers or designs on a schematic drawing.

SYNCHRONOUS: Two or more actions of a similar or related nature occurring at exactly the same rate at the same time.

TABLE: A collection of data displayed in chart form.

TANK CIRCUIT: A parallel resonant circuit.

TAP: A fixed electrical connection on a transformer that is used to provide another voltage level.

TAPER, POTENTIOMETER: The way in which the amount of resistance between the wiper and an end terminal of a potentiometer varies. Common tapers are linear and audio or logarithmic.

TAPPED RESISTOR: A wirewound resistor having additional terminals at points along its length. It is often used for voltage divider circuits.

TELEMETRY: The transmission of analog or digital information recorded on instruments by radio or telephone to another point some distance away.

TELEVISION SIGNAL: The audio, video and sync (timing reference) signals that are broadcast by a television station. These signals are used by a television receiver to produce the picture and sound.

TEMPORARY MAGNET: A magnet that will hold its magnetism for a short period of time after the source of magnetism has been removed.

TERMINAL: A point for electrical connections.

TERMINAL STRIP: An insulated base on which a number of terminals are mounted so that wiring connections will be easy to make.

TEST CLIP: A spring clip used to make temporary connections to circuits, especially when they are being tested.

TEST LEAD: A flexible insulated wire used to connect testing equipment to circuits to be tested.

THERMAL OVERLOAD: A device used to protect electrical circuits from overheating. It will open the circuit when the temperature gets too high because of excessive current.

THERMAL RADIATION: Heat.

THERMAL RUNAWAY: A condition where rising temperature causes the current to increase which in turn, causes the temperature to rise, etc.

THERMISTOR: Thermal resistor. A device that will decrease in resistance with an increase in temperature.

THERMOCOUPLE: A device used to measure temperature by the use of two dissimilar metals which are joined at the point where the measurement is being taken. As heat is applied, the voltage generated can be read on a meter calibrated in degrees.

THERMOSTAT: A device used to control the amount of heat in a building by setting it at some desired temperature level.

THICK FILM CIRCUITS: Films are deposited on a substrate to act as certain components, then transistors and diodes are soldered in place as the active components.

THIN WALLED CONDUIT: Metallic tubing used to hold and protect insulated wires in various circuits.

THREE PHASE MOTOR: An alternating current motor which operates from three phase current.

THREE PHASE SYSTEM: Using three wires to carry current and simultaneously act as the return wires for alternating current with a phase difference of 120 deg. or one third of a cycle.

THREE WAY SWITCH: A switch that connects one conductor to either of two other conductors. It is used in places where a light can be turned on from either the top or bottom of a stairway.

THRESHOLD: The point where something first happens, such as a transistor beginning to conduct.

TIME DELAY RELAY: A relay that can be set so that it will not activate its contacts until after a measured amount of time delay.

TINNED: A coating on wire used to make soldering easier.

TIP: The end of a probe.

TO CAN: One of the shapes used for packaging integrated circuits in which the shape is similar to that normally used with transistors.

TOGGLE SWITCH: A two position switch that is opened and closed (operated) by a lever.

TOLERANCE: The permissable amount from which a value is allowed to vary.

TORQUE: A force that tends to produce rotation.

TRACE: The pattern on the screen of an oscilloscope.

TRANSDUCER: A device that will convert a signal from one form to another, such as converting pressure to electricity.

TRANSFORMER: A device used to step up or step down voltage by induction. Also, an electrical device used to transport energy from one circuit to another circuit at the same frequency.

TRANSIENT: A sudden change or pulse that is not intended in a circuit.

TRANSISTOR: A semiconductor made from germanium or silicon with leads known as emitter, collector and base. It is made to conduct by forward biasing the emitter-base junction.

TRANSISTOR TRANSISTOR LOGIC (TTL): Digital circuits used with computers in which transistors control other transistors.

TRANSMIT: To send a signal from one point to another point.

TRIAC: A three terminal, bi-directional thyristor device that is triggered into conduction by applying a signal to its gate.

TRICKLE CHARGE: A very slow rate of charging, usually taking many hours.

TRIMMER CAPACITOR: A small variable capacitor that is connected in series with another capacitor. It is used to fine adjust a circuit.

TRL: Abbreviation for transistor resistor logic.

TROUBLESHOOT: The process of working with equipment to locate and correct any faults.

TRUTH TABLE: A table used to show the output with all the possible combinations of inputs for various AND, OR and other logic functions.

TTL: Abbreviation for transistor transistor logic.

TUNED: Adjusted to resonate at a specific frequency.

TUNED CIRCUIT: A circuit which consists of capacitance and inductance that is adjustable for resonance at a desired frequency.

TUNGSTEN FILAMENT: The filament used in incandescent lights because of its ability to glow white hot.

TUNNEL DIODE: A special PN junction diode in which initial forward biasing causes the electrons to tunnel through rather than over the barrier. These diodes have special properties that make them suited to high frequencies and use in space.

TURNS RATIO: In a transformer, it is the ratio of the number of turns in the primary to the number of turns in the secondary.

TWEETER: A speaker that is designed to produce high audio frequencies.

UHF: Abbreviation for ultra high frequency.

ULTRA HIGH FREQUENCY: The frequency band from 300 to 3000 megahertz.

ULTRASONIC: Having a frequency above audible sound (average 20,000 Hz).

ULTRASONIC WELDING: Bonding two pieces of metal without reaching melting temperatures.

UNDERCUT: The process where the metal foil on a printed circuit board is etched under the edge of the resist material.

UNGROUNDED SYSTEM: An electrical system in which no point is connected to ground.

UNIDIRECTIONAL: Current which is flowing in only one direction.

UNIJUNCTION TRANSISTOR: A three terminal transistor that has one junction with two base leads and one emitter lead.

V: Symbol for volt or voltmeter.

VA: Abbreviation for volt amperes.

VACUUM TUBE VOLTMETER (VTVM): A voltmeter that uses a vacuum tube to perform the functions of amplification and rectification. It is used to make electrical measurements.

VARIABLE CAPACITOR: A capacitor whose capacitance can be adjusted by turning a shaft to change the distance between its plates or effectively change the area of its plates.

VARIABLE SPEED MOTOR: A motor which, regardless of loading can be adjusted over a range of speeds.

VARISTOR: A two electrode semiconductor device that drops in resistance as the voltage is increased.

VECTOR: A quantity having both magnitude and direction.

VITREOUS: Having the characteristics of glass.

VOLT (V): The unit of measure for EMF or potential difference.

VOLTAGE (E): The electromotive force which will cause current to flow.

VOLTAGE DIVIDER: A method of using a series of resistors across a voltage source so that multiple voltages can be obtained.

VOLTAGE DROP: The difference in voltage between two points in a circuit.

VOLTAIC CELL: Another name for a primary cell.

VOLTMETER: An instrument used for measuring the difference in potential between two points.

VOLT OHM MILLIAMMETER (VOM): A meter that can be used to measure a number of different voltage, resistance and current levels.

VOM: Abbreviation for volt ohm milliammeter.

VTVM: Abbreviation for vacuum tube voltmeter.

W: Symbol for watt or work.

WAFER: A thin slice of semiconductor material used for making integrated circuits.

WALL BOX: A metal or plastic box used in house wiring to enclose a switch or receptacle.

WATT (W): The unit of measure for power (rate of doing work).

WATT HOUR: A unit of measure for electrical power that says you have used one watt for one hour.

WATT HOUR METER: A meter connected into a circuit and used to determine the amount of energy used, usually in kilowatt hours.

WAVEFORM: The graphical shape of a wave which will show the amount of variation in amplitude over some period of time.

WAVELENGTH: Distance between a specific point on a wave and same point on next cycle of the wave.

WET CELL: Cell that has electrolyte in liquid form.

WHEATSONE BRIDGE: A means of measuring an unknown resistance by comparing it to a known value in a special circuit which will be balanced when its value is found.

WHISKER: A thin piece of metal (resembling a cat's whisker) used in making semiconductor devices.

WIDEBAND: Able to pass a broad range of frequencies.

WINDING: In a transformer, it is the length of conductor that is wound in a coil shape to serve as the primary or secondary.

WIPER: The moving contact of a potentiometer that is adjusted to change the resistance.

WIPING CONTACTS: Contacts that rub between two other contacts, usually found in switches.

WIRE: A solid conductor or stranded group of electrical conductors (copper, aluminum, etc.) which serve as one conductor. It is said to have low resistance to the flow of current.

WIREWOUND RESISTOR: A resistor whose resistance element is a piece of resistance wire wound onto some insulating form. Many of them are then coated with an insulating material.

WIRING DIAGRAM: A drawing used to show the circuit and electrical connections needed.

WOOFER: A large speaker specifically designed to produce low frequency sounds.

WORKING VOLTAGE: The maximum voltage at which some device should be operated.

WVDC: Abbreviation for working voltage direct current, which is maximum safe dc operating voltage across a capacitor at normal operating temperature.

WYE CONNECTION: A method of connecting three phase electrical devices so that the end of each winding has a common neutral point. Symbol is Y.

X: Symbol for reactance measured in ohms.

X AXIS: The horizontal axis on a graph.

X_C: Symbol for capacitive reactance.

X_L: Symbol for inductive reactance.

X RAYS: Radiation of very short wavelength generated by bombarding a target with a stream of high speed electrons.

Y AXIS: The vertical axis on a graph.

YOKE: A set of coils used on the neck of a cathode ray tube to deflect the electron beam.

Z: Symbol for impedance.

ZENER BREAKDOWN: A breakdown in a reverse biased semiconductor device. It is caused when electrons in the depletion layer knock other electrons loose until a large reverse current is flowing across a reverse biased junction.

ZENER DIODE: A silicon diode that makes use of the breakdown properties of a PN junction. If reverse voltage across the diode is increased, a point will be reached where the current will greatly increase.

ZERO POTENTIAL: Zero voltage or "earth potential."

INDEX

Index

Index